T0192478

Chapman & Hall/CRC Biostatistics Series

Statistical Design and Analysis of Stability Studies

Chapman & Hall/CRC Biostatistics Series

Editor-in-Chief

Shein-Chung Chow, Ph.D.
Professor
Department of Biostatistics and Bioinformatics
Duke University School of Medicine
Durham, North Carolina, U.S.A.

Series Editors

Byron Jones
Senior Director
Statistical Research and Consulting Centre
(IPC 193)
Pfizer Global Research and Development
Sandwich, Kent, UK

Jen-pei Liu
Professor
Division of Biometry
Department of Agronomy
National Taiwan University
Taipei, Taiwan

Karl E. Peace
Director, Karl E. Peace Center for Biostatistics
Professor of Biostatistics
Georgia Cancer Coalition Distinguished Cancer Scholar
Georgia Southern University, Statesboro, GA

Chapman & Hall/CRC Biostatistics Series

Published Titles

1. *Design and Analysis of Animal Studies in Pharmaceutical Development,*
   Shein-Chung Chow and Jen-pei Liu
2. *Basic Statistics and Pharmaceutical Statistical Applications,*
   James E. De Muth
3. *Design and Analysis of Bioavailability and Bioequivalence Studies,*
   Second Edition, Revised and Expanded, Shein-Chung Chow and Jen-pei Liu
4. *Meta-Analysis in Medicine and Health Policy,* Dalene K. Stangl and
   Donald A. Berry
5. *Generalized Linear Models: A Bayesian Perspective,* Dipak K. Dey,
   Sujit K. Ghosh, and Bani K. Mallick
6. *Difference Equations with Public Health Applications,* Lemuel A. Moyé
   and Asha Seth Kapadia
7. *Medical Biostatistics,* Abhaya Indrayan and Sanjeev B. Sarmukaddam
8. *Statistical Methods for Clinical Trials,* Mark X. Norleans
9. *Causal Analysis in Biomedicine and Epidemiology: Based on Minimal*
   Sufficient Causation, Mikel Aickin
10. *Statistics in Drug Research: Methodologies and Recent Developments,*
    Shein-Chung Chow and Jun Shao
11. *Sample Size Calculations in Clinical Research,* Shein-Chung Chow, Jun Shao, and
    Hansheng Wang
12. *Applied Statistical Design for the Researcher,* Daryl S. Paulson
13. *Advances in Clinical Trial Biostatistics,* Nancy L. Geller
14. *Statistics in the Pharmaceutical Industry, 3rd Edition,* Ralph Buncher
    and Jia-Yeong Tsay
15. *DNA Microarrays and Related Genomics Techniques: Design, Analysis, and Interpretation*
    of Experiments, David B. Allsion, Grier P. Page, T. Mark Beasley, and Jode W. Edwards
16. *Basic Statistics and Pharmaceutical Statistical Applications, Second Edition,* James E.
    De Muth
17. *Adaptive Design Methods in Clinical Trials,* Shein-Chung Chow and
    Mark Chang
18. *Handbook of Regression and Modeling: Applications for the Clinical and Pharmaceutical*
    Industries, Daryl S. Paulson
19. *Statistical Design and Analysis of Stability Studies,* Shein-Chung Chow

Chapman & Hall/CRC Biostatistics Series

Statistical Design and Analysis of Stability Studies

Shein-Chung Chow

Duke University School of Medicine
Duham, North Carolina, U.S.A.

CRC Press
Taylor & Francis Group
Boca Raton London New York

CRC Press is an imprint of the
Taylor & Francis Group, an **informa** business
A CHAPMAN & HALL BOOK

Chapman & Hall/CRC
Taylor & Francis Group
6000 Broken Sound Parkway NW, Suite 300
Boca Raton, FL 33487-2742

First issued in paperback 2020

ISBN-13: 978-0-367-57768-1 (pbk)
ISBN-13: 978-1-58488-905-2 (hbk)

Library of Congress Cataloging-in-Publication Data

Chow, Shein-Chung, 1955-
    Statistical design and analysis of stability studies / Shein-Chung Chow.
        p. ; cm. -- (Biostatistics ; 19)
    "A CRC title."
    Includes bibliographical references and index.
    ISBN-13: 978-1-58488-905-2 (hardcover : alk. paper)
    ISBN-10: 1-58488-905-5 (hardcover : alk. paper)
    1. Drug stability. 2. Drugs--Research--Statistical methods. I. Title. II. Series: Biostatistics (New York, N.Y.) ; 19.
    [DNLM: 1. Drug Stability. 2. Research Design. 3. Pharmaceutical Preparations--standards. 4. Statistics--methods. QV 754 C5595s 2007]

RS424.C46 2007
615'.1901--dc22
                                                  2006100696

Visit the Taylor & Francis Web site at
http://www.taylorandfrancis.com

and the CRC Press Web site at
http://www.crcpress.com

Contents

Preface

In the pharmaceutical industry, labeled shelf-life on the immediate container of a drug product provides the consumer the confidence that the drug product will retain its identity, strength, quality, and purity throughout the expiration period of the drug product. Drug shelf-life is usually supported by stability data collected from stability studies conducted under appropriate storage conditions. The FDA has the authority to issue recalls for drug products due to a problem occurring in the manufacture or distribution of the product that may present a significant risk to public health. As indicated in the FDA Report to the Nation in 2004, stability data not supporting expiration the date is among the top three reasons for drug recalls in the fiscal year of 2004, during which 215 prescription drug products and 71 over-the-counter drug products were recalled. The cost for a recall and possible penalty could be a disaster for the pharmaceutical company. Thus, stability studies play an important role in drug safety and quality assurance.

The purpose of this book is to provide a comprehensive and unified presentation of the principles and methodologies of design and analysis of stability studies. In addition, this book is intended to give a well-balanced summary of current regulatory perspectives and recently developed statistical methods in the area. It is our goal to provide a complete, comprehensive and updated reference and textbook in the area of stability design and analysis in pharmaceutical research and development.

Chapter 1 provides an introduction to basic concepts regarding stability testing in pharmaceutical research and development. Also included in this chapter are regulatory requirements and practical issues for stability studies. Chapter 2 focuses on design and analysis of short-term stability studies such as accelerated testing. Several methods for estimating drug expiration dating periods are reviewed in Chapter 3. Chapter 4 compares several commonly employed study designs including bracketing and matrixing designs. Chapter 5 discusses statistical analysis with fixed batches. Also included in this chapter are several statistical tests for batch-to-batch variability (or batch similarity). Statistical analysis methods with random batches are discussed in Chapter 6. Chapter 7 introduces statistical methods for stability analysis under a linear mixed effects model. Stability analyses with discrete responses, multiple components, and frozen drug products are studied in Chapters 8–10, respectively. Statistical methods of stability testing for dissolution including *United States Pharmacopeia-National Formulary (USP-NF)* dissolution testing and dissolution profile testing are given in Chapter 11. Current issues and recent developments in stability studies such as scale-up and postapproval changes, mean kinetic temperature, and optimality criteria for choosing a stability design are given in the last chapter. Note that some of the chapters such as Chapter 2 (Accelerated Testing), Chapter 4 (Stability Designs), Chapter 5 (Stability Analysis with Fixed Batches), and Chapter 6 (Stability Analysis with Random Batches) are revised based on chapters in *Statistical Design and Analysis in Pharmaceutical Sciences* by Chow, S.C. and Liu, J.P., Marcel Dekker, Inc., New York, New York.

From Taylor & Francis, I would like to thank David Grubbs and Sunil Nair for providing me the opportunity to work on this book and Robert Sims for his editorial support during the book's production. I would like to thank colleagues from the Department of Biostatistics and Bioinformatics and Duke Clinical Research Institute (DCRI) of Duke University School of Medicine for their support during the preparation of this book. I wish to thank Annpey Pong, my fiancée, for her encouragement and support. I also wish to express my gratitude to the following individuals for their encouragement and support: Robert Califf, M.D., Ralph Corey, M.D., and Kerry Lee, Ph.D., of Duke Clinical Research Institute; Steven George, Ph.D., of the Cancer Center at Duke University Medical Center; Yi Tsong, Ph.D., and Greg Campbell, Ph.D., of the U.S. Food and Drug Administration; and many friends from academia, the pharmaceutical industry, and regulatory agencies.

Finally, the views expressed are those of the author and not necessarily those of Duke University School of Medicine. I am solely responsible for the content and errors of this edition. Any comments and suggestions will be very much appreciated.

Shein-Chung Chow, Ph.D.
Department of Biostatistics and Bioinformatics
Duke University School of Medicine, Durham, North Carolina

About the Author

Shein-Chung Chow, Ph.D. is currently a Professor at the Department of Biostatistics and Bioinformatics, Duke University School of Medicine, Durham, North Carolina. Prior to joining Duke University, he was the Director of TCOG (the Taiwan Cooperative Oncology Group) Statistical Center and the Executive Director of the National Clinical Trial Network Coordination Center. Prior to that, Dr. Chow also held various positions in the pharmaceutical industry, such as Vice President, Biostatistics, Data Management, and Medical Writing at Millennium Pharmaceuticals, Inc., Cambridge, Massachusetts; Executive Director, Statistics and Clinical Programming at Covance, Inc.; Director and Department Head at Bristol-Myers Squibb Company, Plainsboro, New Jersey; Senior Statistician and Research Statistician at Parke-Davis Pharmaceutical Division, Warner-Lambert Company, Ann Arbor, Michigan; and Wyeth-Ayerst Laboratories, Rouses Point, New York. Through these positions, Dr. Chow provided technical supervision and guidance to project teams on statistical issues and presentations before partners, regulatory agencies or scientific bodies, defending the appropriateness of statistical methods used in clinical trial design or data analyses or the validity of reported statistical inferences. Dr. Chow identified the best statistical and data management practices, organized and led working parties for development of statistical design, analyses and presentation applications, and participated on Data Safety Monitoring Boards.

Dr. Chow's professional activities include playing key roles in many professional organizations such as officer, Board of Directors member, Advisory Committee member, and Executive Committee member. He has served as program chair, session-chair/moderator, panelist and instructor/faculty at many professional conferences, symposia, workshops, tutorials and short courses. He is the Editor-in-Chief of the *Journal of Biopharmaceutical Statistics*. Dr. Chow is also the Editor-in-Chief of the *Biostatistics Book Series* at Chapman and Hall/CRC Press of Taylor & Francis. He was elected Fellow of the American Statistical Association in 1995 and an elected member of the ISI (International Statistical Institute) in 1999. He was the recipient of the DIA Outstanding Service Award (1996), ICSA Extraordinary Achievement Award (1996), and Chapter Service Recognition Award of the American Statistical Association (1998). Dr. Chow was appointed Scientific Advisor to the Department of Health, Taiwan, Republic of China in 1999–2001 and 2006 to date. Dr. Chow was President of the International Chinese Statistical Association, Chair of the Advisory Committee on Chinese Pharmaceutical Affairs, and a member of the Advisory Committee on Statistics of the DIA.

Dr. Chow is the author or co-author of over 150 methodology papers, and a number of books, which include *Advanced Linear Models, Design and Analysis of*

Bioavailability and Bioequivalence Studies (1st and 2nd Editions), *Statistical Design and Analysis in Pharmaceutical Science, Design and Analysis of Clinical Trials* (1st and 2nd editions), *Design and Analysis of Animal Studies in Pharmaceutical Development, Encyclopedia of Biopharmaceutical Statistics* (1st and 2nd editions), *Sample Size Calculations in Clinical Research,* and *Adaptive Design Methods in Clinical Trials.* Most recently, Dr. Chow is working on two new book projects: *Statistics in Translational Medicine* (with D. Cosmatos) and *Statistics in Medical Devices* (with G. Campbell, L. Yue, and A. Kosinski).

Dr. Chow received a B.S. in mathematics from National Taiwan University, Taiwan, and a Ph.D. in statistics from the University of Wisconsin, Madison, Wisconsin.

Series Introduction

The primary objectives of the *Biostatistics Book Series* are to provide useful reference books for researchers and scientists in academia, industry, and government and to offer textbooks for undergraduate and graduate courses in the area of biostatistics. This book series will provide comprehensive and unified presentations of statistical designs and analyses of important applications in biostatistics, such as those in biopharmaceuticals. A well-balanced summary will be given of current and recently developed statistical methods and interpretations for both statisticians and researchers and scientists with minimal statistical knowledge who are engaged in the field of applied biostatistics. The series is committed to providing easy-to-understand, state-of-the-art references and textbooks. In each volume statistical concepts and methodologies will be illustrated through real-world examples.

For every drug product on the market, the United States Food and Drug Administration (FDA) requires that an expiration dating period (or shelf-life) be indicated on the immediate container label. The expiration date provides the consumer with the confidence that the drug product will retain its identity, strength, quality, and purity throughout the expiration period of the drug product. To provide such assurance, stability studies are often conducted to collect, analyze, and interpret data on the stability of a drug product under study throughout the expiration period. If the drug fails to remain within the approved specifications for the identity, strength, quality, and purity, the drug product is considered unsafe and subject to recall. Statistics plays an important role in the design and analysis of stability studies for a valid and fair assessment of the degradation of a drug product under study and consequently for an accurate and reliable estimation of the expiration dating period of the drug product.

This volume provides a comprehensive and unified presentation of the principles and methodologies in the design and analysis of stability studies. It gives not only a well-balanced summary of current regulatory perspectives and recently developed statistical methods in the area, but also provides a complete, comprehensive, and updated reference and textbook in the areas of stability design and analysis in pharmaceutical research and development. It will be beneficial to biostatisticians and pharmaceutical scientists engaged in the areas of pharmaceutical research and development.

Shein-Chung Chow
Editor-in-Chief
Biostatistics Book Series
Chapman & Hall/CRC

Chapter 1

Introduction

According to a recent report to the nation issued by the Center for Drug Evaluation and Research (CDER) of the United States Food and Drug Administration (FDA), 215 prescription drugs and 71 over-the-counter drugs were recalled in the fiscal year of 2004 (CDER, 2004). The top five reasons for drug recalls in the fiscal year of 2004 were (a) current good manufacturing practice (cGMP) deviations, (b) subpotency, (c) stability data not supporting expiration date, (d) generic drug or new drug application discrepancies, and (e) dissolution failure. As indicated in the CDER Report to the Nation, a drug product must be recalled owing to a problem occurring in the manufacture or distribution of the product that may present a significant risk to public health. These problems usually but not always occur in one or a small number of batches of the drug product. Drug stability plays a very important role in pharmaceutical research and development. For a newly developed drug product, stability analysis not only provides useful information regarding the degradation of the drug product, but also determines an expiration dating period of the drug product. For the purpose of safety and quality assurance, most regulatory agencies such as the FDA require that an expiration dating period be indicated on the immediate container label for every drug product on the market.

In the next section, a brief background regarding stability analysis in pharmaceutical development is provided. Section 1.2 provides an introduction to various aspects of stability testing such as analytical assay development and validation, accelerated testing, preapproval and postapproval stability testing, and regulatory inspection and action. Section 1.3 summarizes current regulatory requirements for conduct of stability studies. Also included in this section is a comparison of FDA stability guidelines and International Conference on Harmonization (ICH) guidelines for stability. Some practical issues that are commonly encountered in design and analysis of stability studies are discussed in Section 1.4. The aim and scope of the book are described in the last section of this chapter.

1.1 Background

In early 1970, although some drug products such as penicillin were known to be unstable, there were no regulations regarding drug stability. Since then it has become a concern that an unstable drug product may not be able to maintain its identity, strength, quality, and purity after being stored over a period of time, especially when the drug product is expected to degrade over time. To ensure the identity, strength, quality, and purity of drug products, in 1975 the United States Pharmacopeia (USP) included a

1

clause regarding the drug expiration dating period. In 1984 the FDA issued the first stability guideline. However, specific requirements on statistical design and analysis of stability studies for human drugs and biologics were not available until the FDA guideline issued in 1987. In 1993 the International Conference on Harmonization issued guideline on stability based on a strong industrial interest in international harmonization of requirements for marketing in the European Union (EU), Japan, and the United States (ICH Q1A, 1993).

The definition of *drug stability* has evolved over time with different meanings by different organizations. For example, the stability in the context of dispensing for pharmacists is defined differently (*USP-NF*, 2000) from that in the context of pharmaceutical dosage forms for manufacturers. Carstensen (1990) gave the following definition: "The term pharmaceutical stability could imply several things. First, it is applied to chemical stability of a drug substance in a dosage form and this is the most common interpretation. However, the performance of a drug when given as a tablet ... depends also on its pharmaceutical properties (dissolution, hardness, etc.) All of these aspects must, therefore, be a part of the stability program." (see also, Connors et al., 1986; Carstensen and Rhodes, 2000; Yoshioka and Stella, 2000). In the FDA guidelines for the stability of human drugs and biologics (FDA, 1987, 1998), stability is defined as "the capacity of a drug product to remain within specifications established to ensure its identity, strength, quality, and purity." In this book, unless otherwise stated, we adopt the definition given in the FDA stability guidelines (FDA, 1987, 1998).

An *active ingredient* is defined as any component that is intended to furnish pharmacological activity or other direct effect in the diagnosis, cure, mitigation, treatment, or prevention of diseases or to affect the structure or any function of the body of man or other animals. An active ingredient includes those components that may undergo chemical change in the manufacture of a drug product and be present in the product in a modified form intended to furnish the specified activity or effect. An *inactive ingredient* is defined as any component other than an active ingredient. A *component* is any ingredient intended for use in the manufacture of a drug product. A *drug product* is considered a finished dosage form, such as tablet, capsule, solution, and so on, that contains an active ingredient generally, but not necessarily, in association with inactive ingredients. A *bulk substance* is defined as the pharmacologically active component of a drug product before formulation. The *strength* of a drug product is defined as either the concentration of the drug substance or the *potency*, which is the therapeutic activity of the drug product as indicated by appropriate laboratory tests or by adequately developed and controlled clinical data. A *batch* of a drug product is defined as a specific quantity of a drug or other material that is intended to have uniform character and quality, within specific limits, and is produced according to a single manufacturing order during the same cycle of manufacture. A *lot* can be interpreted as a batch, or a specific identified portion of a batch, having uniform character and quality within specified limits or a drug product produced in a unit of time or quantity in a manner that ensures its having uniform character and quantity within specified limits. In stability analysis, the 1987 FDA stability guideline indicate that manufacturers should establish stability not only for drug products but also for bulk substances.

As indicated in the FDA stability guidelines (FDA, 1987, 1998), the purpose of a stability study is not only to characterize the degradation of a drug product but also to establish an expiration dating period or shelf-life applicable to all future batches of the drug product. The ICH stability guidelines indicate that the purpose of stability testing is to provide evidence on how the quality of a drug substance or drug product varies with time under the influence of a variety of environmental factors such as temperature, humidity, and light and enables recommended storage conditions, retest periods, and shelf-lives to be established (ICH Q1A, 1993; ICH Q1A [R2], 2003). Some statistical concerns related to the design and analysis of stability studies stated in the guidelines have become popular topics in the pharmaceutical industry. These concerns include, but are not limited to, (a) the use of a significant level of 0.25 for data pooling, (b) the number of batches to be tested, (c) the selection of sampling time points, and (d) search for optimal designs.

1.2 Regulatory Requirements

As indicated earlier, the U.S. FDA issued the first stability guideline in 1984. However, specific requirements on statistical design and analysis of stability studies were not available until 1987 (FDA, 1987). These guidelines were subsequently revised to reflect changes in the regulatory environment for international harmonization (FDA, 1998). In the interest of having international harmonization of stability testing requirements for a registration application within the three areas of the EU, Japan, and the United States, a tripartite guideline for the stability testing of new drug substances and products was developed by the Expert Working Group (EWG) of the ICH and released in 1993. In what follows, regulatory requirements for stability testing as described in the FDA stability guidelines and the ICH guidelines for stability are briefly described.

1.2.1 FDA Stability Guideline

The purpose of the 1987 FDA stability guideline is twofold. One objective is to provide recommendations for the design and analysis of stability studies to establish an appropriate expiration dating period and product requirements. The other objective is to provide recommendations for the submission of stability information and data to the FDA for investigational and new drug applications and product license applications. The 1987 FDA stability guideline indicates that a stability protocol must describe not only how the stability is to be designed and carried out, but also the statistical methods to be used for analysis of the data. As pointed out by the 1987 FDA stability guideline the design of a stability protocol is intended to establish an expiration dating period applicable to all future batches of the drug product manufactured under similar circumstances. Therefore, as indicated in the 1987 FDA stability guideline, the design of a stability study should be able to take into consideration the following variabilities: (a) individual dosage units, (b) containers within a batch, and (c) batches.

The purpose is to ensure that the resulting data for each batch are truly representative of the batch as a whole and to quantify the variability from batch to batch. In addition, the 1987 FDA stability guideline provides a number of requirements for conducting a stability study for determination of an expiration dating period for drug products. Some of these requirements are summarized below.

1.2.1.1   Batch Sampling Consideration

The 1987 FDA guideline indicates that at least three batches and preferably more should be tested to allow for some estimate of batch-to-batch variability and to test the hypothesis that a single expiration dating period for all batches is justifiable. It is a concern that testing a single batch does not permit assessment of batch-to-batch variability and that testing of two batches may not provide a reliable estimate. It should be noted that the specification of at least three batches being tested is a minimum requirement. In general, more precise estimates can be obtained from more batches.

1.2.1.2   Container (Closure) and Drug Product Sampling

To ensure that the samples chosen for stability study can represent the batch as a whole, the 1987 FDA stability guideline suggests that selection of such containers as bottles, packages, and vials from the batches be included in the stability study. Therefore, it is recommended that at least as many containers be sampled as the number of sampling times in the stability study. In any case, sampling of at least two containers for each sampling time is encouraged.

1.2.1.3   Sampling Time Considerations

The 1987 FDA stability guideline suggests that stability testing be done at 3-month intervals during the first year, 6-month intervals during the second year, and annually thereafter. In other words, it is suggested that stability testing be performed at 0, 3, 6, 9, 12, 18, 24, 36, and 48 months for a 4-year duration of a stability study. However, if the drug product is expected to degrade rapidly, more frequent sampling is necessary.

1.2.2   ICH Guidelines for Stability

The ICH Q1A guideline for stability is usually referred to as the parent guideline for stability because (a) it has been revised a couple of times and (b) it is the foundation of subsequent guidelines for stability developed by the ICH EWG since it was issued in 1993 (ICH Q1A, 1993; ICH Q1A [R2], 2003). The ICH Q1A (R2) guideline for stability, given in Appendix A, provides a general indication of the requirements for stability testing but leaves sufficient flexibility to encompass the variety of practical situations required for specific scientific situations and characteristics for the materials being evaluated. The ICH guidelines for stability establish the principle that information on stability generated in any of the three areas of the EU, Japan, and the United States would be mutually acceptable in both of the other two areas provided that it meets the appropriate requirements of the guideline and the

TABLE 1.1: ICH Guidelines Related to Stability Testing

ICH Guideline	Date Issued	Description
Q1A	1993	Stability testing of new drug substances and products
Q1A (R2)	2003	Stability testing of new drug substances and products
Q1B	1996	Photostability testing of new drug substance and products
Q1C	1997	Stability testing of new dosage forms
Q1D	2003	Bracketing and matrixing designs for stability testing of new drug substances and products
Q1E	2004	Evaluation of stability data
Q1F	2004	Stability data package for registration applications in climatic zones III and IV
Q3A	2003	Impurities in new drug substances
Q3B	1996	Impurities in new drug products
Q3B (R)	2003	Impurities in new drug products
Q5C	1995	Stability testing of biotechnological/biological products
Q6A	1999	Specifications: test procedures and acceptance criteria for new drug substances and new drug products: chemical substances
Q6B	1999	Specifications: test procedures and acceptance criteria for new drug substances and new drug products: biotechnological/biological products

labeling is in accordance with national and regional requirements. Table 1.1 lists ICH guidelines related to stability testing issued in the past decade. It should be noted that the choice of test conditions defined in the ICH guidelines is based on an analysis of the effects of climatic conditions in the three areas of the EU, Japan, and the United States. Therefore, the main kinetic temperature in any region of the world can be derived from climatic data (ICH Q1F, 2004).

Basically, the ICH Q1A (R2) guideline for stability is similar to the 1987 FDA stability guideline and the current FDA draft guideline for stability (FDA, 1998). For example, the ICH guidelines suggest that testing under the defined long-term conditions normally be done every 3 months over the first year, every 6 months over the second year, and annually thereafter. It requires that the container to be used in the long-term real-time stability evaluation be the same as or simulate the actual packaging used for storage and distribution. For the selection of batches, it requires that stability information from accelerated and long-term testing be provided on at least three batches and the long-term testing should cover a minimum of 12 months' duration on at least three batches at the time of submission. For the drug product, it is required that the three batches be of the same formulation and dosage form in the containers and closure proposed for marketing. Two of the three batches should be at least pilot scale. The third batch may be smaller (e.g., 25,000 to 50,000 tablets or capsules for solid oral dosage forms). However, the ICH Q1A (R2) guideline for stability also requires that the first three production batches of the drug substances or drug product manufactured postapproval, if not submitted in the original registration application, be placed on the long-term stability studies using the same stability protocol as in the approved drug application. For storage conditions, the ICH Q1A (R2) guideline requires that accelerated testing be carried out at a temperature at least 15° C above the designated long-term storage temperature in conjunction with the appropriate relative humidity conditions for that temperature. The designated long-term testing conditions will be reflected in the labeling and retest date. The retest date is the date when samples of the drug substance should be reexamined to ensure that material is still suitable for use. The ICH Q1A (R2) guideline for stability also indicates that where significant change occurs during six months of storage under conditions of accelerated testing at $40 \pm 2°$ C/$75 \pm 5\%$ relative humidity, additional testing at an intermediate condition (such as $30 \pm 2°$ C/$60 \pm 5\%$ relative humidity) should be conducted for drug substances to be used in the manufacture of dosage forms.

For the evaluation of stability data, the ICH Q1A (R2) guideline for stability indicates that statistical methods should be employed to test goodness of fit of the data on all batches and combined batches (where appropriate) to the assumed degradation line or curve. If it is inappropriate to combine data from several batches, the overall retest period may depend on the minimum time a batch may be expected to remain within acceptable and justified limits. A retest period is defined as the period of time during which the drug substance or drug product can be considered to remain within specifications and therefore acceptable for use in the manufacture of a given drug product, provided that it has been stored under the defined conditions.

1.2.3   Remarks

As indicated earlier, the EU, Japan, and the United States have different but similar stability requirements (see, e.g., Mazzo, 1998). Based on different requirements, pharmaceutical companies may have to conduct stability tests repeatedly for different markets. The ICH guidelines for stability are an attempt to harmonize these requirements so that information generated in any of the three areas of the EU, Japan, and the United States would be acceptable to the other two areas. In what follows, we briefly summarize the differences in requirements regarding stability aspects among the EU, Japan, and the United States, which were discussed in a workshop on stability testing held in Brussels, Belgium, November 5–7, 1991.

1.2.3.1   Minimum Duration of Stability Testing

In the EU it is required to file an application based on the results of stability tests performed after at least 6 months of storage. In the United States, however, the FDA requires that a minimum of 12 months of stability data be provided. The Ministry of Health, Labor, and Welfare (MHLW) of Japan requires 12 months. Statistically, it is undesirable to extrapolate a drug shelf-life too far beyond the sampling intervals under study. Therefore, as a rule of thumb, it is suggested that stability extrapolation not extend beyond 6 months. Stability data should be obtained to cover up to 6 months prior to the desired expiration dating period. In other words, if a desired shelf-life is 18 months, stability testing should cover at least a one year period. However, as indicated in the 1987 FDA stability guideline, although a tentative shelf-life may be granted based on a short-term stability study, the pharmaceutical companies are expected to have commitment to obtain complete data that cover the full expiration dating period.

1.2.3.2 Minimum Number of Batches Required for Stability Testing

Under the current stability guideline, the FDA requires at least three batches, and preferably more should be tested to allow a reasonable estimation of the batch-to-batch variability and to test the hypothesis that a single expiration dating period for all future batches is justifiable. However, the EU requires only that stability data on two batches of the active drug substances be submitted for the evaluation of a drug expiration period. For the number of batches required in stability testing, the MHLW's requirement is consistent with that of the FDA. The 1987 FDA stability guideline provides some justification for the use of a minimum number of three batches for stability testing. The 1987 FDA stability guideline indicates that a single batch does not permit assessment of batch-to-batch variability, and testing two batches provides an unreliable estimate. To provide a more precise estimate of drug shelf-life, it is preferred to have stability testing on more batches. However, practical considerations such as cost, resources, and capacity may prevent the collection of data from more batches. As a result, the specification that at least three batches be tested has become a minimum requirement representing a compromise between statistical and regulatory considerations and actual practice.

1.2.3.3 Definition of Room Temperature

According to the *United States Phamacopeia-National Forumlary (USP-NF)*, the definition of room temperature is between 15 and 30° C in the United States. However, in the EU, the room temperature is defined as being 15 to 25° C, while in Japan, it is defined being 1 to 30° C. If the drug product is sensitive to the temperature range 0 to 30° C, degradation of the drug product may vary from one temperature to another within the range. Therefore, it is important to investigate the stability of the drug product at different ranges of temperatures if the drug product is to be marketed in different regions. In this case harmonization of the definition of room temperature may not be useful. However, if the drug product is not sensitive to this range of temperatures, harmonization of the definition of room temperatures may be needed so that similar stability testing need not be conducted repeatedly to fulfill different requirements.

1.2.3.4 Extension of Shelf-Life

In practice, when a new drug application (NDA) submission is filed, there are usually limited data available on the stability of the drug product. In the United States it is a common practice for the FDA to tentatively grant marketing authorization of the drug product based on limited stability data. However, it is required by the FDA that a pharmaceutical company submit the results of stability studies obtained up to the expiration date granted. However, the EU does not accept an extension of shelf-life beyond real-time data submitted.

1.2.3.5   Least Stable Batch

When there is a batch-to-batch variation, or the batches are not similar (equivalent), the European Health Authorities expect the pharmaceutical industry to consider the least stable batch for the determination of shelf-life and to refrain from averaging the values statistically. When there is batch-to-batch variation, the 1987 FDA stability guideline suggests considering the minimum of individual shelf-lives. It should be noted that the use of the least stable batch for determination of shelf-life is conservative.

1.2.3.6   Least Protective Packaging

The MHLW of Japan prefers to determine the drug shelf-life based on the results of stability testing using the least stable packaging material instead of testing the product in all packages. The 1987 FDA stability guideline, however, encourages sampling of at least two containers of each packaging material for each sampling time in all cases. The idea of testing the least stable packaging material is well taken. However, how to identify the least stable packaging material is an interesting statistical question. To identify the least stable packaging material, a pilot study may be required. As a result, a fractional factorial design may be applied. However, it should be noted that the selected pilot design should be able to avoid any possible confounding and interaction effects.

1.2.3.7   Replicates

In Japan each test must be repeated three times without provision for scientific and statistical justification. The 1987 FDA stability guideline, however, encourages testing an increasing number of replicates at later sampling times, particularly the latest sampling time. The reason for doing this is that it will increase the precision of the estimation of the expiration dating period because the degradation is most likely to occur at later sampling time points than at earlier time points for long-term stability studies. Although the accuracy and precision of the estimated shelf-life based on replicates of test results will be improved, it is not clear how much improvement the test replicates will achieve. Replications at each sampling time point not only increase the precision of the estimated shelf-life, but also provide data on the lack-of-fit test for fitting individual simple linear regressions to each batch. In practice, it is of interest to investigate the impact of replicates at each sampling time point on the accuracy and precision of shelf-life estimation.

1.3   Stability Testing

When a new pharmaceutical compound is discovered, assay and test procedures are necessarily developed for determining the active ingredient of the compound in compliance with *USP-NF* standards for the identity, strength, quality, and purity of the compound (*USP-NF*, 2000). An analytical (or assay) method is usually developed based on instruments such as gas chromatography (GC), high-performance liquid chromatography (HPLC) and liquid chromatography/mass spectrometry/mass spectrometry (LC/MS/MS). cGMP indicates that an instrument must be suitable for its intended purpose and be capable of producing valid results. The instrument is to be calibrated, inspected, and checked routinely according to written procedures. For the development of an assay method, a common approach is to have a number of known standard concentration preparations put through a given instrument (e.g., GC, HPLC, or LC/MS/MS) to obtain the corresponding responses (e.g., absorbance or peak response). On the basis of these standards and their corresponding responses, an estimated *calibration curve* (or *standard curve*) can be obtained by fitting an appropriate statistical model between these standards and their corresponding responses. For a given unknown sample, the concentration can be determined based on the standard curve by replacing the dependent variable with its response. cGMP indicates that, where practical, the calibration standards used for assay development must be in compliance with *USP-NF* standards. If *USP-NF* standards are not practical for the parameter being measured, an independent reproducible standard must be used. If no applicable standards exist, an in-house standard must be developed and used.

1.3.1 Stability-Indicating Assay

In the pharmaceutical industry, stability testing is referred to as stability-indicating assay, which is an analytical method that is employed for the analysis of stability samples collected from stability studies. The ICH Q1A (R2) guideline for stability requires that a stability-indicating assay method be established for stability testing. The ICH Q1A (R2) guideline for stability explicitly requires conduct of forced decomposition studies under a variety of conditions such as pH, light, oxidation, dry heat, and so on, and separation of drug from degradation products. The stability-indicating method is expected to allow analysis of individual degradation products (ICH Q1A, 1993; ICH Q1A (R2), 2003). In the past two decades, ICH has issued a number of guidelines regarding requirements for stability testing such as the ICH Q1A guideline for stability-indicating testing and stress testing methods (ICH Q1A, 1993), the ICH Q1B guideline on photostability testing (ICH Q1B, 1996), the ICH Q3B guideline on impurities and for validation of analytical procedures (ICH Q3B, 1996), the ICH Q6A guideline on specifications (ICH Q6A, 1999), and the ICH Q5C guideline on stability testing of biotechnological and biological products (ICH Q5C, 1995). However, none of the ICH guidelines provides an exact definition of a stability-indicating method.

Elaborate definitions of *stability-indicating method* are given in the 1987 FDA stability guideline and the 1998 FDA draft stability guideline. In this book we will adopt the definition described in the 1998 FDA draft stability guideline. Where *stability-indicating method* is defined as a validated quantitative analytical method that can detect the changes with time in the chemical, physical, or microbiological properties of the drug substance and drug product and that are specific so that the contents of active ingredients, degradation products, and other components of interest can be accurately measured without interference (FDA, 1998). Unlike the definition given in the 1987 FDA stability guideline, the above definition emphasizes the requirement of validation and the requirement of analysis of degradation products and other components, apart from the active ingredients.

Bakshi and Singh (2002) classified the stability-indicating assay method into two types of assay methods, namely the *specific stability-indicating* assay method and the *selective stability-indicating* assay method. The specific stability-indicating assay method is defined as a method that is able to measure unequivocally the drug in the presence of all degradation products, excipients and additives expected to be present in the formulation. The selective stability-indicating assay method is defined as a method that is able to measure unequivocally the drug and all degradation products in the presence of excipients and additives expected to be present in the formulation. By this definition, the selective stability-indicating assay method is a procedure that is selective to the drug as well as its degradation products (separates all of them qualitatively) and is also specific to all of the components (measures them quantitatively).

Bakshi and Singh (2002) proposed a systematic approach for the development of a stability-indicating method by following the following steps:

- Step 1: Critical study of the drug structure to assess the likely decomposition route(s)

- Step 2: Collection of information on physicochemical properties

- Step 3: Stress (forced decomposition) studies

- Step 4: Preliminary separation studies on stress samples

- Step 5: Final method development and optimization

- Step 6: Identification and characterization of degradation products and preparation of standards

- Step 7: Validation of the stability-indicating assay method

In their review article Bakshi and Singh (2002) pointed out that while the current requirement is to subject the drug substance to a variety of stress conditions and then separate the drug from all degradation products, many studies have only shown the separation of the drug from known synthetic impurities and potential degradation products without subjecting it to any type of stress. There are also reports in which the drug has been decomposed by exposing it to one, two, three, four, or more conditions among acidic, neutral, or alkaline hydrolysis, photolysis, oxidation, and thermal stress. Very few studies are truly stability-indicating, where the drug has been exposed to all types of stress conditions and attempts have been made to separate the drug from degradation products.

1.3.2   Analytical Method Validation

Current Good Manufacturing Practices (cGMP) indicates that the suitability of the instrument under actual conditions of use must be verified. If computers are used as part of an automated system, the computer software programs must be validated by adequate, documented testing. cGMP (see, e.g., Part 21 Codes of Federal Regulations [CFR], Section 211.194 [a]) requires that the assay method must meet certain standards of accuracy and reliability. Since assays and specifications of the *USP-NF*

TABLE 1.2: Analytical
Validation Parameters

Accuracy
Precision
Limit of detection
Limit of quantitation
Selectivity
Range
Linearity
Ruggedness

---

constitute the legal standards recognized by the official compendia of the Federal Food, Drug, and Cosmetic Act, the use of assay methods described in the *USP-NF* is not required to validate accuracy and reliability. Any new or revised assay methods proposed for submission to the compendia must be validated and documented with sufficient laboratory data and information according to the requirements stated in the *USP-NF*. The new or revised assay methods are then reviewed for their relative merits and disadvantages by the members of the USP Committee of Revision. The *USP-NF* indicates that the validation of an analytical method is the process by which it is established, by laboratory studies, that the performance characteristics of the method meet the requirements for the intended analytical applications. The performance characteristics of an analytical method or a testing procedure can be assessed through a set of analytical validation parameters. A set of analytical validation parameters suggested by the *USP-NF* is listed in Table 1.2 (see also, Chow, 1997). Note that validation of analytical methods has been extensively covered in the ICH Q2A and Q2B guidelines (ICH Q2A, 1994; ICH Q2B, 1996) and in an FDA guidance (FDA, 2000). Detailed information regarding statistical methods for assessment of these analytical validation parameters can be found in Chow and Liu (1995).

1.3.2.1 An Example

Analytical method development and validation is extremely important in stability testing. An invalidated analytical method could have an impact on stability testing. It may not be able to characterize the degradation of the drug product over time accurately with certain reliability. Moreover, it may not provide an accurate estimate of the drug expiration dating period. One typical example is the stability problem found in levothyroxine sodium products as described in *Federal Register* (Vol, 62, No. 157, 1997). Levothyroxine sodium is the sodium salt of the levo isomer of the thyroid hormone thyroxine (T_4). Thyroid hormones affect protein, lipid, and carbohydrate metabolism and growth and development. They stimulate the oxygen consumption of most cells of the body, resulting in increased energy expenditure and heat production and possess a cardiostimulatory effect that may be the result of a direct action on the heart. Levothyroxine sodium was first introduced into the market before 1962 without an approved NDA. Orally administered levothyroxine sodium is used as replacement therapy in conditions characterized by diminished or absent thyroid function such as cretinism, myxefema, nontoxic goiter, or hypothyroidism. The diminished or absent thyroid function may result from functional deficiency, primary atrophy, partial or complete absence of the thyroid gland, or the effect of surgery, radiation, or antithyroid agents. Until the HPLC method was used, the assay method based on iodine content was used, which is not a stability-indicating assay. Using the HPLC method, there have been numerous reports indicating problems with the stability of orally administered levothyroxine sodium products. As a result, the FDA in conjunction with the United States Pharmacopeial Convention took the initiative in organizing a workshop in 1982 to set the standard for the use of a stability-indicating HPLC assay for the quality control of thyroid hormone drug products (Garnick et al., 1982). In addition to raising concerns about the consistent potency of orally administered levothyroxine sodium products, many reports suggest that the customary 2-year shelf-life may not be appropriate for these products because they are prone to experience accelerated degradation in response to a variety of factors such as light, temperature, air, and humidity (Won, 1992). Won (1992) indicated that stability data of levothyroxine sodium exhibit a biphasic first-order degradation profile with an initial fast degradation rate followed by a slower rate. To compensate for the initial accelerated degradation, some pharmaceutical companies use an overage of active ingredient in their formulation, which can lead to occasional instances of superpotency.

1.3.3 Impurities

The *USP-NF* defines an *impurity* as any component of a drug substance (excluding water) that is not the chemical entity defined as the drug substance. It has been demonstrated that impurities in a finished drug product can cause degradation and lead to stability problems. Further, some adverse reactions in patients have been traced to impurities in the active ingredient. Therefore, the presence or absence of impurities at the time of clinical trial and stability testing is a very important element of drug testing and development, and the appearance of an impurity in scaled-up product that was not present during test stages presents serious questions about the stability of the product and its impact on safety and efficacy.

The FDA expects the manufacturer to establish an appropriate impurity profile for each bulk pharmaceutical chemical (BPC) based on adequate consideration of the process and test results (FDA, 1994). Because different manufacturers synthesize drug substances by different processes and, therefore, will probably have different impurities, the *USP-NF* has developed the *ordinary impurities test* in an effort to establish some specification (*USP-NF*, 2000). Also, in order to protect proprietary information, tests for specific impurities and even solvents are typically not listed in the compendia. The *USP-NF* also notes that the impurity profile of a drug substance is a description of the impurities present in a typical lot of drug substance produced by a given manufacturing process. Such impurities not only should be detected and quantitated, but should also be identified and characterized when this is possible with reasonable effort. Individual limits should be established for all major impurities. The *USP-NF* provides extensive coverage of impurities in the following three sections:

- **Impurities in Official Articles (*USP-NF* Section 1086):** This section defines five types of impurities, both known and unknown including foreign substances, toxic impurities, concomitant components (such as isomers or racemates), signal impurities (which are process related), and ordinary impurities. The *USP-NF* notes that when a specific test and limit is specified for a known impurity, generally a reference standard for that impurity is required. Two of the impurities are singled out for in-depth coverage, ordinary impurities and organic or volatile impurities.

- **Ordinary Impurities (*USP-NF* Section 466):** These are generally specified for each BPC in the individual monograph. The method of detection involves comparison with a *USP-NF* reference standard on a thin-layer chromatographic (TLC) plate, with a review for spots other than the principal spot. The ordinary impurity total should not exceed 2% as a general limit. Be sure to review the extensive *USP-NF* coverage of eight factors that should be considered in setting limits for impurity levels. Related substances are defined as those structurally related to a drug substance such as a degradation product or impurities arising from a manufacturing process or during storage of the BPC. Process contaminants are substances including reagents, inorganics (e.g., heavy metals, chloride, or sulfate), raw materials, and solvents. The *USP-NF* notes that these substances may be introduced during manufacturing or handling procedures. The third and most recent *USP-NF* section regarding impurities is one that appears in the *USP-NF* XXII third supplement.

- **Organic or Volative Impurities (*USP-NF* Section 467):** Several GC methods are given for the detection of specific toxic solvents, and the determination involves use of a standard solution of solvents. There are limits for specified organic volatile impurities present in the BPC unless otherwise noted in the individual monograph. As the USP notes, the setting of limits on impurities in a BPC for use in an approved new drug may be much lower than those levels encountered when the substance was initially synthesized. Further, additional purity data may be obtained by other methods such as gradient HPLC. Be sure to ask for complete impurity profiles. In preparation for a BPC inspection, these sections of the *USP-NF* should be given a detailed review.

1.3.4 Preapproval and Postapproval Testing

Basically, there are two types of stability studies: short-term and long-term studies. A typical short-term stability study is an accelerated stability-testing study under stressed storage conditions. The purpose of an accelerated stability-testing study is not only to determine the rate of chemical and physical reactions, but also to predict a tentative expiration dating period under ambient marketing storage conditions. Information regarding the rate of degradation and tentative expiration dating period are vital and useful for designing long-term stability studies. Long-term studies, which include both preapproval and postapproval stability studies are usually conducted under ambient conditions. A preapproval stability study is also known as an NDA stability study, while a postapproval stability study is usually referred to as a marketing stability study. The purpose of an NDA stability study is to determine (estimate) a drug expiration dating period applicable to all future batches. The objective of a marketing stability study is to make sure the drug product currently on the market can meet *USP-NF* specifications up to the end of the expiration dating period (Chow and Shao, 1990a). The major difference between an NDA stability study and a marketing stability study is that a marketing study is usually conducted on a large number of batches with fewer sampling time points.

1.3.5 Regulatory Inspection and Action

As indicated in the FDA *Guide to Inspection of Bulk Pharmaceutical Chemicals* (FDA, 1994), regulatory inspection is intended to aid agency personnel in determining whether the methods used in, and the facilities and manufacturing controls used for, the production of BPCs are adequate to ensure that they have the quality and purity that they purport or are represented to possess.

1.3.5.1 Inspectional Approach

The FDA notes that the inspectional approach for coverage of a BPC operation is the same whether or not that the BPC is referenced as an active ingredient in a pending application. The purpose, operational limitations, and validation of the critical processing steps of a production process should be examined to determine that the firm adequately controls such steps to ensure that the process works consistently. Overall, the inspection must determine the manufacturer's capability to deliver a product that consistently meets the specifications of the bulk drug substance that the finished dosage form manufacturer listed in the application or the product needed for research purposes. BPC manufacturing plants often produce laboratory scale or *pilot* batches. Scale-up to commercial full-scale (routine) production may involve several stages, and data should be reviewed to demonstrate the adequacy of the scale-up process. Such scale-ups to commercial size production may produce significant problems in consistency among batches. Pilot batches serve as the basis for establishing in-process and finished product purity specifications. Typically, manufacturers will generate reports that discuss the development and limitation of the manufacturing process. Summaries of such reports should be reviewed to determine if the plant is capable of adequately producing the bulk substance. The reports serve as the basis for the validation of the manufacturing and control process and the basic documentation that the process works consistently.

A good starting point for the BPC inspection is a review of product failures evidenced by the rejection of a batch that did not meet specifications, return of a product by a customer, or recall of the product. The cause of the failure should have been determined by the manufacturer, a report of the investigation prepared, and subsequent corrective action initiated and documented. Such records and documents should be reviewed to ensure that such product failures are not the result of a process that has been poorly developed or one that does not perform consistently. In the analytical laboratory, specifications for the presence of unreacted intermediates and solvent residues in the finished BPC should be reviewed. These ranges should be at or near irreducible levels. An inspectional team consisting of investigators and engineers, laboratory analysts, or computer experts should participate in the inspection, as appropriate, when resources permit.

1.3.5.2 Stability Testing

Most BPC manufacturers conduct stability-testing programs for their products; however, such programs may be less comprehensive than the programs now required for finished pharmaceuticals. Undetected changes in raw materials specifications, or subtle changes in manufacturing procedures, may affect the stability of BPCs. This, together with the generally widespread existence of stability-testing programs, make it reasonable to require such programs for BPCs. The FDA notes that a stability-testing program for BPCs should contain the following features: (a) the program should be formalized in writing, (b) stability samples should be stored in containers that approximate the market container, (c) the program should include samples from the first three commercial-size batches, (d) a minimum of one batch a year, if there is one, should be entered in the program, (e) the samples should be stored under conditions specified on the label for the marketed product, (f) it is recommended that additional samples be stored under stressful conditions (e.g., elevated temperature, light, humidity, or freezing) if such conditions can be reasonably anticipated, and (g) stability-indicating methods should be used. In addition, the FDA notes that conducting a stability-testing program does not usually lead to a requirement to employ expiration dates. If testing does not indicate a reasonable shelf-life, for example, two years or more, under anticipated storage conditions, then the BPC can be labeled with an expiration date or should be reevaluated at appropriate intervals. If the need for special storage conditions exists, for example, protection from light, such restrictions should be placed on the labeling.

It should be noted that reserve samples of the released BPCs should be retained for one year after distribution is complete or for one year after the expiration or reevaluation date. In addition, documentation of the BPC manufacturing process should include a written description of the process and production records similar to those required for dosage form production. However, it is likely that computer systems will be associated with BPC production. Computer systems are increasingly used to initiate, monitor, adjust, and otherwise control both fermentations and syntheses. These operations may be accompanied by recording charts that show key parameters (e.g., temperature) at suitable intervals or even continuously throughout the process. In other cases key measurements (e.g., pH) may be displayed on a television screen for that moment in time but are not available in hard copy. In both cases conventional hard-copy batch-production records may be missing. In other words, records showing addition of ingredients, actual performance of operations by identifiable individuals, and other information usually seen in conventional records may be missing.

1.3.5.3 Drug Recalls Owing to Stability Problems

As indicated earlier, 215 prescription drugs and 71 over-the-counter drugs were recalled in the fiscal year of 2004 (CDER, 2004). The top 10 reasons for drug recalls in the fiscal year of 2004 are listed in Table 1.3. As can be seen, "stability data do not support expiration date" is listed as the number 3 reason for drug recalls. Most recently, three drug products were recalled owing to problems in stability. The reasons include: (a) the incorrect stability test method was used to ensure that the product meets its specifications throughout its shelf-life (Carbidopa, Mylan), (b) the incorrect

TABLE 1.3:   Top 10 Reasons for Drug Recalls
in Fiscal Year 2004

---

cGMP deviations
Subpotency
Stability data does not support expiration date
Generic drug or new drug application discrepancies
Dissolution failure
Label mix-ups
Content uniformity failure
Presence of foreign substance
pH failures
Microbial contamination of nonsterile products

---

expiration date was on the label (Citalopram Hydrobromide, Ivax), and (c) stability data do not support expiration date (FiberCon, Wyeth).

1.3.6   Other Related Tests

In addition to potency testing for stability, related tests including weight variation testing, content uniformity testing, dissolution testing, and disintegration testing are usually performed at various stages of the manufacturing process of a drug product to ensure that the product meets the *USP-NF* standards for identity, strength, quality, and purity of the drug product. These tests are usually referred to as USP tests. As indicated in the FDA Report to the Nation: 2004, stability data not supporting expiration date, dissolution failure, and content uniformity are the top 3, 5, and 7 reasons, respectively, for drug recalls in fiscal year 2004. The *USP-NF* requires that a specific sampling plan for the individual USP test be employed and that specific acceptance criteria be met to pass the test. Although these USP tests are mainly performed for the purpose of quality assurance and quality control (QA/QC), results of these USP tests may or may not directly or indirectly have an impact on the stability of the drug product. Note that the uniformity of dosage units is usually demonstrated either by weight variation testing or content uniformity testing. Requirements for testing the uniformity of dosage units are described in the *USP-NF*. The requirements apply both to dosage forms containing a single active ingredient and to dosage forms containing two or more active ingredients. Specific sampling plans, acceptance criteria, and procedures of these USP tests are briefly outlined below (see also Chow and Liu, 1995).

1.3.6.1   Weight Variation Testing

For the determination of dosage uniformity by weight variation, accurately weigh ten dosage units individually and calculate the average weight. From the result of the assay, as directed in the individual monograph of the *USP-NF*, calculate the content of the active ingredient in each of the 10 units, assuming homogenous distribution of the active ingredient. The requirements for dosage uniformity are met if the amount of the active ingredient in each of the 10 dosage units lies within the range of 85 to 115% of label claim and the relative standard deviation (or coefficient of variation) is less than 6.0%. If one unit is outside the range of 85 to 115% of label claim and no unit is outside the range of 75 to 125% of label claim, or if the relative standard deviation is greater than 6.0%, or if both conditions prevail, test 20 additional units. The requirements are met if not more than one unit of the 30 is outside the range of 85 to 115% of label claim and no unit is outside the range of 75 to 125% of label claim and the relative standard deviation of the 30 dosage units does not exceed 7.8%.

1.3.6.2 Content Uniformity Testing

For the determination of dosage uniformity by assay of individual units, the *USP-NF* requires that the following be done. First, assay 10 units individually, as described in the assay in the individual monograph, unless specified otherwise in the test for content uniformity. Where a special procedure is specified in the test for content uniformity in the individual monograph, the results should be adjusted. Note that the requirements as described for weight variation apply only if the average of the limits specified in the potency definition in the individual monograph is 100% or less. In the case where the average value of the dosage units tested is greater than or equal to the average of the limits specified in the potency definition in the individual monograph, replace "label claim" with "label claim multiplied by the average of the limits specified in the potency definition in the monograph divided by 100."

1.3.6.3 Dissolution Testing

The *USP-NF* contains an explanation of the test for acceptability of dissolution rates. The requirements are met if the quantities of active ingredient dissolved from the units conform to the *USP-NF* acceptance criteria. Let Q be the amount of dissolved active ingredient specified in the individual monograph, which is usually expressed as a percentage of label claim. The *USP-NF* dissolution acceptance criteria is composed of a three-stage sampling plan. For the first stage (S_1), six dosage units are to be tested. The requirement for the first stage is met if each unit is not less than $Q + 5\%$. If the product fails to pass S_1, an additional six units will be tested at the second stage (S_2). The product is considered to have passed if the average of the 12 units from S_1 and S_2 is equal to or greater than Q and if no unit is less than $Q - 15\%$. If the product fails to pass both S_1 and S_2, an additional 12 units will be tested at a third stage (S_3). If the average of all 24 units from S_1, S_2 and S_3 is equal to or greater than Q, no more than two units are less than $Q - 15\%$, and no unit is less than $Q - 25\%$, the product has passed the *USP-NF* dissolution test. Under this sampling plan and acceptance, Chow, Shao, and Wang (2002b) derived the probability lower bound for the *USP-NF* dissolution test.

1.3.6.4 Disintegration Testing

Similar to dissolution testing, disintegration testing has a two-stage sample plan. In the first stage (S_1) of disintegration testing, six dosage units are tested. The requirements are met if all six units disintegrate completely. Complete disintegration is defined as that state in which any residual of the unit, except fragments of an insoluble coating or capsule shell, that may remain on the test apparatus screen is a short mass with no palpably firm core. If one or two units fail to disintegrate completely, repeat the test on 12 additional units at the second stage (S_2). The requirements are met if no fewer than 16 units of the total of 18 units tested disintegrate completely.

1.4 Practical Issues

1.4.1 Accelerated Testing

At a very early stage of drug development, the primary stability data usually are not available to characterize the degradation of the drug product. To determine the rates of chemical and physical reactions and their relationships with storage conditions such as temperature, moisture, light (Tonnesen, 2004), and others, *accelerated stability testing* is usually conducted. An accelerated stability test is a short-term stability study conducted under exaggerated (or stressed) conditions to increase the rate of chemical or physical degradation of a drug substance or drug product. Thus, accelerated testing is also known as *stressed testing*. As indicated in the 1987 FDA stability guideline, exaggerated storage conditions may include temperature (e.g., 5, 50 or 75° C), humidity of 75% or greater, exposure to various wavelengths of electromagnetic radiation, and storage in an open container. Exaggerated storage conditions are used to accelerate the reaction rate so that significant degradation of the drug product can be observed in a relatively short period of time (e.g., a few months). Based on the degradation data observed, the kinetic parameters of the reaction rate can be estimated. A predicted shelf-life under marketplace storage conditions can then be obtained by extrapolation. Note that the extrapolation from stressed testing conditions to ambient conditions is usually done based on established relationships between kinetic parameters and storage conditions. In the pharmaceutical industry the Arrhenius equation is often employed to relate the degradation reaction rate and the corresponding temperature. Other models such as the Eyring equation (Kirkwood, 1977) can also be used.

The primary objective of accelerated stability testing is to provide an accurate and reliable estimate of the tentative expiration dating period. To achieve this objective, it is important to select an efficient design that will provide the maximum information to an accelerated stability test. Since the amount of information provided by a design is a function of the inverse of the variance, in practice it is recommended that the design generate an estimate of time on the logarithmic scale with the smallest possible variance. More details regarding chemical kinetic reaction, statistical analysis, and verification of model assumptions for an accelerated testing are given in the next chapter.

1.4.2 Batch Similarity

As indicated in the FDA stability guideline, a minimum requirement for a stability study is to test at least three batches. If batch-to-batch variability is small, it would be advantageous to combine the data into one overall estimate with high precision and a large degree of freedom for mean squared error. However, combining the data should be supported by preliminary testing of batch similarity. The similarity of the degradation curves for each batch tested should be assessed by applying statistical tests of the equality of slopes and of zero-time intercepts at a significant level of $\alpha = 0.25\%$ (Bancroft, 1964). Chow and Shao (1989) proposed several tests for batch-to-batch variability under a normality assumption. If tests for equality of slopes and for equality of intercepts do not result in rejection at the 25% level of significance, the data from the batches would be pooled. However, if tests result in p-values less than 0.25, a judgment would be made by the FDA reviewers as to whether pooling would be permitted.

It should be noted that there are some criticisms regarding the use of a significance level of 0.25. Among these criticisms, the following are probably the most common:

- Acceptance of rejection of the null hypothesis if there is no difference in slopes among batches does not guarantee that the batches have similar degradation rates. This is because that problem of similarity is incorrectly formulated by the wrong hypothesis of difference.

- It is not a common practice to increase test power by increasing the level of significance.

In addition, Lin and Tsong (1991) pointed out that the level of significance required for a given minimum relative efficiency of the estimate based on results of the pooling test depends on sample size, time points measured, mean slope of all batches, and the tightness of the stability data. Hence, for a test of a fixed level of significance, the pooling test will have low power for a given batch-to-batch difference when either absolute magnitude of mean slope or within-batch variability is large.

1.4.3 Matrixing and Bracketing Designs

Suppose we are interested in conducting a stability study under ambient conditions (e.g., 60% relative humidity and 25° C room temperature). A complete (full) factorial stability study consisting of three batches for each combination of three strengths (e.g., 15 mg, 30 mg, and 60 mg) and three package types (e.g., bottle, blister, and tube) is considered. As a result, a full factorial design consists of $3^3 = 27$ combinations. If each combination is to be tested at the time intervals of 0, 3, 6, 9, 12, 18, 24, 36, and 48 months for a 4-year stability study, there are a total of 270 ($3 \times 3 \times 3 \times 10$) assays. In practice, if every batch by strength-by-package combination is tested (i.e., a complete factorial design is used), a substantial expense is involved. Besides, it is in the best interest of the pharmaceutical companies that a longer shelf-life can be claimed by testing fewer batches for strength-by-package combinations within a short period of time. Therefore, for considerations of time and cost, a fractional factorial design is often used to reduce the total number of tests (or assays).

Although a fractional factorial design is preferred in the interest of reducing the number of tests (i.e., cost), it suffers the following disadvantages:

- If there are interactions such as a strength-by-package interaction, the data cannot be pooled to establish a single shelf-life. In this case it is recommended that individual shelf-lives be obtained for each combination of strength and package. However, we may not have three batches for each combination of strength and package for a fractional factorial design.

- We may not have sufficient precision for the estimated drug shelf-life.

Generally, a reduction of stability tests could be achieved if we apply a different method such as a matrixing design or a bracketing design (see, e.g., Barron, 1994; Lin, 1994; Nordbrock, 1994), which are also special cases of fractional factorial designs. More details regarding matrixing and bracketing designs are provided in Chapter 4.

1.4.4   Stability Analysis with Random Batches

To establish an expiration dating period, the 1987 FDA stability guideline requires that at least three batches, and preferably more, be tested in stability analysis to account for batch-to-batch variation so that a single shelf-life is applicable to all *future* batches manufactured under similar circumstances. If there is no documented evidence for batch-to-batch variation (i.e., all batches have the same shelf-life), the single shelf-life can be determined based on the ordinary least squares method as the time point at which the 95% confidence bound for the mean degradation curve of the drug characteristics intersects the approved lower specification limit. When there is a significant batch-to-batch variation, a typical approach is to consider so-called stability analysis with fixed effects model (Chow and Liu, 1995). This fixed effects model may not be appropriate because statistical inference about the expiration dating period obtained can only be made to the batches under study and cannot be applied to future batches. As indicated in the FDA stability guidelines, the batches used in long-term stability studies for establishment of drug shelf-life should constitute a *random* sample from the population of future production batches. As a result, *batch* should be considered a random variable in stability analysis (see, e.g., Brandt and Collings, 1989; Murphy and Weisman, 1990; Ruberg and Hsu, 1990; Chow and Shao, 1991; Ho, Liu, and Chow 1993; Grimes and Foust, 1994; Lee and Gagnon, 1994; Silverberg, 1997) . For this purpose, Chow and Shao (1991) and Shao and Chow (1994) proposed statistical methods for stability analysis with random batches.

Note that the difference between a random effects model and a fixed effects model is that the batches used in a random effects model for stability analysis are considered a random sample drawn from the population of all future production batches. As a result, the intercepts and slopes, which are often used to characterize the degradation of a drug product, are no longer fixed unknown parameters but random variables. More details regarding stability analysis with fixed batches and random batches can be found in Chapter 5 and Chapter 6, respectively.

1.4.5 Stability Analysis with Discrete Responses

For solid oral dosage forms such as tablets and capsules, the 1987 FDA stability guidelines indicate the characteristics of appearance, friability, hardness, color, odor, moisture, strength, and dissolution for tablets and the characteristics of strength, moisture, color, appearance, shape, brittleness, and dissolution for capsules should be studied in stability studies. Some of these characteristics are measured based on a discrete rating scale. The responses obtained from a discrete rating scale may be classified into acceptable (pass) and not acceptable (failure) categories, which results in binary stability data. Although in most stability studies, continuous responses such as potency are the primary concern, discrete responses such as appearance, color, and odor should be considered for quality assurance or safety. For establishment of drug shelf-life based on discrete responses, however, there is little discussion in either the FDA or the ICH stability guidelines. Chow and Shao (2003) considered the estimation of the shelf-life of a drug product when the stability data are discrete. When there is no batch-to-batch variation, their proposed shelf-life is an approximate 95% lower confidence bound of the true shelf-life. In the presence of batch-to-batch variation, their proposed shelf-life is an approximate 95% lower prediction bound of the shelf-life of future batches. As a result, their proposed shelf-life is applicable to all future batches of the same drug product. More details regarding Chow and Shao's proposed method for discrete responses will be provided in Chapter 8.

1.4.6 Stability Analysis with Multiple Components

As most drug products contain a single active ingredient, the 1987 FDA stability guidelines and the ICH guidelines for stability are developed for drug products with a single active ingredient. Many drug products consist of multiple ingredients (components). For example, Premarin (conjugated estrogens, USP) is known to contain at least five active ingredients: estrone, equilin, 17α-dihydroequilin, 17α-estradiol, and 17β-dihydroequilin. Other examples include combinational drug products such as most traditional Chinese medicine (TCM). For determination of the shelf-life of a drug product with multiple ingredients, an ingredient-by-ingredient stability analysis may not be appropriate, since multiple ingredients may have some unknown interactions. In this case Chow and Shao (2007) proposed a statistical method assuming that ingredients are linear combinations of some factors. Their proposed method was found to be efficient and useful. Details of Chow and Shao's method are provided in Chapter 9 of this book.

1.4.7 Stability Analysis with Frozen Drug Products

Unlike most drug products, some products must be stored at several temperatures, such as $-20°$ C, $5°$ C, and $25°$ C, in order to maintain the stability of the drug products. Drug products of this kind are usually referred to as *frozen products*. The determination of shelf-life for frozen drug products involves the estimation of drug shelf-lives at different temperatures, which requires multiple-phase linear regression. Mellon

(1991) suggested obtaining a combined shelf-life by determining shelf-lives based on data available at different temperatures. This method, however, does not account for the fact that the shelf-life at the second phase would depend on the shelf-life at the first phase. As an alternative, Shao and Chow (2001a) proposed a method for determination of drug shelf-lives for the two phases using a two-phase regression analysis based on the statistical principle as described in the 1987 FDA stability guidelines and the ICH Q1A (R2) guidelines for stability. Their proposed method was shown to be quite satisfactory. More details can be found in Chapter 10.

1.4.8   Stability Testing for Dissolution

As indicated in the FDA Report to the Nation: 2004, dissolution failure is the top 5 reason for drug recalls in fiscal year 2004. In addition to dissolution testing as described in Section 1.3, the FDA recommends the comparison of dissolution profiles at initial and at time $t = x$ in order to characterize the degradation in dissolution of the drug product. The FDA suggests that two dissolution profiles be compared based on a so-called f_2 similarity factor proposed by Moore and Flanner (1996). The use of the f_2 factor, however, has been criticized by many authors. See, for example, Chow and Liu (1995), Liu, Ma, and Chow (1997), Shah et al. (1998), Tsong et al. (1996), Ma et al., (1999), and Chow and Shao (2002b). Ma, et al. (2000) studied the size and power of the f_2 similarity factor for assessment of similarity based on the method of moments and the method of bootstrap. More details regarding the statistical methods for assessment of similarity between dissolution profiles in terms of f_2 similarity factor are given in the last chapter of this book.

1.4.9   Mean Kinetic Temperature

Haynes (1971) pointed out that changes in the actual field storage temperature could cause the reaction rate constant of some drug products to change according to the Arrhenius relationship. Since drug products stored in pharmacies and warehouses for extended periods of time are exposed to a range of temperatures, the exact determination of drug shelf-life becomes almost impossible. The degradation curve of a given drug product may not be consistent at different times under different environmental conditions. This will definitely have an impact on stability testing of the drug product. Haynes (1971) proposed establishing the mean kinetic temperature for a defined period. The mean kinetic temperature is a single derived temperature that affords the same thermal challenge to a drug substance or drug product as would be experienced over a range of both higher and lower temperatures for an equivalent defined period. Based on the mean kinetic temperature, the whole world is divided into four zones that are distinguished by their characteristic prevalent annual climatic conditions (Grimm, 1985, 1986). At different climatic zones, slightly different requirements for storage conditions for accelerated testing and long-term stability testing are imposed. More details are given in the last chapter of this book.

1.5 Aim and Scope of the Book

The goal of this book is to provide a comprehensive and unified presentation of statistical methods employed for design and analysis of stability studies in pharmaceutical research and development. In addition, it is intended to give a well balanced summary of current and recently developed statistical methods in stability analysis. It is our goal to provide a useful reference book for chemical scientists, pharmaceutical scientists, development pharmacists, and biostatisticians in the pharmaceutical industry, regulatory agencies, and academia. This book can also serve as a textbook for graduate courses in the areas of pharmacy, pharmaceutical development, stability studies, and biostatistics. Our primary emphasis is on the application of stability studies in pharmaceutical research and development. All statistical methods and their interpretations regarding design and analysis of stability studies are illustrated through real examples whenever possible.

The scope of this book is restricted to statistical design and analysis of stability studies. This book consists of 12 chapters. Chapter 1 introduces basic concepts of stability testing in pharmaceutical research and development. Also included in this chapter are regulatory requirements and practical issues for stability studies. Chapter 2 focuses on design and analysis of short-term stability studies such as accelerated testing. Several methods for estimating drug expiration dating periods are reviewed in Chapter 3. Chapter 4 compares several commonly employed study designs including matrixing and bracketing designs. Chapter 5 discusses statistical analysis with fixed batches. Also included in this chapter are several statistical tests for batch-to-batch variability (or batch similarity). Statistical analysis with random batches are discussed in Chapter 6. Chapter 7 introduces statistical methods for stability analysis under a linear mixed effects model. Stability analyses with discrete responses, multiple components, and frozen drug products are studied in Chapters 8, 9 and 10, respectively. Statistical methods of stability testing for dissolution including *USP-NF* dissolution testing and dissolution profile testing are given in Chapter 11. Current issues and recent developments in stability studies such as scale-up and postapproval changes, mean kinetic temperature, and optimality criteria for choosing a stability design are given in the last chapter.

For each chapter, real examples are given to illustrate the concepts, application, and limitations of statistical methods whenever possible. The comparisons of statistical methods available in terms of their relative merits and disadvantages are also discussed. When applicable, topics for possible future research development are provided.

Chapter 2

Accelerated Testing

As defined in the 1987 FDA stability guideline, the expiration dating period, or shelf-life, of a drug product is the time interval that the drug product is expected to remain within specifications after manufacture. The shelf-life of a drug product is usually established based on the primary stability data. The primary stability data are obtained from long-term stability studies conducted under approved stability protocols with ambient storage conditions. Drug products under ambient storage conditions in long-term stability studies, however, usually degrade very slowly over time. Therefore, for most drug products, it may take more than a year to observe significant degradation.

As is well known, the development of a drug product is a lengthy process that usually involves many stages. The goals and meanings of drug stability functions may be different from stage to stage. For example, the purpose of stability studies before filing an investigational new drug application (IND) is twofold. First, it is to verify that the stability of the drug product will be maintained within specifications for animal studies such as toxicological trials. Second, it is to provide useful stability information for modification of the formulation of the drug product. At the very early stages of drug development, however, the primary goal of stability functions is to determine the rates of chemical and physical reactions and their relationship with storage conditions such as temperature, moisture, and light. To achieve this goal, accelerated stability testing is usually conducted. As defined in the 1987 FDA stability guideline, an accelerated stability test is a short-term stability study conducted under exaggerated (or stressed) conditions to increase the rate of chemical or physical degradation of a drug substance or drug product. As stated in the 1987 FDA stability guideline, exaggerated storage conditions may include temperature, humidity of 75% or greater, exposure to various wavelengths of electromagnetic radiation, and storage in an open container. The use of exaggerated storage conditions to accelerate the reaction rate so that significant degradation of the drug product can be observed allows the kinetic parameters of the reaction rate to be estimated. A predicted shelf-life under marketplace storage conditions can then be obtained by extrapolation. Note that extrapolation from stressed testing conditions to ambient conditions is usually done based on established relationships between the kinetic parameters and storage conditions. In the pharmaceutical industry, the Arrhenius equation is often employed to relate the degradation reaction rate and the corresponding temperature. Other models such as the Eyring equation (Kirkwood, 1977) can also be used. In this chapter, however, we focus on application of the Arrhenius equation to the prediction of drug product shelf-life.

For a given drug product, the estimated shelf-life based on the date obtained from accelerated stability testing is usually referred to as the tentative expiration period. As indicated by the 1987 FDA stability guideline, this period is a provisional expiration dating period determined by projecting results from less-than-full-term data using the drug product to be marketed in its proposed container closure. Therefore, the 1987 FDA stability guideline indicates that the results obtained from accelerated testing can be used only as supportive stability data. However, accelerated testing is useful in the following ways: (a) the results provide estimates of the kinetic parameters for the rates of reactions, (b) the results can be used to characterize the relationship between degradation and storage conditions, and (c) the results supply critical information in the design and analysis of long-term stability studies under ambient conditions at the planning stage.

In this chapter our efforts are directed to a discussion of accelerated stability testing in terms of application of the Arrhenius equation for the relationship between degradation and temperature. In the next section we describe briefly some deterministic chemical kinetic models. The Arrhenius equation is also described in this section. Applications of statistical methods for estimation of kinetic parameters and a tentative expiration dating period using the Arrhenius equation are given in Section 2.2. Section 2.3 covers determination of the order of a reaction by selection of adequate models. Also included in this section are issues that often occur in the design of accelerated stability testing. A numerical example using the data set given in Carstenson (1990) is given in Section 2.4. A brief discussion regarding other methods and possible future research topics can be found in Section 2.5.

2.1 Chemical Kinetic Reaction

In this section we describe some functional relationships of chemical kinetic reactions. Unless otherwise stated, all quantities and equations considered in this chapter will be deterministic. In other words, there are no random error terms. Carstensen (1990) indicated that if a reaction with two entities A and B is

$$A + B \rightarrow C,$$

the reaction rate is given by

$$\frac{dY(t)}{dt} = -K_{r_A+r_B}[A]^{r_A}[B]^{r_B} \tag{2.1}$$

The reaction is said to be of order $r_A + r_B$, where $dY(t)/dt$ is a differential quotient between concentration and time, $Y(t)$ is the concentration of the species being studied at time t, $[A]$ and $[B]$ represent the concentrations of A and B, and K is a rate constant. Note that in the pharmaceutical industry we are only interested in the integer orders of a reaction (i.e., 0, 1, and 2). The differential equation for a zero-order reaction is given by

$$\frac{dY(t)}{dt} = -K_0 \tag{2.2}$$

where K_0 is the zero-order rate constant, which is expressed as concentration unit per time unit. Integrating both sides of Equation 2.2 with respect to time gives

$$\int \frac{dY(t)}{dt} = - \int K_0 dt,$$

or

$$Y(t) = C_0 - K_0 t, \tag{2.3}$$

where C_0 is a constant. Let $Y(0)$ be the initial concentration at time 0. Then Equation 2.3 becomes

$$Y(t) = Y(0) - K_0 t,$$

or

$$Y(t) - Y(0) = -K_0 t, \tag{2.4}$$

It can be seen from Equation 2.4 that a drug product based on the zero-order reaction degrades at a constant rate over time that is independent of both concentrations at the initial and at time t.

However, for the first-order reaction, since the amount of degradation is proportional to the concentration at time t, the corresponding differential equation is given by

$$\frac{dY(t)}{dt} = -K_1 Y(t), \tag{2.5}$$

where K_1 is the first-order rate of constant. Equation 2.5 can be rewritten as

$$\int \frac{dY(t)}{dt} = - \int K_0 dt. \tag{2.6}$$

Similarly, integrating both sides of Equation 2.6 with respect to time yields

$$\int \frac{dY(t)}{Y(t)} dt = - \int K_1 dt,$$

or

$$\ln Y(t) = C_1 - K_1 t, \tag{2.7}$$

where ln denotes the natural logarithm and C_1 is a constant. If the initial concentration at time 0 is $Y(0)$, it can be shown that

$$\ln Y(t) = \ln Y(0) - K_1 t,$$

or

$$\ln Y(t) - \ln Y(0) = -K_1 t, \tag{2.8}$$

or

$$\ln \frac{Y(t)}{Y(0)} = -K_1 t.$$

Hence, it can be seen from Equation 2.8 that the degradation of a drug product described by a first order reaction can be characterized by a rate constant for the logarithm of concentrations.

For the second-order reaction, since it proceeds at a constant rate that is proportional to the square of the concentration, that is,

$$\frac{dY(t)}{Y(t)} = -K_2 Y^2(t), \tag{2.9}$$

or

$$\frac{dY(t)}{d(t)} \Big/ Y^2(t) = -K_2,$$

we have

$$\int \frac{dY(t)}{Y^2(t)} dt = -\int K_2 dt,$$

or

$$\frac{1}{Y(t)} = \frac{1}{Y(0)} - K_2 t. \tag{2.10}$$

As a result, the second-order reaction describes a degradation characterized by a rate constant for the inverse of concentration.

In the pharmaceutical industry the first-order reaction is probably the most commonly employed model for describing the decomposition and the degradation of active ingredients of a drug product. The zero-order reaction is used occasionally, but the second-order reaction is rarely adopted. It however, should be noted, that the discussion above is based on the assumption that there is only one decomposed end chemical entity for each active ingredient contained in the drug product. In many situations the active ingredient of a drug product may be decomposed into more than one end chemical entity. The description and determination of orders for such a reaction are usually very complicated. In this chapter, for simplicity, we focus on the zero- and first-order reactions for a single end chemical entity decomposed from one active ingredient of a drug product.

For thermal stability accelerated testing, samples of drug products are usually stored over time at different temperatures (e.g., $5°$ C, $50°$ C, or $75°$ C). There are, in general, two methods for determining the strength (i.e., concentration or potency) of a sample that has been stored over time t at temperature T (Davies and Hudson, 1981, 1993). The time is usually expressed in terms of either days or months. Absolute temperature is used for the study of the relationship between rate constant and temperature. For the first method, each sample stored at an elevated temperature over time t is assayed side by side with a sample stored for the same period of time but at a lower temperature in which no appreciable degradation can occur. Then the strength of the sample stored at the elevated temperature is expressed as a percentage of the sample at the lower temperature. The second method assays the sample stored at temperature T and time t and a sample prior to storage together against an independent standard. The strength of the sample at temperature T and time t is scaled to that of the sample prior to storage, which is made to be 100.

The difference between these two methods is the determination of the strength for the sample at time 0. The strength at time 0 obtained by the second method is an observation subject to random error. The first method provides an initial strength by definition of exactly 100 without error. Because of this advantage in determining the initial strength at time 0, the first method has become popular in practice. Therefore, in this chapter, we focus on the projection of a tentative expiration dating period based on the data obtained from the first method. It should be noted that there are other sources of variations in the first method that are not discussed in this chapter. For more details, see Davies (1980).

Based on the first method, since the initial strength at time 0 is 100% of label claim without error, that is, $Y(0) = 100$, the equations for the zero- and first-order reactions given in Equations 2.4 and 2.8 become

$$Y(t) = 100 - K_0 t, \qquad (2.11)$$

and

$$\ln[Y(t)] = \ln(100) - K_1 t, \qquad (2.12)$$

respectively. At the early stage of drug development, it is necessary to project the degradation rate at marketing storage temperature based on the data collected from thermal stability accelerated testing at elevated temperatures. To achieve this goal, we first need to establish the relationship between rate constant and temperature. The relationship between the rate constant (or reaction rate) and absolute temperature can be expressed by the following Arrhenius equation (see, e.g., Bohidar and Peace, 1988; Davies and Hudson, 1981; Carstensen, 1990):

$$\frac{d \ln K}{dT} = \frac{E}{RT^2}, \qquad (2.13)$$

where T is the absolute temperature, E is the activation energy, and R is the gas constant. Integrating both sides of Equation 2.13 gives

$$\ln K = -\frac{E}{R} \cdot \frac{1}{T} + \ln A,$$

or

$$K = A \exp\left(-\frac{E}{RT}\right), \qquad (2.14)$$

where A is a frequency factor. Substituting Equation 2.14 into Equations 2.11 and 2.12 yields the following equations:

$$Y(t) = 100 - A \exp\left(-\frac{E}{RT}\right) t, \tag{2.15}$$

$$\ln[Y(t)] = \ln(100) - A \exp\left(-\frac{E}{RT}\right) t. \tag{2.16}$$

Rearranging the terms in Equations 2.15 and 2.16 gives

$$\frac{Y(t) - 100}{t} = -A \exp\left(-\frac{E}{RT}\right), \tag{2.17}$$

and

$$\frac{\ln[Y(t)/100]}{t} = -A \exp\left(-\frac{E}{RT}\right). \tag{2.18}$$

The quantities on the left-hand side of Equations 2.17 and 2.18 are the degradation per time unit based on either the original scale of the strength for the zero-order reaction or the log scale of the strength for the first-order reaction. $[Y(t) - 100]/t$ or $\ln[Y(t)/100]/t$ can be interpreted as the observed reaction rates that can be used for estimation of the unknown parameters A and $-E/R$ in Equation 2.14. For the purpose of estimating the parameters A and $-E/R$, it is preferable to use the following negative observed reaction rates:

$$\frac{100 - Y(t)}{t} = A \exp\left(-\frac{E}{RT}\right), \tag{2.19}$$

$$\frac{\ln[100/Y(t)]}{t} = A \exp\left(-\frac{E}{RT}\right). \tag{2.20}$$

The estimates obtained from Equations 2.19 and 2.20 are the same as those from Equations 2.17 and 2.18.

2.2   Statistical Analysis and Prediction

Let Y_{ij} be the strength of a sample stored after time t_j at temperature T_i, where $i = 1, 2, \ldots, I$ and $j = 1, \ldots, J$. Since the first method for determination of strength is used throughout the rest of this chapter, Y_{ij} will not represent the initial strength at time 0, which is 100 without error, and all Y_{ij} are expressed in terms of percentage. Define the degradation for the zero- and first-order reactions, respectively, as

$$D_{ij}(0) = Y_{ij} - 100$$

and

$$D_{ij}(1) = \ln(Y_{ij}/100)$$
$$= \ln Y_{ij} - \ln 100, \tag{2.21}$$

where $i = 1, 2, \ldots, I$ and $j = 1, 2, \ldots, J$. The negative reaction rates or the negative degradation rates per time unit are given, respectively, as follows:

$$K_{ij}(0) = -\frac{D_{ij}(0)}{t_j} = \frac{100 - Y_{ij}}{t_j},$$

$$K_{ij}(1) = -\frac{D_{ij}(1)}{t_j} = \frac{\ln(100/Y_{ij})}{t_j}, \tag{2.22}$$

where $i = 1, 2, \ldots, I$ and $j = 1, 2, \ldots, J$.

In this section we illustrate the application of some well-established statistical methods for the estimation of rate constants and the unknown parameters in the Arrhenius equation. Based on estimates of the unknown parameters and their estimated standard deviations and covariance, a projected tentative expiration dating period can be obtained using the relationship given in Equations 2.15 and 2.16. From Equations 2.4 and 2.8, the degradation at time t_j can be expressed by the following linear regression model through the origin:

$$D_{ij}(h) = \beta_i(h)t_j + e_{ij}, \quad i = 1, \ldots I, \quad j = 1, \ldots, J, \tag{2.23}$$

where

$$\beta_i(h) = -K_{hi},$$

and

$$h = \begin{cases} 0 & \text{for the zero-order reaction} \\ 1 & \text{for the first-order reaction.} \end{cases}$$

It is assumed that random errors e_{ij} in Model 2.23 follow a normal distribution with mean 0 and variance $\sigma_i^2(h)$. The least-squares estimator for $\beta_i(h)$ is then given by (Draper and Smith, 1981):

$$b_i(h) = \frac{\sum\limits_{j=1}^{J} t_j D_{ij}(h)}{\sum\limits_{j=1}^{J} t_j^2}. \tag{2.24}$$

Note that the above estimator is not the same as that obtained from the linear regression model with intercept. However, they are expressed in a similar form. Under the normality assumption, $b_i(h)$ is the minimum variance unbiased estimator (MVUE) for $-K_{hi}$. An unbiased estimator for the error variance can also be obtained as

$$\hat{\sigma}_i^2(h) = \frac{SSE_i(h)}{J - 1}, \quad i = 1, \ldots, I, \quad h = 0, 1, \tag{2.25}$$

where $SSE_i(h)$ is the sum of squares of residuals for the hth order of reaction at temperature i, that is,

$$SSE_i(h) = SST_i(h) - SSR_i(h), \qquad (2.26)$$

where $SST_i(h)$ and $SSR_i(h)$ are the total uncorrected sum of squares and the sum of squares due to regression; that is

$$SST_i(h) = \sum_{j=1}^{J} D_{ij}^2(h), \qquad (2.27)$$

$$SSR_i(h) = \frac{\left[\sum_{j=1}^{J} D_{ij}(h)t_j\right]^2}{\sum_{j=1}^{J} t_j^2}. \qquad (2.28)$$

The corresponding degrees of freedom for $SST_i(h)$, $SSR_i(h)$, and $SSE_i(h)$ are J, 1, and $J - 1$, respectively. The relationship among the total uncorrected sum of squares, the sum of squares due to regression, and the sum of squares of residuals is the same as that for the linear regression model with intercept. This relationship is summarized in the ANOVA table (Table 2.1). The variance of $b_i(h)$ can be estimated by

$$\widehat{Var}[b_i(h)] = \frac{\hat{\sigma}_i^2(h)}{\sum_{j=1}^{J} t_j^2}, \quad i = 1, .., I, \quad h = 0, 1. \qquad (2.29)$$

Thus, the standard error of $b_i(h)$, denoted by $SE[b_i(h)]$, is given by

$$SE[b_i(h)] = \sqrt{\widehat{Var}[b_i(h)]} \qquad (2.30)$$

Under the hypothesis that $\beta_i(h) = 0$, the statistic

$$T_b = \frac{b_i(h)}{SE[b_i(h)]} \qquad (2.31)$$

TABLE 2.1: ANOVA Table for Simple Linear Regression Without Intercept

Source of Variation	df	Sum of Squares	Mean Squares	F-Value
Regression	1	$SSR_i(h)$	$MSR_i(h) = SSR_i(h)$	$F = \frac{MSR_i(h)}{MSE_i(h)}$
Residual	$J - 1$	$SSE_i(h)$	$MSE_i(h) = SSE_i(h)/(J - 1)$	
Total	J	$SST_i(h)$		

$SSR_i(h) = \frac{[\sum D_{ij}(h)t_j]^2}{\sum t_j^2}$

$SSE_i(h) = \sum D_{ij}^2(h) - \frac{[\sum D_{ij}(h)t_j]^2}{\sum t_j^2}$

$SST_i(h) = D_{ij}^2(h)$

Source: Chow, S.C. and Liu, J.P. (1995). *Statistical Design and Analysis in Pharmaceutical Science.* Marcel Dekker, New York.

follows a central t distribution with $J - 1$ degrees of freedom. Based on Equation 2.31, a $(1 - \alpha)100\%$ confidence interval for $\beta_i(h)$ can be obtained as follows:

$$b_i(h) \pm t_{\alpha/2, J-1} SE[b_i(h)], \tag{2.32}$$

where $t_{\alpha/2, J-1}$ is the $(\alpha/2)$th upper quantile of a central t distribution with $J - 1$ degrees of freedom. To test for a negative reaction after time t_j at temperature T_i, we may consider the following hypotheses:

$$H_0 : \beta_i(h) = 0 \text{ vs } H_0 : \beta_i(h) < 0. \tag{2.33}$$

The null hypothesis of Equation 2.33 is rejected at the α level of significance if

$$T_b < -t_{\alpha, J-1}, \tag{2.34}$$

where $t_{\alpha, J-1}$ is the αth upper quantile of a central t distribution with $J - 1$ degrees of freedom. If one is interested in examining whether the reaction rate is different from zero, the following hypotheses should be tested:

$$H_0 : \beta_i(h) = 0 \text{ vs } H_0 : \beta_i(h) \neq 0. \tag{2.35}$$

The above null hypothesis of zero reaction rate is rejected at the α level of significance if

$$F_i = \frac{MSR_i(h)}{MSE_i(h)} > F_{\alpha, 1, J-1}, \tag{2.36}$$

where $F_{\alpha, 1, J-1}$ is the αth upper quantile of a central F distribution with 1 and $J - 1$ degrees of freedom.

We have illustrated the application of a simple linear regression model without intercept through the least-squares method for the estimation of a rate constant for each combination of time points and temperature. At the early stage of drug development, however, strength data are usually available at only a few time points for each elevated temperature. In this case the estimates of error variances may not be reliable owing to insufficient degrees of freedom. To enhance the precision of the estimates of error variances, we may consider the following regression model:

$$D_{ij}(h) = \sum_{i=1}^{I} \beta_i(h) X_{ij}(h) + e_{ij}, \tag{2.37}$$

where $i = 1, \ldots, I$, $j = 1, \ldots, J$, $h = 0, 1$, and the value of $X_{ij}(h)$ for $D_{ij}(h)$ is t_j if the temperature is T_i and is 0 otherwise; that is,

$$X_{ij}(h) = \begin{cases} t_j & \text{if } i = j \text{ for } D_{ij}(h) \\ 0 & \text{otherwise.} \end{cases} \tag{2.38}$$

If the random errors e_{ij} in Model 2.37 are independent and identically distributed as a normal distribution with mean 0 and variance σ^2, then $b_i(h)$, given in Equation 2.24, is also the MVUE of $\beta_i(h)$ in Model 2.37. The total uncorrected sum of squares, the sum of squares due to regression, and the sum of squares of residuals under Model 2.37 are given, respectively, as in Model 2.24, which is also the MVUE of $\beta_i(h)$ in Model 2.37. The total uncorrected sum of squares, the sum of squares due to regression, and the sum of squares of residuals under Model 2.37 are given, respectively, as

$$SST(h) = \sum_{i=1}^{I} SST_i(h) = \sum_{i=1}^{I} \sum_{j=1}^{J} D_{ij}^2(h), \tag{2.39}$$

$$SSR(h) = \sum_{i=1}^{I} SSR_i(h) = \sum_{i=1}^{I} \left\{ \frac{\left[\sum_{j=1}^{J} t_j D_{ij}(h)\right]^2}{\sum_{j=1}^{J} t_j^2} \right\},$$

$$SSE(h) = \sum_{i=1}^{I} SSE_i(h), \quad h = 0, 1.$$

The corresponding degrees of freedom for $SST(h)$, $SSR(h)$, and $SSE(h)$ are IJ, I, and $I(J-1)$, respectively. Table 2.2 gives the analysis of the variance table for Model 2.37. Since samples obtained at different temperatures are independent of each other, the sum of squares due to regression under Model 2.37 can be partitioned into I independent sums of squares, that is, $SSR_i(h), i = 1, \ldots, I$, which can be obtained separately under Model 2.23. Under Model 2.37, statistical inference for the reaction rate can be obtained based on the following estimates for the error variances

$$\hat{\sigma}^2(h) = MSE(h) = \frac{SSE(h)}{I(J-1)}, \tag{2.40}$$

which has $I(J-1)$ degrees of freedom. It should be noted that under Equation 2.37, $\sigma^2(h)$ and the upper quantiles of a central t distribution with $I(J-1)$ degrees of freedom should be substituted for the estimated standard error, $(1-\alpha)100\%$

TABLE 2.2: ANOVA Table for Simple Regression Without Intercept, Corresponding to Equation 2.37

Source of Variation	df	Sum of Squares[a]	Mean Squares	F-Value
Regression	I	SSR(h)	MSR(h) = SSR(h)/f	F = MSR(h)/MSE(h)
Temp. 1	1	SSR$_1$(h)	MSR$_1$(h) = SSR$_1$(h)	F$_1$ = MSR$_1$(h)/MSE(h)
.	.	.	.	.
.	.	.	.	.
.	.	.	.	.
Temp. I	I	SSR$_I$(h)	MSR$_I$(h) = SSR$_I$(h)	F$_I$ = MSR$_I$(h)/MSE(h)
Residual	I(J − 1)	SSE(h)	MSE(h) = SSE(h)/I(j−1)	
Total	IJ	SST(h)		

[a]SST(h) = $\sum$ SST$_i$(h); SSR(h) = $\sum$ SSR$_i$(h); SSE(h) = $\sum$ SSE$_i$(h).

Source: Chow, S.C. and Liu, J. P. (1995). *Statistical Design and Analysis in Pharmaceutical Science*, Marcel Dekker, New York.

confidence interval, and hypothesis testing of reaction rates $\beta_i(h)$ given in Equations 2.30 to 2.34 and 2.36, respectively. Although the discussion above assumes the same time points for all temperatures, the methodology described above for estimation and inference about rate constants can easily be applied to the situation where there are different time points at different temperatures without modification.

Once the rate constants are estimated at each temperature, Bohidar and Peace (1988) suggested obtaining estimates of the unknown parameters in the Arrhenius Equation 2.14 by fitting a linear regression (or weighted) model to the logarithm of the estimated rate constants $\ln[b_i(h)]$ with temperature as the independent variable. Since typical thermal accelerated stability testing is usually conducted at three or four different elevated temperatures, statistical inference about the unknown parameters in the Arrhenius equation is then based on only one or two degrees of freedom. To overcome this drawback, one may consider utilizing the observed negative degradation rate per time unit or observed negative reaction rates $K_{ij}(h)$, $h = 0, 1, i = 1, \ldots, I$, and $j = 1, \ldots, J$. As a result, all observations are used to estimate the two unknown parameters in the Arrhenius equation, and consequently, statistical inference regarding the tentative expiration dating period can be obtained based on an estimate of the error variance with $IJ - 2$ degrees of freedom.

Recall that the negative reaction rates obtained are defined as

$$K_{ij}(h) = -\frac{D_{ij}(h)}{t_j} = \begin{cases} \frac{100 - Y_{ij}}{t_j} & \text{if } h = 0 \\ \frac{\ln(100/Y_{ij})}{t_j} & \text{if } h = 1 \end{cases}. \tag{2.41}$$

However, the Arrhenius equation states that the relationship between the reaction rate and absolute temperature is

$$K(h) = A \exp\left(-\frac{E}{RT}\right).$$

Let $\alpha(h) = A$, $\beta(h) = -E/R$, and $X = 1/T$. Then the Arrhenius equation can be rewritten as

$$K(h) = \exp[\alpha(h) + \beta(h)X], \tag{2.42}$$

or

$$\ln[K(h)] = \alpha(h) + \beta(h)X. \tag{2.43}$$

For the unknown parameters in Equation 2.42, we may apply the following two methods to obtain estimates of $\alpha(h)$ and $\beta(h)$. The first method is to apply ordinary least squares in a simple linear regression model to the logarithm of the rate constants. The other method is to simply fit a nonlinear regression model directly to the original data of the reaction rates.

Let $K_{ij}^*(h)$ be the logarithm of the observed negative reaction rate, that is

$$K_{ij}^*(h) = \ln[K_{ij}(h)], \quad i = 1, \ldots, I, \quad j = 1, \ldots, J.$$

Then, according to Equation 2.14, the model for the logarithm of $K_{ij}(h)$ is given by

$$K_{ij}^*(h) = \alpha(h) + \beta(h)X_i + e_{ij}, \qquad (2.44)$$

where $i = 1, \ldots, I$ and $j = 1, \ldots, J$. Let $a(h)$ and $b(h)$ be the MVUE of $\alpha(h)$ and $\beta(h)$. Also, let $\widehat{Var}[a(h)]$ and $\widehat{Var}[b(h)]$ be the estimates of the variances of $a(h)$ and $b(h)$, respectively. The covariance between $a(h)$ and $b(h)$ can be estimated by

$$\widehat{Cov}[a(h), b(h)] = -\hat{\sigma}^2(h)\frac{\bar{X}}{J S_{xx}}, \qquad (2.45)$$

where

$$\bar{X} = \frac{1}{I}\sum_{i=1}^{I} X_i,$$

$$S_{xx} = \sum_{i=1}^{I}(X_i - \bar{X})^2,$$

and $\hat{\sigma}^2(h)$ is an estimate of the error variance.

To obtain an estimate of the tentative expiration dating period, we need to obtain the predicted mean reaction rate at the marketing storage temperature T for reaction order h. This can be done by considering the following linear regression model:

$$\hat{K}^*(h) = a(h) + b(h)X \qquad (2.46)$$

where $X = 1/T$. An estimate of the variance of $\hat{K}^*(h)$ can be obtained as

$$\widehat{Var}[\hat{K}^*(h)] = \hat{\sigma}^2(h)\{\widehat{Var}[a(h)] + X^2\widehat{Var}[b(h)] \qquad (2.47)$$
$$+ 2X\widehat{Cov}[a(h), b(h)]\}.$$

From Equation 2.41, it can be verified that the predicted degradation after time t at the marketing storage temperature T is given by

$$\hat{D}(h) = -\exp[\hat{K}^*(h)]t. \qquad (2.48)$$

Let $G(h)$ be the minimum strength required for a drug product to maintain under reaction order h, that is

$$G(h) = \begin{cases} 100 - P(0) & if\ h = 0 \\ \ln\left[\dfrac{100}{100 - P(1)}\right] & if\ h = 1 \end{cases}, \qquad (2.49)$$

where $P(h)$ is the amount of maximum degradation allowed for reaction order h, where $h = 0$ or 1. For the zero-order reaction (i.e., $h = 0$),

$$D(0) = G(0) - 100 = -P(0)$$

It follows that

$$P(0) = \exp[\hat{K}^*(0)]t,$$

or

$$\ln[P(0)] = \hat{K}^*(0) + \ln(t)$$
$$= a(0) + b(0)X + \ln(t).$$

Consequently,

$$\ln(\hat{t}) = \ln[P(0)] - [a(0) + b(0)X]. \qquad (2.50)$$

Since $P(0)$ is a predetermined fixed constant, for example $P(0) = 10\%$, the estimate of the variance of $\ln(\hat{t})$ is the same as that for $\hat{K}^*(h)$ given in Equation 2.47. The $(1 - \alpha)100\%$ lower confidence limit for the time based on the logarithm scale is then given by

$$L_t(0) = \ln(\hat{t}) - t_{\alpha, IJ-2} SE[\hat{K}^*(0)], \qquad (2.51)$$

where

$$SE[\hat{K}^*(0)] = \sqrt{\widehat{Var}[\hat{K}^*(0)]},$$

and $t_{\alpha, IJ-2}$ is the αth upper quantile of a central t distribution with $IJ - 2$ degrees of freedom. Thus, an estimate of the tentative expiration dating period for a maximum allowable degradation of $P(0)$ at the marketing temperature T under the zero-order reaction is obtained as

$$t_T(0) = \exp[L_t(0)]. \qquad (2.52)$$

Note that the estimated tentative expiration dating period given in Equation 2.52 is not derived from the mean degradation but is based on the 95% lower confidence limit for degradation. Thus, the obtained tentative expiration dating period ensures that 95% of future samples at marketing storage temperature T are expected to remain above the specified minimum strength $G(0)$. For the first-order reaction, we have

$$G(1) = \ln\left[\frac{100}{100 - P(1)}\right]$$
$$= -\ln\left[\frac{100 - P(1)}{100}\right]$$
$$= -\ln\left[\frac{Y(t)}{100}\right] = -D(1).$$

Thus, a similar method can be applied to obtain an estimate of the tentative expiration dating period based on the fact that

$$G(1) = \exp[\hat{K}^*(1)]t.$$

Let $a(1)$ and $b(1)$ be the MVUE of $\alpha(1)$ and $\beta(1)$. Then

$$\ln[G(1)] = \hat{K}^*(1) + \ln(t)$$
$$= a(1) + b(1)X + \ln(t).$$

Following the same arguments, an estimate of the tentative expiration dating period for a maximum allowable degradation of $P(1)$ at the marketing temperature T under the first-order reaction is given by

$$t_T(1) = \exp[L_t(1)], \tag{2.53}$$

where $L_t(1)$ is the $(1 - \alpha)100\%$ lower confidence limit for the time based on the logarithmic scale at temperature T, that is,

$$L_t(1) = \ln(\hat{t}) - t_{\alpha,IJ-2}SE[\hat{K}^*(1)],$$
$$\ln(\hat{t}) = \ln[G(1)] - [a(1) + b(1)X],$$

and $SE[\hat{K}^*(1)]$ is defined similarly.

The other method is to fit a nonlinear regression model directly to obtain estimates of $\alpha(h)$ and $\beta(h)$ in the Arrhenius equation. Once estimates of $\alpha(h)$ and $\beta(h)$ and their variances and covariance are obtained, an estimate of the tentative expiration dating period at marketing storage temperature T can be obtained similarly using Equations 2.52 and 2.53.

As suggested by the functional relationship stated in Equation 2.42, consider the following nonlinear regression model for the observed negative reaction rate:

$$K_{ij} = \exp(\alpha + \beta X_i) + e_{ij}, \quad i = 1, \ldots, I, \quad j = 1, \ldots, J. \tag{2.54}$$

Note that for simplicity, the index for reaction order was dropped in the model above. Similarly, we assume that random errors e_{ij} are independent and identically distributed as a normal distribution with mean zero and variance σ^2.

To obtain estimates of α and β, we consider a Taylor series expansion of K_{ij} around K_{ij}^0 with respect to α and β up to the first derivative, where K_{ij}^0 is the value of K_{ij} evaluated at a^0 and b^0, and a^0 and b^0 are some selected initial values of α and β. Thus, Model 2.54 can be approximated by the following model:

$$K_{ij} = K_{ij}^0 + Z_{\alpha ij}^0(\alpha - a^0) + Z_{\beta ij}^0(\beta - b^0) + e_{ij} \tag{2.55}$$
$$i = 1, \ldots, I, \quad \text{and} \quad j = 1, \ldots, J,$$

where

$$Z_{\alpha ij}^0 = \left. \frac{\partial K_{ij}}{\partial \alpha} \right|_{\alpha=a^0} = \exp(a^0 + b^0 X_i), \tag{2.56}$$

$$Z_{\beta ij}^0 = \left. \frac{\partial K_{ij}}{\partial \beta} \right|_{\beta=b^0} = X_i \exp(a^0 + b^0 X_i).$$

Equation 2.56 can be rewritten as

$$k_{ij} = K_{ij} - K_{ij}^0 = Z_{\alpha ij}^0 \delta_0^0 + Z_{\beta ij}^0 \delta_1^0 + e_{ij} \tag{2.57}$$
$$i = 1, \ldots, I, \quad \text{and} \quad j = 1, \ldots, J,$$

where

$$\delta_0^0 = \alpha - a^0,$$
$$\delta_1^0 = \beta - b^0.$$

Let $K^0 = (K_{11} - K_{11}^0, \ldots, K_{IJ} - K_{IJ}^0)$,

$$Z^0 = (Z_\alpha^0, Z_\beta^0) = \begin{bmatrix} Z_{\alpha 11}^0 & Z_{\beta 11}^0 \\ \vdots & \vdots \\ Z_{\alpha IJ}^0 & Z_{\beta IJ}^0 \end{bmatrix},$$

$$\delta^0 = (\delta_0^0, \delta_1^0) = (\alpha - a^0, \beta - b^0).$$

The ordinary least-squares estimator of δ_0, which minimizes the sum of squares

$$\sum_{i=1}^{I} \sum_{j=1}^{J} \left(K_{ij} - K_{ij}^0 - Z_{\alpha ij}^0 \delta_0^0 - Z_{\beta ij}^0 \delta_1^0 \right)^2$$

is given by

$$\hat{\delta}^0 = (Z^{0\prime} Z^0)^{-1} Z^{0\prime} K^0. \tag{2.58}$$

After $\hat{\delta}^0$ is obtained, we can repeat the steps from Equations 2.56 to 2.58 to improve the linear approximation. Denote $\beta = (\alpha, \beta)'$ and let $b = (a, b)'$ be an estimator of β. At the uth iteration, the resulting estimator of β is given by

$$b_u = b_{u-1} + \hat{\delta}_{u-1} \tag{2.59}$$
$$= b_{u-1} + (Z_{u-1}' Z_{u-1})^{-1} Z_{u-1}' K_{u-1},$$

where

$$b_u = (a_u, b_u)',$$
$$Z_{u-1} = \left(Z_\alpha^{u-1}, Z_\beta^{u-1} \right)',$$
$$K^{u-1} = \left(K_{11} - K_{11}^{u-1}, \ldots, K_{IJ} - K_{IJ}^{u-1} \right)'. \tag{2.60}$$

Let λ be some prespecified small number (e.g., 10^{-5}). Then the iterative procedure continues until the following criteria are met:

$$\left| \frac{a_u - a_{u-1}}{a_{u-1}} \right| < \lambda, \tag{2.61}$$

$$\left| \frac{b_u - b_{u-1}}{b_{u-1}} \right| < \lambda.$$

Many statistical software packages for nonlinear regression analysis are available. For example, the PROC NLIN in SAS provides estimates for the unknown parameters and their variances and covariance and 95% confidence intervals for the unknown parameters. Once estimates of the unknown parameters in the Arrhenius equation are obtained from the nonlinear regression model, an estimate of the tentative expiration dating period at the marketing storage temperature can be obtained using Equations 2.52 and 2.53 for the zero- and first-order reactions, respectively.

2.3   Examinations of Model Assumptions

In the previous section we demonstrated how to apply statistical models, including a simple linear regression model and a nonlinear regression model for obtaining an estimate of the tentative expiration dating period under the assumption that the reaction is either zero or first order. The order of the reaction has an impact on the estimate of the tentative expiration dating period. One of the primary objectives of accelerated stability testing at the early stage of drug development is to empirically determine the order of reaction. As indicated earlier, the number of elevated temperatures examined in an accelerated stability testing study is usually between three and five. Thus, at each elevated temperature, the degradation of the drug product is evaluated at three to five time points, including time zero. Based on these few observations, it is difficult to ensure the accuracy and precision of the empirically determined order of reactions for the drug product. After the reaction order of the degradation of the drug product is determined, the next step is to apply the Arrhenius equation. It is then important to evaluate whether the Arrhenius equation can adequately describe the relationship between degradation and temperature. In practice, it is suggested that the adequacy of the two postulated models be examined. It is necessary to check whether the models of the zero- or first-order reaction can adequately describe the relationship between degradation and time. In addition, it is of interest to determine which model provides a better description of the relationship.

At each temperature, a zero-order reaction is generally used to describe a linear relationship between strength and time based on the original scale, while a first-order reaction dictates a linear relationship between log(strength) and time. To provide a visual inspection of the linear relationship, scatter plots of the strength and log(strength) against time points by temperatures are often employed as a useful graphical presentation of reaction orders. If the scatter plots reveal that linearity exists for the strength on the original (or log) scale, the reaction for the degradation of an ingredient of the drug product may be of order zero (or one). In practice, however, if the degradation is not sufficient, it is very difficult to determine the order of reaction either by graphical or by other sophisticated statistical methods. Carstensen (1990) showed that if degradation is less than 15%, we may not be able to distinguish a first-order reaction from a zero-order reaction. Let $P(t)$ denote the amount of strength that has been decomposed after time t. The remaining strength at time t is then given by $Y(t) = 100 - P(t)$, or equivalently,

$$\frac{Y(t)}{100} = 1 - \frac{P(t)}{100}.$$

If the reaction is of first order, we have

$$\ln\left[\frac{Y(t)}{100}\right] = -K_1 t,$$

or

$$\ln\left[1 - \frac{P(t)}{100}\right] = -K_1 t.$$

If $P(t)/100 < 15\%$, we can approximate

$$\ln\left[1 - \frac{P(t)}{100}\right]$$

by $-P(t)/100$, that is,

$$\ln\left[1 - \frac{P(t)}{100}\right] \approx -\frac{P(t)}{100}. \tag{2.62}$$

Consequently,

$$\frac{P(t)}{100} = 1 - \frac{Y(t)}{100} = K_1 t. \tag{2.63}$$

For the zero order reaction, we have

$$Y(t) - 100 = -K_0 t, \tag{2.64}$$

or

$$1 - \frac{Y(t)}{100} = \frac{K_0}{100} t.$$

By comparing Equations 2.63 and 2.64, it appears that the first-order reaction is similar to a zero-order reaction with a rate constant equal to that of the first-order reaction normalized by the initial strength at time zero. As a result, when the degradation is small, it is very difficult to differentiate these two orders.

As indicated earlier, Model 2.37 is employed to describe the relationship between degradation and time. Suppose there are r_{ij} replicates at time t_j and temperature T_i. The test statistic for lack of fit can be applied to Model 2.37. The sum of squares of pure error, denoted by $SSPE(h)$ can then be obtained by subtracting the sum of squares of residuals from $SSPE(h)$. The degrees of freedom for $SSPE(h)$ and $SSLF(h)$ are given by

$$df(SSPE(h)) = \sum_{i=1}^{I}\sum_{j=1}^{J}(r_{ij} - 1) = N - IJ,$$

$$df(SSLF(h)) = (N - I) - (N - IJ) = I(J - 1),$$

where

$$N = \sum_{i=1}^{I}\sum_{j=1}^{J} r_{ij}.$$

Similarly, the mean squares of pure error and lack of fit can be obtained by dividing the respective sums of squares by their corresponding degrees of freedom as follows:

$$MSPE(h) = \frac{SSPE(h)}{N - IJ},$$

$$MSLF(h) = \frac{SSLF(h)}{I(J - 1)}.$$

TABLE 2.3:   ANOVA Table for Lack of Fit for Equation (2.37)

Source of Variation	df	Sum of Squares	Mean Squares	*F*-value
Regression	I	SSR(h)	MSR(h) = SSR(h)/I	$F = \frac{\text{MSR}(h)}{\text{MSE}(h)}$
Residual	$N - I$	SSE(h)	MSE(h) = SSE(h)/($N - 1$)	
Lack of fit	$I(J - 1)$	SSLF(h)	MSLF(h) = SSLF(h)/$I(J - 1)$	$F_{LF} = \frac{\text{MSLF}(h)}{\text{MSPE}(h)}$
Pure error	$N - IJ$	SSPE(h)	MSPE(h) = SSPE(h)/($N - IJ$)	
Total	N	SST(h)		

Source: Chow, S.C. and Liu, J.P. (1995). *Statistical Design and Analysis in Pharmaceutical Science,* Marcel Dekker, New York.

Table 2.3 provides the analysis of variance table, which partitions the residual sum of squares into *SSPE* and *SSLF*. Model 2.37 is considered adequate for a description of the relationship between degradation and time if we fail to reject the null hypothesis of no lack of fit. The null hypothesis of no lack of fit is rejected at the α level of significance if

$$F_{LF} = \frac{MSLF(h)}{MSPE(h)} > F_{\alpha, I(J-1), N-IJ}, \qquad (2.65)$$

where $F_{\alpha, I(J-1), N-IJ}$ is the αth upper quantile of a central F distribution with $I(J - 1)$ and $N - IJ$ degrees of freedom. If we fail to reject the null hypothesis of no lack of fit at the α level of significance, then $\hat{\sigma}^2(h)$ given in Model 2.40 provides an unbiased estimate for the error variance. However, if the null hypothesis of no lack of fit is rejected, Model 2.37 is considered inadequate. In this case one needs to carefully examine residual plots for possible outliers. Residual plots may also provide useful information for alternative models.

If Model 2.37 is considered adequate for a linear relationship between degradation and time, the next step is to investigate whether the Arrhenius equation can provide a satisfactory description for the relationship between degradation and temperature. Note that there are I unknown parameters in Model 2.37, while the Arrhenius Equations 2.42 and 2.43 consist of only two unknown parameters. Therefore, if the Arrhenius equation is adequate, no statistically significant increase in sum of squares of residuals would occur. The sum of squares of residuals under the Arrhenius equation is given by

$$SSE_A(h) = \sum_{i=1}^{I} \sum_{j=1}^{J} \left[D_{ij}(h) - \hat{D}_{ij}(h) \right]^2, \qquad (2.66)$$

where $\hat{D}_{ij}(h)$ is as given in Equation 2.48.

Let $SSE_L(h)$ be the residual sum of squares obtained from Model 2.37. The sum of squares due to the lack of fit under the Arrhenius equation is then given by

$$SSLF_A(h) = SSE_A(h) - SSE_L(h). \qquad (2.67)$$

It follows that the null hypothesis of no lack of fit for the Arrhenius equation is rejected at the α level of significance if

$$F_A = \frac{SSLF_A(h)/(I - 2)}{MSL_L(h)} > F_{\alpha, I-2, I(J-1)}, \qquad (2.68)$$

where $MSE_L(h)$ is as defined in Equation 2.40 and $F_{\alpha, I-2, I(J-1)}$ is the αth upper quantile for a central F distribution with $I - 2$ and $I(J - 1)$ degrees of freedom.

Note that it is suggested that the residuals from fitting that Arrhenius equation be examined thoroughly for special patterns. When the null hypothesis of no lack of fit is rejected, it is useful to examine the nature of inadequacy and the departure from the Arrhenius equation by plotting the logarithm of the estimates of the rate constants obtained from Model 2.37 versus the inverse of the absolute temperature.

2.4 An Example

In this section we use the data set adopted from Carstensen (1990) to illustrate statistical methods discussed in previous sections for determination of a tentative expiration dating period. This data set was obtained from an accelerated stability testing study that consists of three temperatures: $35°$ C, $45°$ C, and $55°$ C. Different time points were used for different temperatures: 0, 1, 2, and 3 months at $35°$ C; 0, 1, and 3 months at $45°$ C; and 0, 0.5, and 2 month at $55°$ C. All data except those at initial time points (which are 100%) are reproduced in Table 2.4 along with log(strength), $D_{ij}(0)$, $D_{ij}(1)$, $K_{ij}^*(0)$, and $K_{ij}^*(1)$. The scatter plot of assay results versus time by temperature is provided in Figure 2.1 for the zero-order reaction. The scatter plot of the logarithm of assay results versus time given in Figure 2.2 is used to examine a possible first-order reaction. Both plots reveal that a simple linear regression model can provide a better description of the relationship between strengths and time points at higher temperature than at lower temperature. This can be explained in part by the fact that at most 3% was decomposed for the first three months at $35°$ C.

TABLE 2.4: Strength of an Accelerated Stability Testing Study

Temperature (°C)	Time (months)	Strength (mg/mL)	ln(strength)	$D_{ij}(0)$	$D_{ij}(1)$	$K_{ij}^*(0)$	$K_{ij}^*(1)$
35	1	99.5	4.600	−0.5	−0.005	−0.693	−5.296
35	2	98.0	4.585	−2.0	−0.020	0.000	−4.595
35	3	97.0	4.575	−3.0	−0.031	0.000	−4.590
45	1	98.0	4.585	−2.0	−0.020	0.693	−3.902
45	3	95.2	4.556	−4.8	−0.049	0.470	−4.111
55	0.5	97.5	4.580	−2.5	−0.025	1.609	−2.983
55	1	95.1	4.555	−4.9	−0.050	1.589	−2.991
55	2	90.4	4.504	−9.6	−0.101	1.569	−2.987

Source: Carstensen, J. T. (1990), *Drug Stability*. Marcel Dekker, New York.

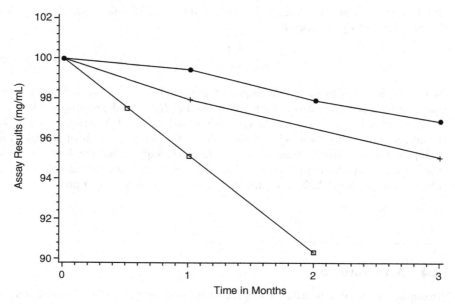

Figure 2.1:   Assay result (mg/mL) versus time in months. Circle 35°C; plus, 45°C; square, 55°C.

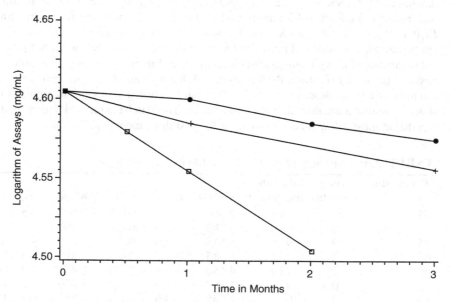

Figure 2.2:   Logarithm of assay result (mg/mL) versus time in months. Circle. 35°C; plus, 45°C; square, 55°C.

TABLE 2.5: Results of Estimation of Rate Constants by Equation 2.38

Order of Reaction	Temperature (°C)	Estimate	SE	T_b	p-Value
Zero	35	−0.9643	0.0748	−12.888	<0.0001
	45	−1.6400	0.0885	−18.525	<0.0001
	55	−4.8286	0.1222	−39.520	<0.0001
First	35	−0.0098	0.0007	−13.353	<0.0001
	45	−0.0168	0.0009	−19.378	<0.0001
	55	−0.0504	0.0012	−42.203	<0.0001

Test results for estimating rate constants under Model 2.37 for both orders are summarized in Table 2.5. Also included in the table are their standard errors, t statistics, and p-values for the null hypothesis of a negative reaction rate. It can be seen from the table that all rate constants are negative and all p-values are less than 0.0001. Hence, the null hypothesis of Equation 2.33 for each rate constant is rejected at the 5% level of significance. As a result, we conclude that all rate constants are statistically significantly smaller than 0. Table 2.6 also gives the ANOVA table for both orders. Note that although both orders under Model 2.37 yield a R^2 value greater than 99%, this does not guarantee that they gives a good fit. It can be easily verified that the studentized residual at time point 1 month and temperature 35° C exceeds 1.7 for both orders. Since no replicate tests were conducted at all combinations between time point and temperature, we are unable to perform a lack-of-fit test for this data set. However, to demonstrate the techniques, three strengths after 1 month at each of the three temperatures were added artificially to the data set. The values are 99.2, 97.8, and 96.1 for 35, 45, and 55° C, respectively. The ANOVA table for this modified data set of one data point under a zero-order reaction and Model 2.37 is given in Table 2.4. The residual sum of squares is given by 1.426 with 8 degrees of freedom, and the sum of squares for pure error is 0.565. Since there are only two replicates available at three combinations of time point and temperature, the degrees of freedom for the sum of squares of pure error is 3. Therefore, the sum of squares for lack of fit is equal to 0.861 with 5 degrees of freedom. As a result, the F-value for the lack-of-fit test is 0.914, with a p-value of 0.433. Hence, we conclude that there is no evidence of inadequacy for Equation 2.37 with respect to the modified data set.

TABLE 2.6: ANOVA Table Under Model 2.37

Order of Reaction	Source of Variation	df	Sum of Squares	Mean Squares	F-value	p-value	R^2
Zero	Regression	3	162.318	54.106	690.380	<0.0001	99.76%
	Residual	5	0.392	0.078			
	Total	8	162.710				
First	Regression	3	0.0175	0.0058	778.296	<0.0001	99.79%
	Residual	5	3.75×10^{-5}	7.496×10^{-6}			
	Total	8	0.0175				

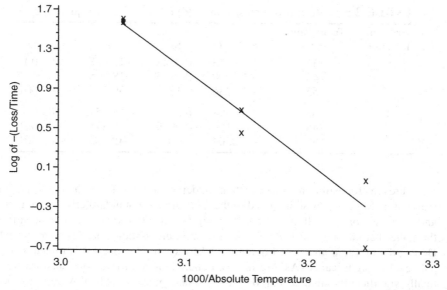

Figure 2.3:   Regression of logarithm of minus loss divided by time versus temperature. Log of $-$ (loss/time) $= \log[-(y - 100)/\text{time}]$.

The original data set is used to estimate the unknown parameters in the Arrhenius equation by a nonlinear regression technique. The estimates of $\alpha(h)$ and $\beta(h)$ obtained from simple linear regression Equation 2.44 can be used as initial values for nonlinear regression. These values are 29.57 and -9193.75 for the zero-order reaction and 25.30 and -9294.54 for the first-order reaction. Figures 2.2 and 2.3 display the fitted simple regression lines for the zero- and the first-order reactions, respectively. Estimates for $\alpha(h)$, $\beta(h)$, and their standard errors for Model 2.54 are presented in Table 2.7. The resulting estimates, which were obtained by fewer than seven iterations, are quite close to those obtained from a simple linear regression.

Suppose one wishes to establish a tentative expiration dating period at a marketing temperature 25° C such that the strength of the drug product is at least 90% of the label claim within the tentative expiration dating period. We may extrapolate temperature in the Arrhenius equation to a marketing temperature of 25° C with estimates provided in Table 2.8. Table 2.9 provides 95% lower confidence limits for the time on the logarithmic scale and hence the tentative expiration dating period at 25° C. Note that both orders produce a similar tentative expiration dating period. Under the zero-order reaction, the tentative expiration dating period is 26.6 months, while it is 28.6 months for the first-order reaction. The adequacy of the Arrhenius equation can be examined by testing for lack of fit. The residuals from the Arrhenius equation are given in Table 2.10 for both orders of reaction. Table 2.11 provides the results for the tests' of lack of fit. It can be seen from Table 2.11 that the p-values of the tests for lack of fit are less than 0.05 for both orders. Hence, at the 5% level of significance, the null hypothesis of no lack of fit is rejected for both orders. As a result, for this data set,

TABLE 2.7: ANOVA Table for the Modified Data for a Zero-Order Reaction under Model 2.37

Source of Variation	df	Sum of Squares	Mean Squares	F-value	p-value
Regression	3	181.974	60.658	340.195[a]	<0.0001
Residual	8	1.426	0.178		
Lack of fit	5	0.861	0.172		
Pure error	3	0.565	0.188	0.914	0.433
Total	11	183.400			

[a]Use the sum of squares of residuals as the error term.

TABLE 2.8: Summary of Estimates of Parameters in the Arrhenius Equation

Order of Reaction	Parameter	Estimate	Standard Error	Correlation
Zero	α	31.091	2.140	−0.9999
	β	−9682.682	699.439	
	α	26.883	2.071	−0.9999
	β	−9803.517	676.885	

TABLE 2.9: Summary of the Predicted Tentative Expiration Dating Period at a Marketing Storage Temperature

Order of Reaction	$\ln(\hat{t})$	$SE(\ln(\hat{t}))$	L_t	$\exp(L_t)$
Zero	3.6856	0.2071	3.2832	26.66
First	3.7458	0.2008	3.3556	28.66

TABLE 2.10: Residuals from the Arrhenius Equation

Temp (°C)	Time (months)	$D_{IJ}(0)$	$\hat{D}_{IJ}(0)$	$R(0)^*$	$D_{IJ}(1)$	$\hat{D}_{IJ}(1)$	$R(1)^*$
35	1	−0.5	−0.719	0.219	−0.0050	−0.0072	0.0022
35	2	−2.0	−1.439	−0.561	−0.0202	−0.0145	−0.0057
35	3	−3.0	−2.158	−0.842	−0.0305	−0.0217	−0.0088
45	1	−2.0	−1.931	−0.069	−0.0202	−0.0197	−0.0005
45	3	−4.8	−5.794	0.994	−0.0492	−0.0590	0.0098
55	0.5	−2.5	−2.441	−0.059	−0.0253	−0.0251	0.0002
55	1	−4.9	−4.882	−0.018	−0.0502	−0.0503	0.0001
55	2	−9.6	−9.764	0.164	−0.1010	−0.1005	0.0004

*$R(0)$ and $R(1)$ are residuals for reactions of order 0 and 1, respectively.

Accelerated Testing

TABLE 2.11: Summary of Residuals for Lack-of-Fit for the Arrhenius Equation

Order of Reaction	Source of Variation	df	Sum of Squares	Mean Squares	F-value	p-value
Zero	Residual from the Arrhenius equation	6	2.0953			
	Residual from Equation 2.38	5	0.3919	0.0784		
	Lack of fit	1	1.7034		21.73	0.0055
First	Residual from the Arrhenius equation	6	2.1074×10^{-4}			
	Residual from Equation 2.38	5	3.7481×10^{-5}	7.4961×10^{-6}		
	Lack of fit	1	1.7326×10^{-4}		23.11	0.0049

the Arrhenius equation is inadequate for describing the relationship between reaction rates and temperature. Failure of the Arrhenius equation for both orders becomes clear in Figures 2.5 and 2.6 when the logarithm of rate constants is plotted against the inverse of the absolute temperature. Figures 2.5 and 2.6 indicate that the relationship between rate constant and inverse of the absolute temperature is not linear.

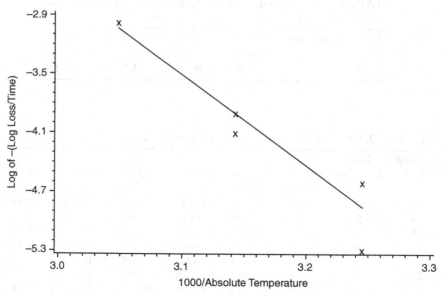

Figure 2.4: Regression of logarithm of minus log loss divided by time versus temperature. Log of $-$ (log loss/time) $= \log[-\log(y/100)/\text{time}]$.

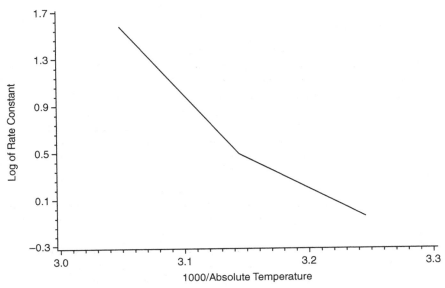

Figure 2.5:   Regression of logarithm of zero-order rate constant by time versus temperature.

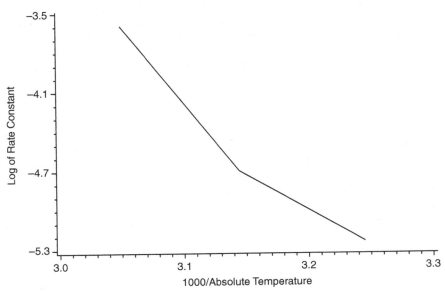

Figure 2.6:   Regression of logarithm of first-order rate constant versus temperature.

2.5 Discussion

The FDA stability guidelines provide some general guidance on conditions for stress testing. For example, it is suggested that the condition of H_2O_2 from 3% to 35% be considered for testing for minutes up to days at ambient temperature. For the conditions of heat and humidity, we may consider heat from 50° C to greater than 100° C for several days and humidity greater than 90% for up to several days, respectively. For acid and base, we may consider acid from 0.1 to 1 N HCl as low as pH $= 1$ and base from 0.1 to 1 N N_aOH as high as pH $= 13$. For light, the ICH Q1B guideline for stability recommendations to other wavelengths may be considered (ICH Q1B, 1996).

For the numerical example presented in the previous section, the null hypothesis of no lack of fit for the Arrhenius equation is rejected at the 5% level of significance for both reaction orders. One possible explanation for failure of the Arrhenius equation is that the study was not designed properly. This may be because no proper prior information regarding the rate of degradation was available for the selection of time points and temperature. For example, the maximum degradation in this data set is only 9.6% after a storage period of 2 months at 55° C. As indicated earlier, if the proportion of degradation is less than 15% of the initial strength, it is difficult to distinguish a zero-order reaction from a first-order reaction. Davies and Hudson (1981) presented an example that showed that the departure from linearity of strengths versus time for a zero-order reaction does not become apparent until the degradation exceeds at least 40%. Therefore, unless the strength has decreased by more than 40%, one is unable to distinguish the first order from the zero order, and consequently, is unable to describe adequately the relationship among degradation, time, and temperature.

As indicated earlier, the primary objective of an accelerated stability study is to provide an accurate and reliable estimate of the tentative expiration dating period. To achieve this objective, it is important to select an efficient design that will provide the maximum information for accelerated stability testing. Since the amount of information provided by a design is a function of the inverse of the variance, it is recommended that the design generate an estimate of time on the logarithmic scale based on Equation 2.40 with the smallest possible variance.

The relationship among the amount of degradation, temperature, and time stated in Equations 2.15 and 2.16 implies that the difference between the degradation and time on the logarithmic scale is a linear function of the absolute temperature. Hence, the time on the logarithmic scale can be viewed as the residual between the prespecified maximum allowable degradation and that predicted by the Arrhenius equation. Consequently, the information from an accelerated stability testing study may be defined as the inverse of the variance for the estimate of the time on the logarithmic scale, which is a function of the variances of the estimates for the unknown parameters in the Arrhenius equation.

In general, the principles and sample size determination for designing a calibration experiment can be applied separately to the selection of time points and temperatures in an accelerated stability testing study. For example, to examine the possible departure from linearity, at least three design points should be used for both time and temperature. The range of time and temperature should be chosen for maximum degradation to determine the correct order of reaction. However, since the technique for determining the tentative expiration dating period is the extrapolation of temperature, it is preferable to select the lowest temperature in a stressed test: either lower than or as close to the marketing storage temperature as possible. In the following we provide some recommendations for temperatures and time points in the design selection of an accelerated stability testing program:

- Use at least three design points separately for time and temperature.

- Select the highest temperature such that accurate discrimination of the degradation of a first-order reaction at this temperature from a zero-order reaction can be reached in a short period of time, for examples, within three months.

- Select the lowest and highest temperature possible that are close to the marketing storage temperature, which will be used in long-term stability studies under ambient conditions.

- The lowest and highest temperatures should be selected as far apart as scientifically and physically possible. This defines the range of temperatures.

- The lower limit for time points is the initial point, at zero. The upper limit of the range for time points is suggested as the time point at which the amount of degradation for an adequate determination of reaction rates can be reached at the highest temperature selected in the second bullet point.

- Select at least one more data point within the range of temperatures and time points such that at least three data points are available for an investigation of a departure from linearity with respect to both time and temperature. Techniques such as factorial or fractional designs and designs for a response surface (e.g., central composite design) can also be applied here, provided that the time and temperature ranges have been determined.

- Replicates should be obtained at various combinations to test lack-of-fit. The number of replicates can be determined based on $\beta(h)$ in Model 2.54.

Equations 2.15 and 2.16 are a combination of linear and nonlinear functions. For example, degradation relates to time points in a linear fashion but depends on temperature in a nonlinear fashion. The recommendations discussed above fall under the assumption that the relationship between degradation and temperature is also linear. As a result, the recommendations above will not provide optimal designs. However, if one only considers the nonlinear relationship between reaction rate and temperature in Model 2.54, the method for construction of a design for nonlinear regression proposed by Box and Lucas (1959) might be useful. Let I be the number of temperatures employed in an accelerated stability study. Define a matrix Z of order $I \times 2$ as

$$Z = (Z_\alpha, Z_\beta),$$

TABLE 2.12: Means and Variances of
Logarithmic Strength by Temperature

Temperature (°C)	Mean	Variance
35	4.5866	0.00016
45	4.5705	0.00042
55	4.5463	0.01485

where

$$Z_\alpha = (Z_{\alpha 1}, \ldots, Z_{\alpha I})',$$
$$Z_\beta = (Z_{\beta 1}, \ldots, Z_{\beta I})',$$

and $Z_{\alpha i}$ and $Z_{\beta i}$ are as defined in Equation 2.56. Box and Lucas (1959) suggested selecting temperatures that maximize the determinant of matrix $Z'Z$. When $I = 2$, it can be shown that the temperatures that maximize $|Z'Z|$ are lower and upper limits of the temperature range. However, more research is needed for the selection of optimal designs that can accommodate both linear and nonlinear parts of Equations 2.15 and 2.16.

The means and variances of the logarithm of strengths for the data set given in Table 2.4 are presented in Table 2.12 by temperature. Table 2.12 reveals that the variance of the logarithm of strengths increases as the temperature increases. This indicates that the assumption of a constant error variance might be violated. In addition, it can be seen from Table 2.8 that the correlation between the estimates of the two unknown parameters in the Arrhenius equation is close to -1. Therefore, the nonlinear technique may not converge to obtain estimates owing to the almost perfect correlation. To alleviate these problems, Davies and Hudson (1981) suggested that the strengths be weighted by the inverse of sample variance at each temperature. In addition, the inverse of the absolute temperature in Equations 2.42 and 2.43 needs to be centralized to avoid high correlation. For more details, see Davies and Hudson (1981).

Chapter 3

Expiration Dating Period

As indicated earlier, for every drug product on the market, the FDA requires that an expiration dating period (or shelf-life) be indicated on the immediate container label. The expiration date provides the consumer with the confidence that the drug product will retain its identity, strength, quality, and purity throughout the expiration period of the drug product. If the drug fails to remain within the approved specifications for the identity, strength, quality, and purity, the drug product is considered unsafe and subject to recall. To provide such assurance, pharmaceutical companies usually conduct stability studies to collect, analyze, and interpret data on the stability of their drug products throughout the expiration period. According to the FDA stability guidelines, the time at which the average drug characteristic (e.g., potency) remains within an approved specification after manufacture is recommended as the shelf-life of the drug product (FDA, 1987, 1998).

Since the true shelf-life of a drug product is usually unknown, it is typically estimated based on assay results of the drug characteristics from a stability study conducted during the process of drug development. In the next section basic concepts for determining an expiration dating period as given in the FDA stability guidelines are briefly described. Following this concept, several methods including the method recommended by the FDA, the direct method, and the inverse method are introduced in Section 3.2. Also included in this section is a comparison of these methods. Other methods such as a nonparametric method based on ranks, the slope approach, and the method of interval estimate are discussed in Section 3.4. Some concluding remarks are given in the last section of this chapter.

3.1 Basic Concepts

According to the FDA stability guidelines, the expiration dating period or shelf-life of a drug product can be determined as the time at which the average drug characteristic (e.g., potency) remains within an approved specification after manufacture (FDA, 1987, 1998). It is suggested that an expiration dating period or shelf-life of a drug product be determined as the time point at which the 95% lower confidence bound of the mean drug characteristic (e.g., potency) intersects the approved lower specification of the drug product. The use of the one-sided 95% lower confidence bound of the mean degradation of the drug product is to assure that the drug product will remain within the approved specifications for the identity, strength, quality, and purity prior to the expiration date.

3.1.1   Drug Characteristics

Generally, there are different criteria for acceptable levels of stability with respect to chemical, physical, microbiological, therapeutic, and toxicological characteristics of drug products (*USP-NF*, 2000). The requirements of stability on these characteristics are also different from dosage form to dosage form. Table 3.1 lists drug characteristics for different dosage forms, which should be evaluated in a stability study. As indicated earlier, the objective of stability studies is to characterize the degradation of drug products in terms of some essential drug characteristics and consequently, to establish an expiration dating period. The approach suggested in the FDA stability guidelines for determining drug shelf-life is based primarily on a single drug characteristic such as strength. The strength of a drug product is defined as either: (a) the concentration of the drug substance or (b) the potency, that is, the therapeutic activity of the drug product, which can be determined by an appropriate analytical method (laboratory test) or by adequately developed and controlled clinical data. However, as indicated in the FDA stability guidelines, the strength of a drug product is interpreted as a quantitative measure of the active ingredient of a drug product as well as other ingredients requiring quantitation, such as alcohol and preservatives. For an analysis of stability data, the FDA requires that percent of label claim, not percent of initial average value, be used as the primary variable for strength.

TABLE 3.1:   Drug Characteristics for Different Dosage Forms

Dosage Form	Drug Characteristics
Tablets	appearance, friability, hardness, color, odor, moisture, strength, dissolution
Capsules	strength, moisture, color appearance, shape, brittleness, dissolution
Emulsions	appearance, color, odor, pH, viscosity strength
Oral solution and suspensions	appearance, strength, pH, color, odor, redispersibility, dissolution, clarity
Oral powder	appearance, pH, dispersibility , strength
Metered dose ophthalmic preparations	strength delivered dose per actuation, number of metered doses, color, clarity, particle size, loss of propellant, pressure, valve corrosion, spray pattern
Topical and ophthalmic preparations	appearance, clarity, color, homogeneity, odor, pH, redispersibility, consistency, particle size distribution, strength, weight loss
Small-volume parenterals	strength, appearance, color, particulate matter, pH, starility, pyrogenicity
Large-volume parenterals	strength, appearance, color, clarity, particulate matter, pH, volume, extractables, sterility, pyrogenicity
Suppositories	strength, softening, range, appearance, dissolution

As can be seen from Table 3.1, for a given dosage form, the FDA guidelines require that a number of drug characteristics be evaluated for determination of drug shelf-life. However, in most stability studies, shelf-life is usually determined based on the primary drug characteristic of interest such as strength (or potency) of the drug product rather than all of the drug characteristics. On the other hand, a drug product may have more than one active ingredient. In practice, to fulfill the FDA requirements, we may determine shelf-lives for each drug characteristic of each active ingredient and consider the minimum shelf-life if different drug characteristics of different active ingredients have different shelf-lives of the drug product. Chow and Shao (2007) proposed an alternative approach for determining drug shelf-life for drug products with multiple ingredients (or components). Their method will be discussed in detail in Chapter 9.

3.1.2 Model and Assumptions

For a given batch, let y_j be the assay result (percent of label claim) of a pharmaceutical compound at time x_j, $j = 1, \ldots, n$. The following simple linear regression model is usually assumed:

$$y_j = \alpha + \beta x_j + e_j, \quad j = 1, \ldots, n, \tag{3.1}$$

where α and β are unknown parameters, x_j's are deterministic time points selected in the stability study, and e_j's are measurement errors independently and identically distributed as a normal random variable with mean 0 and variance σ^2, denoted by $N(0, \sigma^2)$. Under model (3.1), the average drug characteristic at time x is $\alpha + \beta x$. Assuming that the drug characteristic decreases as time increases, that is, β in Model 3.1 is negative, the drug product expires if its average characteristic is below a given specification limit η. Thus, the true shelf-life, denoted by θ, is the solution of

$$\eta = \alpha + \beta x.$$

Hence, we have

$$\theta = \frac{\eta - \alpha}{\beta}.$$

Note that α and β are the intercept and slope of the degradation curve, where α is the average drug characteristic at the time of manufacture (i.e., $x = 0$), which is usually larger than η. Thus, $\theta > 0$. The slope β is also known as stability loss in the drug characteristic over time.

Let $\hat{\theta}$ be an estimator of the true shelf-life θ based on (y_j, x_j)'s. It is desirable that $\hat{\theta} \leq \theta$ be statistically evident; that is, $\hat{\theta}$ is a conservative estimator. According to the 1987 FDA stability guideline, the probability of $\hat{\theta} \leq \theta$ should be nearly 95%, that is, $\hat{\theta}$ is approximately a 95% lower confidence bound for θ. Thus, $\hat{\theta}$ has a negative bias of the same order of magnitude as the standard deviation of $\hat{\theta}$. Studying the magnitude of the bias of $\hat{\theta}$ is particularly important for pharmaceutical companies, because the closeness of $\hat{\theta}$ to θ is directly related to the bias of $\hat{\theta}$, and a less biased shelf-life estimator is preferred.

3.2 Shelf-Life Estimation

Following the concept of determination of a drug shelf-life as suggested by the FDA, several methods have been proposed in the literature (see, e.g., Pong (2000); Shao and Chow, 2001b). These methods are described below.

3.2.1 The FDA's Approach

Let $(\hat{\alpha}, \hat{\beta})$ be the least squares estimator of (α, β) based on stability data $(y_j, x_j)'s$ under Model 3.1. For any fixed time x, a 95% lower confidence bound for $\alpha + \beta x$ is given by

$$L(x) = \hat{\alpha} + \hat{\beta}x - \hat{\sigma}t_{n-2}\sqrt{\frac{1}{n} + \frac{(x - \bar{x})^2}{S_{xx}}}, \qquad (3.2)$$

where t_{n-2} is the 95th percentile of the t-distribution with $n - 2$ degrees of freedom, $\bar{x}$ is the average of x_j's,

$$\hat{\sigma}^2 = \frac{1}{n-2}\left(\frac{S_{yy} - S_{xy}^2}{S_{xx}}\right)$$

in which

$$S_{yy} = \sum_{j=1}^{n}(y_j - \bar{y})^2,$$

$$S_{xx} = \sum_{j=1}^{n}(x_j - \bar{x})^2,$$

$$S_{xy} = \sum_{j=1}^{n}(x_j - \bar{x})(y_j - \bar{y}),$$

and $\bar{y}$ is the average of y_j's. The FDA recommends the following as the shelf-life estimator:

$$\hat{\theta}_F = \inf\{x \geq 0 : L(x) \leq \eta\},$$

the smallest $x \geq 0$ satisfying $L(x) = \eta$. Thus, the points where Equation 3.2 intersects the acceptable lower specification limit η (if it exists) are the two roots of the following quadratic equation:

$$[\eta - (\hat{\alpha} + \hat{\beta}x)]^2 = \hat{\sigma}^2 t_{n-2}^2 \left[\frac{1}{n} + \frac{(x - \bar{x})^2}{S_{xx}}\right]. \qquad (3.3)$$

The two roots, denoted by x_L and x_U, of Equation 3.3 constitute the lower and upper limits of the 90% confidence interval of $(\eta - \alpha)/\beta$. Denote

$$SE(x) = \left\{\hat{\sigma}^2\left[\frac{1}{n} + \frac{(x - \bar{x})^2}{S_{xx}}\right]\right\}^{1/2},$$

and let

$$T_{\hat{\alpha}} = \frac{\hat{\alpha} - \eta}{SE(\hat{\alpha})} \quad \text{and} \quad T_{\hat{\beta}} = \frac{\hat{\beta}}{SE(\hat{\beta})}. \tag{3.4}$$

If the slope is statistically significantly smaller than zero and the intercept is statistically significantly larger than η, which is the acceptable lower specification limit at the 5% level of significance, that is,

$$\text{(a)} \; T_{\hat{\beta}} = \frac{\hat{\beta}}{SE(\hat{\beta})} < -t_{0.05,n-2}, \tag{3.5}$$

$$\text{(b)} \; T_{\hat{\alpha}} = \frac{\hat{\alpha} - \eta}{SE(\hat{\alpha})} > t_{0.05,n-2}, \tag{3.6}$$

then the 90% confidence interval for $(\eta - \alpha)/\beta$ is an inclusive and close interval $[x_L, x_U]$. In this case the expiration dating period of a single batch is defined as $\hat{\theta}_F = x_L$ (Kohberger, 1988). However, in other cases the 90% confidence intervals for $(\eta - \alpha)/\beta$ is either the entire real line or two disjoint open intervals; consequently, the expiration dating period is not defined.

Note that the above approach is based on the 95% lower confidence limit for mean degradation line. Its interpretation in terms of hypothesis testing is given below (Easterling, 1969). Under Model 3.1, the pth upper quantile of the distribution for the percent of label claim at a given time point $t = x$ is $\alpha + \beta x + z_p \sigma$, where z_p is the pth upper quantile of a standard normal distribution. The null hypothesis that the pth upper quantile of the distribution for the percent of label claim at time point $t = x_0$ is larger than the acceptable lower specification η can be stated as

$$H_0 : \alpha + \beta x_0 + \sigma z_p \geq \eta \quad \text{vs} \quad H_a : \alpha + \beta x_0 + \sigma z_p < \eta,$$

which can be rewritten as

$$H_0 : \eta - (\alpha + \beta x_0 + \sigma z_p) \leq 0 \quad \text{vs} \quad H_a : \eta - (\alpha + \beta x_0 + \sigma z_p) > 0.$$

Furthermore, the hypotheses can be expressed in terms of time point $t = x_0$ as follows (with $\beta < 0$):

$$H_0 : \frac{\eta - \alpha - \sigma z_p}{\beta} \leq x_0 \quad \text{vs} \quad H_a : \frac{\eta - \alpha - \sigma z_p}{\beta} > x_0. \tag{3.7}$$

For estimation of the shelf-life of a drug product, as specified in the 1987 FDA guideline, the mean degradation line is to be used. Hence, $p = 0.5$ is chosen. When $p = 0.5$ and $z_p = 0$, the hypotheses of 3.7 reduces to

$$H_0 : \frac{\eta - \alpha}{\beta} \leq x_0 \quad \text{vs} \quad H_a : \frac{\eta - \alpha}{\beta} > x_0. \tag{3.8}$$

Note that $(\eta - \alpha)/\beta$ is the time point at which the mean degradation line intersects the acceptable lower specification limit η. The null hypothesis of 3.7 is to test whether this time point is less than $t = x_0$. Therefore, if the null hypothesis is tested at the

5% level of significance, the corresponding set of x for which the null hypothesis of Equation 3.7 is not rejected at the 5% level of significance constitutes the 95% one-sided confidence interval, that is, (x_L, ∞) for $(\eta - \alpha)/\beta$. The lower limit of the 95% one-sided confidence interval for $(\eta - \alpha)/\beta$ is x_L, which can be obtained as the smaller root of the quadratic Equation 3.3.

3.2.1.1 Asymptotic Bias

By definition, $\hat{\theta}_F > \theta$ implies $L(\theta) > \eta$ and

$$P(\hat{\theta}_F > \theta) \le P(L(\theta) > \eta) = 5\%$$

since $L(\theta)$ is a 95% lower confidence bound for $\alpha + \beta\theta = \eta$. This means $\hat{\theta}_F$ is a conservative 95% lower confidence bound for θ. Shao and Chow (2001b) studied the asymptotic bias and asymptotic mean squared error of the above estimator suggested by the FDA. Define

$$A_n = \hat{\sigma}^2 t_{n-2}^2 \left(\frac{1}{n} + \frac{\bar{x}^2}{S_{xx}} \right), \quad B_n = -\frac{\bar{x}\hat{\sigma}^2 t_{n-2}^2}{S_{xx}}, \tag{3.9}$$

and

$$C_n = \frac{\hat{\sigma}^2 t_{n-2}^2}{S_{xx}}.$$

Without loss of generality, assume that S_{xx} is exactly of order n. Then A_n, B_n, and C_n are exactly of order n^{-1}. Thus, asymptotically, $\hat{\theta}_F$ is the unique solution of $L(x) = \eta$. A straightforward calculation shows that the solution is given by

$$\frac{1}{\hat{\beta}^2 - C_n} \{ (\eta - \hat{\alpha})\hat{\beta} + B_n - \sqrt{[(\eta - \hat{\alpha})\hat{\beta} + B_n]^2 - (\hat{\beta}^2 - C_n)[(\eta - \hat{\alpha})^2 - A_n]} \}.$$

Removing terms of order n^{-1}, we have

$$\hat{\theta}_F = \frac{\eta - \hat{\alpha}}{\hat{\beta}} - \frac{\sqrt{A_n \hat{\beta}^2 + 2B_n(\eta - \hat{\alpha})\hat{\beta} + C_n(\eta - \hat{\alpha})^2}}{\hat{\beta}^2} + o_p(n^{-1/2}). \tag{3.10}$$

From the asymptotic theory for the least squares estimators, and Taylor's expansion, we have

$$\left(\frac{\eta - \hat{\alpha}}{\hat{\beta}} - \frac{\eta - \alpha}{\beta} \right) \Big/ \frac{\alpha}{|\beta|} \sqrt{\frac{1}{n} + \frac{(\theta - \bar{x})^2}{S_{xx}}} \longrightarrow N(0, 1). \tag{3.11}$$

Since $\theta = (\eta - \alpha)/\beta$, the asymptotic expectation of $\hat{\theta} - \theta$ (i.e., $(\eta - \hat{\alpha})/\hat{\beta} - (\eta - \alpha)/\beta$) is 0. Since

$$\hat{\alpha} \longrightarrow_p \alpha, \ \hat{\beta} \longrightarrow_p \beta, \ \text{and} \ \hat{\sigma} \longrightarrow_p \sigma$$

the asymptotic expectation of the second term on the right-hand side of Equation 3.10 is given by

$$-\frac{\sigma t_{n-2}}{\beta^2}\sqrt{\left(\frac{1}{n}+\frac{\bar{x}}{S_{xx}}\right)\beta^2 - \frac{2\bar{x}(\eta-\alpha)\beta}{S_{xx}} + \frac{(\eta-\alpha)^2}{S_{xx}}}$$

$$= -\frac{\sigma t_{n-2}}{|\beta|}\sqrt{\frac{1}{n}+\frac{(\theta-\bar{x})^2}{S_{xx}}}. \tag{3.12}$$

This is the asymptotic bias of $\hat{\theta}_F$ as $n \longrightarrow \infty$ and is of order $n^{-1/2}$. Furthermore, it follows from Equations 3.11 and 3.12 that the asymptotic mean squared error of $\hat{\theta}_F$ is given by

$$\frac{\sigma^2(1+t_{n-2}^2)}{\beta^2}\left[\frac{1}{n}+\frac{(\theta-\bar{x})^2}{S_{xx}}\right]. \tag{3.13}$$

Note that stability studies are often conducted under controlled conditions so that the assay measurement error variance σ^2 is usually very small. This leads to the study of the *small error asymptotics*. When n is fixed and $\sigma \longrightarrow 0$, we have

$$\hat{\beta} = \frac{S_{xy}}{S_{xx}} = \frac{\sum_{j=1}^{n}(x_j-\bar{x})y_j}{S_{xx}}$$

$$= \beta + \frac{\sum_{j=1}^{n}(x_j-\bar{x})e_j}{S_{xx}}$$

$$= \beta + O_p(\sigma) \xrightarrow{p} \beta,$$

where $O_p(\sigma)$ denotes a random variable of order σ as $\sigma \longrightarrow 0$. This result holds because e_j/σ is $N(0, 1)$. Similarly, we have

$$\hat{\alpha} = \bar{y} - \hat{\beta}\bar{x}$$

$$= \alpha + O_p(\sigma) \xrightarrow{p} \alpha,$$

Furthermore, $(n-2)\hat{\sigma}^2/\sigma^2$ has the chi-square distribution with $(n-2)$ degrees of freedom. Thus Equation 3.10 holds with $o_p(n^{-1/2})$ replaced by $o_p(\sigma)$. The asymptotic $(\sigma \longrightarrow 0)$ bias of the second term on the right-hand side of Equation 3.10 is given by Equation 3.12, which is now of order σ. Using Taylor's expansion and the fact that $\hat{\alpha} - \alpha$ and $\hat{\beta} - \beta$ are jointly normal with mean 0 and covariance matrix

$$\frac{\sigma^2}{S_{xx}}\begin{pmatrix} \bar{x}^2+n^{-1}S_{xx} & -\bar{x} \\ -\bar{x} & 1 \end{pmatrix},$$

we conclude that Equation 3.11 holds when $\sigma \longrightarrow 0$ and n is fixed. Hence, the asymptotic bias and mean squared error of $\hat{\theta}_F$, in the case of $\sigma \longrightarrow 0$, are the same as those for the case of $n \longrightarrow \infty$, given by Equations 3.12 and 3.13, respectively.

Note that Equations 3.12 and 3.13 indicate that, when n and x_j's are fixed, the asymptotic bias and mean squared error of $\hat{\theta}_F$ depend mainly upon the noise-to-signal ratio $\sigma/|\beta|$. If $\sigma/|\beta|$ cannot be controlled to a desirable level, then an increase of sample size n is necessary to reduce bias and mean squared error.

3.2.2   The Direct Method

As discussed above, the labeled shelf-life is determined as the time point at which the 95% one-sided lower confidence limit for the mean degradation curve intersects the lower acceptable specification limit. Alternatively, we may consider the 95% lower bound for the time point at which the mean degradation curve intersects the lower acceptable specification limit as the labeled shelf-life. Shao and Chow (2001b) referred the above method as the *direct* method for obtaining a shelf-life estimator.

From the discussion above, by the asymptotic theory (either $n \longrightarrow \infty$ or $\sigma \longrightarrow 0$), we have

$$\left(\frac{\eta - \hat{\alpha}}{\hat{\beta}} - \theta\right) \bigg/ \frac{\hat{\sigma}}{|\hat{\beta}|} \sqrt{\frac{1}{n} + \frac{1}{S_{xx}}\left(\frac{\eta - \hat{\alpha}}{\hat{\beta}} - \bar{x}\right)^2} \longrightarrow N(0, 1).$$

Let z be the 95th percentile of the standard normal distribution. Then, an approximate (large n or small σ) 95% lower confidence bound for θ is given by

$$\hat{\theta}_D = \frac{\eta - \hat{\alpha}}{\hat{\beta}} - \frac{\hat{\sigma} z}{|\hat{\beta}|} \sqrt{\frac{1}{n} + \frac{1}{S_{xx}}\left(\frac{\eta - \hat{\alpha}}{\hat{\beta}} - \bar{x}\right)^2}.$$

Using A_n, B_n, and C_n given in Equation 3.2, the above estimator becomes

$$\hat{\theta}_D = \frac{\eta - \hat{\alpha}}{\hat{\beta}} - \frac{z}{t_{n-2}} \frac{\sqrt{A_n \hat{\beta}^2 + 2B_n(\eta - \hat{\alpha})\hat{\beta} + C_n(\eta - \hat{\alpha})^2}}{\hat{\beta}^2}. \tag{3.14}$$

Note that when $n \longrightarrow \infty$, $z/t_{n-2} \longrightarrow 1$. It follows from Equations 3.10 and 3.14 that

$$\hat{\theta}_D - \hat{\theta}_F = o_p(n^{-1/2}).$$

Hence, the shelf-life estimators obtained by using the FDA's method and the direct method are asymptotically equivalent, and their large-sample asymptotic bias and mean squared error agree. The small error asymptotic bias and mean squared error of $\hat{\theta}_D$ are given by

$$-\frac{\sigma z}{|\beta|} \sqrt{\frac{1}{n} + \frac{(\theta - \bar{x})^2}{S_{xx}}},$$

and

$$\frac{\sigma^2(1 + z^2)}{\beta^2} \left[\frac{1}{n} + \frac{(\theta - \bar{x})^2}{S_{xx}}\right],$$

respectively. When n is fixed, z/t_{n-2} is a fixed constant less than 1. Hence, $\hat{\theta}_D > \hat{\theta}_F$ holds asymptotically as $\sigma \longrightarrow 0$, that is, $\hat{\theta}_D$ is less conservative than $\hat{\theta}_F$. This result indicates that, when σ^2 is small, $\hat{\theta}_D$ is preferred.

3.2.3 The Inverse Method

Another shelf-life estimator can be obtained using so-called inverse regression method (Krutchkoff, 1967; Halperin, 1970). To derive the inverse estimator, we start with

$$x_j = \alpha^* + \beta^* y_j + e_j^*, \quad j = 1, \ldots, n \qquad (3.15)$$

which is the same as Model 3.1 except that x_j and y_j are switched. In a stability study, however, the x_j's are deterministic time points and the y_j's are assay results. Therefore, the error term e_j^* is not independent of y_j. Nevertheless, suppose that we fit Model 3.15 based on (x_j, y_j)'s. Since the true shelf-life is the x-value when the mean of y is η, the shelf-life estimator, denoted by $\hat{\theta}_I$, based on the inverse method is the 95% lower confidence bound for $\alpha^* + \beta^* \eta$. Treating Model 3.15 as an ordinary linear regression model, we obtain the least squares estimators of α^* and β^* as follows

$$\hat{\alpha}^* = \bar{x} - \bar{y} \frac{S_{xy}}{S_{yy}},$$

$$\hat{\beta}^* = \frac{S_{xy}}{S_{yy}}.$$

As a result, an unbiased estimator of the variance of $\hat{\alpha}^* + \hat{\beta}^* \eta$ can be obtained as

$$\frac{1}{n-2} \left(S_{xx} - \frac{S_{xy}^2}{S_{yy}} \right) \left[\frac{1}{n} + \frac{(\eta - \bar{y})^2}{S_{yy}} \right]$$

$$= \hat{\sigma}^2 \frac{S_{xx}}{S_{yy}} \left[\frac{1}{n} + \frac{(\eta - \bar{y})^2}{S_{yy}} \right].$$

Consequently, the shelf-life estimator based on the inverse method is given by

$$\hat{\theta}_I = \bar{x} + \frac{S_{xy}}{S_{yy}}(\eta - \bar{y}) \qquad (3.16)$$

$$- \hat{\sigma} t_{n-2} \sqrt{\frac{S_{xx}}{S_{yy}} \left[\frac{1}{n} + \frac{(\eta - \bar{y})^2}{S_{yy}} \right]}.$$

Under Equation 3.1 with S_{xx} having order n^{-1} as $n \longrightarrow \infty$, $\hat{\theta}_I$ has the same limit as

$$\bar{x} + \frac{\beta}{\beta^2 + \frac{\sigma^2 n}{S_{xx}}}(\eta - \alpha - \beta \bar{x}) = \frac{\frac{\sigma^2 n}{S_{xx}}}{\beta^2 + \frac{\sigma^2 n}{S_{xx}}} + \frac{\beta^2}{\beta^2 + \frac{\sigma^2 n}{S_{xx}}} \theta,$$

which is a convex combination of $\bar{x}$ and θ. Unless $\bar{x} = \theta$, $\hat{\theta}_I$ has a nonzero limiting bias as $n \longrightarrow \infty$. Since $\bar{x}$ is the average of the time values used in the stability study, it is usually much smaller than the true shelf-life θ. Hence, the limiting bias of $\hat{\theta}_I$ is negative, that is, $\hat{\theta}_I$ can be too conservative. If $\bar{x} < \theta$ for all n, then

$$\lim_{n \longrightarrow \infty} P(\hat{\theta}_I < \theta) = 1.$$

When n is fixed but $\sigma \longrightarrow 0$, the difference between the last term on the right-hand side of Equation 3.16 and the quantity on the right-hand side of Equation 3.12 is of the order $o_p(\sigma)$. Thus,

$$\hat{\theta}_I - \hat{\theta}_F = o_p(\sigma)$$

and the small error asymptotic properties of $\hat{\theta}_I$ are the same as those of $\hat{\theta}_F$. The inverse method has a better asymptotic performance when $\sigma \longrightarrow 0$ than if $n \longrightarrow \infty$ since Model 3.15 and 3.1 are asymptotically the same as $\sigma \longrightarrow 0$ but asymptotically different from $n \longrightarrow \infty$. Note that Models 3.1 and 3.15 are the same if and only if $\sigma = 0$, regardless of how large n is. The inverse method is appealing because of its simplicity. However, it is not valid unless $\sigma \longrightarrow 0$.

Chow and Shao (1990b) compared the two estimators (i.e., the FDA's approach and the inverse method) in a different setting for calibration in assay development, which is similar to the case where the two estimates of the time point $t = x$ at which the mean degradation curve intersects the lower acceptable specification limit. Denote the two estimates by $\hat{x}_F$ and $\hat{x}_I$, respectively. Chow and Shao (1990b) assessed the closeness between the two estimators in terms of their ratio (i.e., $\hat{x}_I / \hat{x}_F$) and relative ratio (i.e., $[\hat{x}_I - \bar{x}]/[\hat{x}_F - \bar{x}]$). In other words, we have

$$\frac{\hat{x}_I}{\hat{x}_F} = 1 + (1 - R^2)\left(\frac{\bar{x}}{\hat{x}_F} - 1\right) \tag{3.17}$$

and

$$\frac{\hat{x}_I - \bar{x}}{\hat{x}_F - \bar{x}} = R^2, \tag{3.18}$$

where

$$R^2 = \frac{S_{xy}^2}{S_{xx} S_{yy}}.$$

From Equations 3.17 and 3.18, we have the following observations:

- The difference between the two estimates is zero if and only if $R^2 = 1$ (i.e., there is a perfect fit between x and y or $\eta = \bar{y}$).

- Since $R^2 \leq 1$, the inverse estimate is always closer to $\bar{x}$ than $\hat{x}_F$.

- The distribution of the relative ratio is independent of the unknown shelf-life.

Chow and Shao (1990b) also evaluated the closeness between $\hat{x}_F$ and $\hat{x}_I$ in terms of the probabilities that $(\hat{x}_I - \bar{x})/(\hat{x}_F - \bar{x})$ and $\hat{x}_I / \hat{x}_F$ differ from unity by the small amount δ. Let $\lambda = \beta^2/\sigma^2$. Then, the probability that $(\hat{x}_I - \bar{x})/(\hat{x}_F - \bar{x})$ differs from unity by δ can be expressed as

$$p(\lambda) = P\left\{1 - \frac{\hat{x}_I - \bar{x}}{\hat{x}_F - \bar{x}} > \delta\right\} \tag{3.19}$$

$$= P\left\{F < \frac{(n-2)(1-\delta)}{\delta}\right\},$$

where

$$F = \frac{(n-2)R^2}{1-R^2}$$

has a noncentral F distribution with 1 and $n-2$ degrees of freedom and noncentrality parameter $S_{xx}\lambda$. Note that $\hat{x}_F$ is derived from the regression of y and x, where x is assumed to be fixed, while $\hat{x}_I$ can be viewed as it can be obtained from the regression of x on y, where x is assumed to be random and y is fixed. It can be verified that

$$\frac{R^2}{1-R^2} = \frac{SSR}{SSE},$$

where

$$SSR = \frac{S_{xy}^2}{S_{xx}},$$

and

$$SSE = S_{xx} - SSR.$$

Thus, the probability given in Equation 3.19 is a decreasing function of λ. Therefore, the probability is small if the ration $|\beta|/\sigma$ is large. To provide a better understanding, Chow and Shao (1990b) provided a table of $p(\lambda)$ with respect to λ (with $n=12$) for various values of δ (see Table 3.2). From Table 3.2, it can be seen that the difference between $\hat{x}_I$ and $\hat{x}_F$ is not appreciable when $\lambda \geq 152$ for $\delta = 1\%$. Chow and Shao (1990b) also examined the closeness between $\hat{x}_I$ and $\hat{x}_F$ in terms of the following probability:

$$P\left\{\left|\frac{\hat{x}_I}{\hat{x}_F} - 1\right| \geq \delta\right\}.$$

Since there exists no closed form for the distribution of $\hat{x}_I/\hat{x}_F$, Chow and Shao (1990b) suggested using the following approximation:

$$P\left\{\left|\frac{\hat{x}_I}{\hat{x}_F} - 1\right| \geq \delta\right\} \approx \begin{cases} t(\lambda, x_\eta) + s(\lambda, x_\eta) & \text{if } \delta < 1 - r \\ t(\lambda, x_\eta) & \text{if } \delta = 1 - r \text{ ,} \\ t(\lambda, x_\eta) + s(\lambda, x_\eta) - 1 & \text{if } \delta > 1 - r \end{cases} \quad (3.20)$$

where x_η is the true x value corresponding to $y = \eta$,

$$r = \frac{1}{(1 + \lambda^{-1}\Sigma^{-1})},$$

$$t(\lambda, x_\eta) = \Phi\left\{\left[(1 - r + \delta)^{-1}(1 - r)\bar{x} - x_\eta\right]\lambda^{1/2}\right\},$$

$$s(\lambda, x_\eta) = \Phi\left\{\left[x_\eta - (1 - r + \delta)^{-1}(1 - r)\bar{x}\right]\lambda^{1/2}\right\},$$

and $\Phi(x)$ is the standard normal distribution function. Note that for the above approximation, we assume that as $n \to \infty$, $\bar{x} \to \mu_x$ and $n^{-1}S_{xx} \to \Sigma$, where Σ is some

TABLE 3.2:   Values of $p(\theta)$, δ and $\lambda = S_{xx}\theta/n(n=12)$

	Values of λ for the following values of δ:									
$p(\theta)$	0.01	0.02	0.03	0.04	0.05	0.06	0.07	0.08	0.09	0.10
1.00	3	1	1	0	0	0	0	0	0	0
0.95	32	16	10	8	6	5	4	3	3	3
0.90	40	20	13	9	7	6	5	4	4	3
0.85	46	23	15	11	9	7	6	5	4	4
0.80	51	25	16	12	10	8	6	6	5	4
0.75	55	27	18	13	11	9	8	7	6	5
0.70	60	30	19	14	11	9	8	7	6	6
0.65	64	32	21	16	12	10	9	7	6	6
0.60	68	34	23	17	13	11	9	8	7	6
0.55	73	36	24	18	14	12	10	8	7	7
0.50	77	38	25	19	15	12	10	9	8	7
0.45	82	40	27	20	16	13	11	10	8	8
0.40	87	43	28	21	17	14	12	10	9	8
0.35	92	45	30	23	18	15	12	11	10	9
0.30	98	48	32	24	19	16	13	12	10	9
0.25	104	52	34	25	20	17	14	12	11	10
0.20	111	55	37	27	22	18	15	13	12	10
0.15	120	60	40	30	23	19	17	14	13	11
0.10	133	66	44	33	26	21	18	16	14	13
0.05	152	75	50	38	30	25	21	18	16	15
0.01	193	96	64	48	38	31	27	23	21	19

Source: Chow, S.C. and Shao, J. (1990b). *Journal of the Royal Statistics Society*, C, 39, 219–228.

positive number. Let $q(\lambda, x_\eta)$ be the function given on the right-hand side of Equation 3.13. Then, $q(\lambda, x_\eta)$ is a continuous function of δ since $s(\lambda, x_\eta) \to 0$ as $\delta \uparrow 1 - r$ and $s(\lambda, x_\eta) \to 1$ as $\delta \downarrow 1 - r$. Furthermore, since $r \to 1$ as $\lambda \to \infty$ (Σ is fixed), $t(\lambda, x_\eta) \to 0$ and $s(\lambda, x_\eta) \to 1$ for any $\delta > 0$. Hence, $q(\lambda, x_\eta)$ is small if λ is large. Note that probability Equation 3.20 can be approximated by

$$q(\hat{\lambda}, x_\eta) = \begin{cases} t(\hat{\lambda}, x_\eta) + s(\hat{\lambda}, x_\eta) & \text{if } \delta < 1 - R^2 \\ t(\hat{\lambda}, x_\eta) & \text{if } \delta = 1 - R^2 \\ t(\hat{\lambda}, x_\eta) + s(\hat{\lambda}, x_\eta) - 1 & \text{if } \delta > 1 - R^2, \end{cases}$$

where $\hat{\lambda}$ is an estimator of λ. It can be seen that $q(\hat{\lambda}, x_\eta)$ also depends on the unknown x_η.

3.2.4   Comparison of Methods

Shao and Chow (2001b) compared the finite sample performance of the FDA's approach ($\hat{\theta}_F$), the direct method ($\hat{\theta}_D$), and the inverse method ($\hat{\theta}_I$) for estimation of the expiration dating period of a drug product through a simulation study. They also studied whether the asymptotic bias and mean squared error formulas for Equations 3.13 and 3.14 described above are close to the bias and mean squared error given by simulation.

In their simulation study Shao and Chow (2001b) considered a typical study design:

$$x_j = 0, 3, 6, 9, 12, 18, \text{ and } 24 \text{ months}$$

with three replicates at each x_j. Thus, $n = 21$. Values of α (initial assay at month 0, or manufacture), β (rate of degradation), and η (product specification limit) were chosen to be 105, -0.5 and 90, respectively, so that $\theta = 30$. To see the asymptotic effect, they considered values of σ ranging from 0.1 to 2.0. Based on 2000 simulations, Table 3.3 lists (a) the bias (BIAS) and mean squared error (MSE) of $\hat{\theta}_F$, $\hat{\theta}_D$, and $\hat{\theta}_I$, (b) the asymptotic bias (ABIAS) and asymptotic mean squared error (AMSE) computed using Equations 3.13 and 3.14, and (c) the coverage probability (CP) when $\hat{\theta}_F$, $\hat{\theta}_D$, and $\hat{\theta}_I$ are considered to be 95% lower confidence bounds for θ. The results are summarized below:

- The performances of $\hat{\theta}_F$ and $\hat{\theta}_D$ are quite satisfactory, especially when σ is small. The coverage probabilities for $\hat{\theta}_F$ and $\hat{\theta}_D$ are close to 95% and never below 94%. Comparing $\hat{\theta}_F$ and $\hat{\theta}_D$, we found that $\hat{\theta}_D$ is slightly better when σ is small, whereas $\hat{\theta}_F$ is slightly better when σ is large.

- For $\hat{\theta}_F$ or $\hat{\theta}_D$, asymptotic bias and mean squared error from Equations 3.13 and 3.14 are very close to exact bias and mean squared error when σ is small. For large σ, the asymptotic bias (or the asymptotic mean squared error) is quite different from the exact bias (or the exact mean squared error).

- In general, $\hat{\theta}_I$ is too conservative unless σ is very small, which supports the results described in the previous section. Even when $\sigma = 0.2$, the bias and mean squared error of $\hat{\theta}_I$ are still much larger than those of $\hat{\theta}_F$ (or $\hat{\theta}_D$), and the coverage probability of $\hat{\theta}_I$ is over the nominal level by more than 2%.

3.2.5   Remarks

According to current regulatory practice of the FDA for approving shelf-life, the labeled shelf-life is limited to be within 12 months beyond the last stability testing date when the 95 to 105% of label claim is used as the specification limit. It is limited to 6 months beyond the last testing date when 90 to 110% of label claim is used as the specification limit. It should be noted, however, that the above comparison of methods did not take this regulatory practice into consideration. As indicated earlier, the asymptotic bias of $\hat{\theta}_F$ is proportional to $\sigma/|\beta|$ and could be large if $\sigma/|\beta|$ is not well controlled. Under the current regulatory practice that the labeled shelf-life must extrapolate to no more than 12 months beyond the last testing date, the bias can be better controlled. In this situation the difference between $\hat{\theta}_D$ and $\hat{\theta}_F$ is often very small for regulatory shelf-life.

TABLE 3.3: Simulation Averages of Bias (ABIAS), Mean
Squared Error (MSE), and Coverage Probability (CP)
of Shelf-Life Estimators

σ	Estimator	Bias	ABIAS	MSE	AMSE	CP
0.1	$\hat{\theta}_F$	−0.2002	−0.2044	0.0545	0.0557	0.9510
	$\hat{\theta}_D$	−0.1922	−0.1944	0.0513	0.0518	0.9460
	$\hat{\theta}_I$	−0.2136	−0.2044	0.0603	0.0557	0.9615
0.2	$\hat{\theta}_F$	−0.4042	−0.4088	0.2193	0.2230	0.9585
	$\hat{\theta}_D$	−0.3917	−0.3889	0.2093	0.2071	0.9525
	$\hat{\theta}_I$	−0.4571	−0.4088	0.2667	0.2230	0.9720
0.3	$\hat{\theta}_F$	−0.5850	−0.6132	0.4617	0.5017	0.9475
	$\hat{\theta}_D$	−0.5715	−0.5833	0.4462	0.4660	0.9430
	$\hat{\theta}_I$	−0.7020	−0.6132	0.6170	0.5017	0.9760
0.4	$\hat{\theta}_F$	−0.7757	−0.8176	0.8138	0.8920	0.9445
	$\hat{\theta}_D$	−0.7646	−0.7777	0.7972	0.8284	0.9425
	$\hat{\theta}_I$	−0.9824	−0.8176	1.1865	0.8920	0.9755
0.5	$\hat{\theta}_F$	−0.9437	−1.0219	1.2001	1.3937	0.9505
	$\hat{\theta}_D$	−0.9382	−0.9721	1.1909	1.2944	0.9485
	$\hat{\theta}_I$	−1.2599	−1.0219	1.9117	1.3937	0.9820
0.6	$\hat{\theta}_F$	−1.1588	−1.2263	1.7660	2.0069	0.9580
	$\hat{\theta}_D$	−1.1623	−1.1666	1.7742	1.8639	0.9580
	$\hat{\theta}_I$	−1.6088	−1.2263	3.0227	2.0069	0.9915
0.8	$\hat{\theta}_F$	−1.4868	−1.6351	2.9398	3.5678	0.9495
	$\hat{\theta}_D$	−1.5169	−1.5554	3.0291	3.3135	0.9535
	$\hat{\theta}_I$	−2.2552	−1.6351	5.8382	3.5678	0.9915
1.0	$\hat{\theta}_F$	−1.8407	−2.0439	4.4785	5.5747	0.9555
	$\hat{\theta}_D$	−1.9111	−1.9443	4.7305	5.1774	0.9615
	$\hat{\theta}_I$	−3.0114	−2.0439	10.128	5.5747	0.9970
1.5	$\hat{\theta}_F$	−2.5670	−3.0658	8.8510	12.543	0.9490
	$\hat{\theta}_D$	−2.7854	−2.9164	9.8985	11.649	0.9630
	$\hat{\theta}_I$	−4.9163	−3.0658	26.042	12.543	0.9990
2.0	$\hat{\theta}_F$	−3.2363	−4.0878	13.940	22.299	0.9475
	$\hat{\theta}_D$	−3.6868	−3.8885	16.659	20.710	0.9710
	$\hat{\theta}_I$	−6.9898	−4.0878	51.426	22.299	0.9990

Source: Shao, J. and Chow, S.C. (2001b). *Statistica Sinica,* 11,737–745.

3.3 Other Methods

In addition to the method suggested by the FDA, the direct method, and the inverse method, other methods for determining drug shelf-life are also available. These methods include a nonparametic method, a slope approach, and an interval estimate (see, e.g., Rahman, 1992; Chen et al. 2003; Kiermeier et al. 2004), which are briefly described below.

3.3.1 Nonparametric Method

Chen et al. (2003) considered a nonparametric method for determining drug shelf-life based on ranks (see also Min, 2004). For a single batch, consider the model

$$y_j = \alpha + \beta x_j + e_j, \quad j = 1, \ldots, n.$$

Under the null hypothesis of $\beta = 0$, a standard rank test statistic is given by

$$U = \sum_{j=1}^{n} (x_j - \bar{x}) R(y_j),$$

where $\bar{x}$ is the mean of x_j's and $R(y_1)$, $R(y_2)$, $\ldots$, and $R(y_n)$ are the ranks of y_1, y_2, $\ldots$, y_n, respectively. When β is not necessarily zero, Chen et al. (2003) suggested considering the residuals, that is, $e_j = y_j - (\alpha + \beta x_j)$:

$$U(\beta) = \sum_{j=1}^{n} (x_j - \bar{x}) R(e_j).$$

Note that $E(U(\beta)|\beta) = 0$. As a result, it is reasonable to solve the unbiased estimating equation of $U(\beta)$ to obtain an estimate of β, that is,

$$U(\hat{\beta}) = 0.$$

Since $U(\beta)$ is a monotonically decreasing step function in β, there may not exist values of β that satisfy $U(\beta) = 0$. Alternatively, as indicated by Chen et al. (2003), one may obtain the rank estimate $\hat{\beta}$ by minimizing

$$D(\beta) = \sum_{j=1}^{n} w_j[R(e_j)]e_j,$$

where $w_j(.)$ is the nonconstant sequence of scores such that $w_j + w_{n-j+1} = 0$ (Jaeckel, 1972). Thus, $D(\beta)$ is a nonnegative, continuous, and convex function of β. Specifically, we may consider the Wilcoxon scores of

$$w(j) = \phi[j/n + 1],$$

where

$$\phi(u) = \sqrt{12}(u - 1/2).$$

By Theorem 5.2.3 of Hettmansperger (1984), we have

$$n^{1/2}(\hat{\beta} - \beta) \xrightarrow{D} N(0, \tau^2 \Sigma^{-1}),$$

where

$$\tau = \left(\sqrt{12} \int_0^\infty f^2(u) du\right)^{-1},$$

and

$$\Sigma = \lim_{n \to \infty} n^{-1} X_c' X_c,$$

which is the centered X (and is assumed to be positive definite) by

$$X_c = X - 1(\bar{x}_1, \bar{x}_2, \ldots, \bar{x}_p),$$

where $\bar{x}_i$ is the mean of ith column in X.

For the estimation of α, it should be noted that α is not estimable from $U(\beta)$ because the ranks are invariant to the constant intercept. In practice, we may consider the median of $(y_j - \hat{\beta} x_j)$, $j = 1, \ldots, n$ as an estimate of α. However, when the distribution of error terms is symmetric, α can be estimated by the median of the Walsh average of residuals of $(e_i + e_j)/2$, $1 \leq i \leq j \leq n$ (see, e.g., Hettmansperger, 1984; Hettmansperger and McKean, 1998). When $\hat{\alpha}$ is the median of $\hat{e}_j = y_j - \hat{\beta} x_j$, it can be shown that

$$n^{1/2}[(\hat{\alpha}, \hat{\beta})' - (\alpha, \beta)']$$

has an asymptotic distribution of multivariate normal with mean zero and variance-covariance

$$V = \tau^2 \begin{bmatrix} \{2f(0)\tau\}^{-2} + \mu' \Sigma^{-1} \mu & -\mu' \Sigma^{-1} \\ -\mu' \Sigma^{-1} & \Sigma^{-1} \end{bmatrix},$$

where μ is the mean vector of y_j's and $f(.)$ is the density function of e_j's. Note that when $\hat{\alpha}$ is the Walsh average, $n^{1/2}[(\hat{\alpha}, \hat{\beta})' - (\alpha, \beta)']$ has the asymptotic distribution of multivariate normal with mean zero and variance–covariance $V = \tau^2 \Lambda^{-1}$, where

$$\Lambda^{-1} = \lim_{n \to \infty} [1, X]'[1, X]$$

where 1 is the unit vector and X is the design matrix. As a result, an approximately one-sided 95% lower confidence bound for $E(y)$ is given by

$$L(x) = \hat{\alpha} + \hat{\beta} x + z_{0.05} \left[(1, X) \hat{V} (1, X)'\right]^{1/2}.$$

Thus, an estimate of shelf-life can be obtained at which the above 95% lower confidence bound for the mean degradation curve intersects the lower approved specification limit η.

3.3.2 Slope Approach

Rahman (1992) indicated that some pharmaceutical companies consider the following alternative method for estimation of drug shelf-life based on the 95% confidence interval of the slope of the true mean degradation line.

$$\hat{\beta}_L = \hat{\beta} - t_{0.05, n-2} SE(\hat{\beta})$$

Thus, a new estimated degradation line can be constructed from the same estimated intercept and $\hat{\beta}_L$ as follows:

$$L(x) = \hat{\alpha} + \hat{\beta}_L x$$

The expiration dating period can then be estimated as the time interval at which the new estimated degradation line intersects the acceptable lower specification limit η. We will refer to this approach as the slope approach. In other words, the slope approach yields an estimated shelf-life, denoted by $\hat{\theta}_S$, as the time point $t = x'_L$ such that

$$
\begin{aligned}
L(x'_L) &= \hat{\alpha} + \hat{\beta}_L x'_L \\
&= \hat{\alpha} + [\hat{\beta} - t_{0.05, n-2} SE(\hat{\beta})] x'_L \\
&= \eta.
\end{aligned}
$$

The above equation is a linear function of x'_L. Hence, the solution for x'_L always exists and is given by

$$
\begin{aligned}
\hat{\theta}_S &= x'_L \\
&= \frac{\eta - \hat{\alpha}}{\hat{\beta} - t_{0.05, n-2} SE(\hat{\beta})}.
\end{aligned}
$$

Rahman (1992) showed that there exists one and only one point of intersection between $L(x)$ and $L(x'_L)$, which is given as

$$x^* = \frac{\sum_{j=1}^{n} x_j^2}{2 \sum_{j=1}^{n} x_j}.$$

If the sampling time points in a stability study are those recommended in the FDA guideline for up to five years (i.e., 0, 3, 6, 9, 1, 18, 24, 36, 48, and 60 months), x^* is about 19.4 months, which is about one-third of the entire length of the study period. Rahman (1992) also pointed out that when $x \le x^*$,

$$P\{L(x) > L(x'_L)\} < 0.95$$

and when $x > x^*$,

$$P\{L(x) > L(x'_L)\} > 0.95$$

Furthermore, under usual practical situations, x_L is always longer than x'_L. Hence, x'_L is much more conservative than x'_L. Note that the estimated expiration dating period derived from the 95% lower confidence interval for the degradation rate is not based on the mean degradation line as suggested in the FDA guideline. Moreover, the probability statement about the relationship among time points, observed percent of label claim, and the acceptable lower specification limit cannot be made by this method. Therefore, this approach may not be appropriate for estimation of the expiration dating period following the basic concept as described in the FDA stability guidelines.

3.3.3   Interval Estimates

As suggested by the FDA stability guidelines, drug shelf-life can be determined as the time at which the 95% one-sided lower confidence limit for the mean degradation curve intersects the approved specification limit, which is often used as the labeled shelf-life. The labeled shelf-life provides the consumer with confidence that the drug product will retain its identity, strength, quality, and purity throughout the expiration period. Although there is no assurance that the drug product will retain its identity, strength, quality, and purity or that the drug product will be safe beyond the expiration dating period, the expired drug product may still be used by the consumer. It should be noted, however, that point estimates of shelf-life may overestimate or underestimate the true shelf-life. If the labeled shelf-life underestimates the true shelf-life, the drug product will retain its identity, strength, quality, and purity beyond the expiration period. If the labeled shelf-life overestimates the true shelf-life, the expired drug product is no longer safe. According to Lyon et al. (2006), of 119 drug products tested, all except for four or five were stable beyond their original expiration dates. Some were stable for as long as 10 years beyond their expiration dates. As a result, it is suggested that consumers follow expiration dates or beyond-use dates very carefully (see also Kellicker, 2006). The California Board of Pharmacy (CBP 2001) also advised that the pharmacy is not to dispense expired drugs. During an inspection, the inspector will randomly select some filled prescriptions to compare the expiration dates to the manufacturer's container. If the expiration date on the prescription label exceeds the manufacturer's date, this is a violation.

In general, it is believed that the drug product beyond the expiration period will maintain its identity, strength, quality, and purity for a short period of time. We will refer to this short period of time as the *safety margin* for the drug product. Pharmaceutical companies often receive queries regarding the safety of newly expired drug products. It is a common practice for them to suggest that consumers not take any expired drug products. However, it may be of interest for the pharmaceutical companies to establish an interval estimate rather than a point estimate for the drug shelf-life. An interval estimate may provide useful information regarding drug safety beyond labeled shelf-life. To establish an interval estimate, a conservative approach is to consider $(\hat{\theta}_L, \hat{\theta}_F)$, where $\hat{\theta}_L = \hat{\theta}_F - \Delta$, and Δ is the safety margin. In other words, we consider $\hat{\theta}_L$ (instead of $\hat{\theta}_F$) the labeled shelf-life.

3.4 Concluding Remarks

In the previous sections, we considered shelf-life estimation for a single batch. In practice, drug products are usually manufactured in multiple batches. Thus, the values of α and β in Model 3.1 may be different for different batches. As a result, there are four possible scenarios: (a) common intercept and common slope, (b) common intercept but different slopes, (c) different intercepts but common slope, and (d) different intercepts and different slopes. Different intercepts or slopes may indicate that there is batch-to-batch variation in intercepts or slopes. If there is no batch-to-batch variation, then the results described in the previous sections can be applied after combining data from different batches. If there is batch-to-batch variation, statistical methods for single batches described above cannot be applied based on pooled stability data to justify a single shelf-life. In this case two approaches are typically applied. The first approach is so-called stability analysis with fixed batches, and the second approach is known as stability analysis with random batches. These two approaches will be introduced in Chapter 5 and Chapter 6, respectively.

For determining drug shelf-life, it is often assumed that the primary drug characteristic such as potency for stability testing will decrease linearly over time. In some cases, however, this is not the case. A typical example is drug products containing levothyroxine sodium. In a stability study Won (1992) reported that levothyroxine sodium exhibits a biphasic first-order degradation profile with an initial fast degradation rate followed by a slower rate. This observation suggests a time-dependent degradation for drug products containing levothyroxine sodium. In this case, the usual approach for determining drug shelf-life is not appropriate.

For determining drug shelf-life, the FDA stability guideline recommends that Equation 3.1, be used. In Model 3.1 the response variable y is assumed to be a continuous variable. Under normality assumptions, the shelf-life can be determined. In practice, however, some drug characteristics, such as particle size, odor, color, and hardness, are discrete rather than continuous variables. For discrete or categorical response variables, Chow and Shao (2003) proposed a method for determining drug shelf-life following a concept similar to that described in the previous sections. More details regarding stability analysis with discrete response variables are given in Chapter 8.

The methods introduced in the previous sections are mainly for drug products with a single active ingredient. These methods are not applicable for drug products with multiple ingredients (components). For determining the shelf-life of a drug product with multiple ingredients, an ingredient-by-ingredient stability analysis may not be appropriate since multiple ingredients may have unknown interactions. In this case Chow and Shao (2007) proposed a statistical method assuming that ingredients are linear combinations of some factors. Their proposed method was found to be efficient and useful. Details of this method are provided in Chapter 9.

As indicated earlier, frozen drug products must be stored at several different temperatures, such as $-20°$ C, $5°$ C, and $25°$ C, to maintain their stability. Thus, the determination of shelf-life for frozen drug product involves the estimation of drug shelf-lives at different temperatures, which requires multiple-phase linear regression. Although we may obtain a combined shelf-life by applying the methods described in the previous sections based on stability data available at different temperatures (Mellon, 1991), this method, does not account for the fact that the shelf-life at the second phase would depend on the shelf-life at the first phase. As an alternative, Shao and Chow (2001a) proposed a method of determining drug shelf-lives for the two phases using a two-phase regression analysis following a similar concept and statistical principle as described in the previous sections. Their proposed method was shown to be quite satisfactory. More details can be found in Chapter 10.

Chapter 4

Stability Designs

In the pharmaceutical industry stability programs are usually applied at various stages of drug development. For example, at an early stage of drug development, a stability program is necessarily carried out to study the stability of bulk drug substances. The purpose is to evaluate excipient compatibility under various storage factors, such as heat, humidity, light, and container type. At a later stage it is required to conduct a stability program for the formulations used in preclinical and clinical studies to make sure the drug product is within *USP-NF* specifications during the entire study. For the proposed market formulation, a stability program is required to establish an expiration dating period applicable to all future batches of the drug product. For the production batches, it is a common practice to have a stability monitoring program in place to ensure that all drug characteristics remain within *USP-NF* specifications prior to the established expiration date. The success of a stability program requires an approved stability study protocol in which reasons for choosing an appropriate stability design should be described in detail. In addition, the 1987 FDA stability guideline indicates that a stability study protocol must describe: (a) how the stability study is to be designed and carried out, and (b) statistical methods to be used in analyzing the data. Since the design of a stability study is intended to establish an expiration dating period, the design should be chosen so that it can reduce bias and at the same time identify and control any expected or unexpected sources of variations. The goal for selection of an appropriate stability design is to improve the accuracy and precision of the established shelf-life.

The remainder of this chapter is organized as follows. In the next section, we provide some basic design considerations at the planning stage of a stability program. In Section 4.2 various long-term stability designs, including factorial design and fractional factorial designs, are introduced. Also included in this section are some useful new drug application (NDA) stability designs proposed by Nordbrock (1992). Section 4.3 focuses on commonly used matrixing and bracketing designs. Criteria proposed in the literature for comparing stability study designs are discussed in Section 4.4. A brief discussion is given in the last section of this chapter.

4.1 Basic Design Considerations

As pointed out by Chow and Liu (1995) and Lin and Chen (2003), a good stability study design is the key to a successful stability program. The program should start with a stability protocol that clearly specifies the study objective, the study design,

technical details of the drug substance and excipients, batch and packaging information, specification, time points, storage conditions, sampling plan, and planned
statistical analysis. The protocol should be well designed and followed rigorously,
and data collection should be complete and in accordance with the protocol. The
planned statistical analysis should be described in the protocol to produce the most
desirable outcome at the time of data analysis. Any deviations from the design makes
it difficult to interpret the resulting data. Any changes made to the design or analysis
plan without modification to the protocol or after the examination of the data collected should be clearly documented. Pogany (2006) provided the following list of
basic considerations for inclusion in a stability protocol:

- Study objectives and design

- Batch tested

- Container and closure system

- Literature and supporting data

- Testing plan

- Test parameters

- Test results

- Other requirements

- Conclusions

For the batch tested, Pogany (2006) suggested that: (a) batch number, (b) date of
manufacture, (c) site of manufacture, (d) batch size (kg) and units, (e) primary packing
materials, and (f) date of initial analysis be clearly specified in the study protocol. In
addition, the test results must be dated and signed by the responsible personnel with
a signature indicating Quality Assurance approval. In this section, some basic design
considerations as described in the 1987 FDA stability guideline are discussed (see
also Chow and Liu, 1995; Lin and Chen, 2003). These baisc design considerations
include background information, regulatory considerations, design factors, sampling
time considerations, sample size, statistical analysis, and other issues, which are
described below.

4.1.1 Background Information

When choosing an appropriate stability study design, some background information
must be obtained to ensure the success of the study. In practice, it is helpful to have
some knowledge regarding marketing requirements, manufacturing practice, previous
formulation study results, and the variability of the assay method. For marketing
requirements, the following questions may affect the direction of a stability study:

- What package types will be used for the drug product?

- What are the *USP-NF* specifications for the drug product?

- What is the desired shelf-life?

- When will the stability data be filed with the regulatory agency?

- Where is the drug product to be marketed?

In practice it is recognized that degradation of a drug product may differ from one package type to another. The specifications for the characteristics of the drug product can usually be found in the *USP-NF* (*USP-NF*, 2000). The information regarding which package type is to be used for the drug product is useful in determining whether the desired shelf-life can be achieved. Moreover, if a pharmaceutical company wants to file a submission in a short period of time, a short-term stability test such as an accelerated test may be desirable to establish a tentative shelf-life. Therefore, knowledge of the lead time prior to regulatory submission is critical in planning an appropriate stability study. Finally, different regulatory agencies in different countries (e.g., the European Union [EU] and Japan) may have different requirements on stability. Information on where the drug product is to be marketed is useful in devising a strategy at an early stage of planning a stability study.

In addition, knowledge of manufacturing practices is usually helpful in the design of a stability study. For example, answers to the following questions provide crucial information for the selection of an appropriate stability design.

- How many strengths of the same formulation will be manufactured?

- Will multiple strengths be made out of a common granulation batch?

- Will a common encapsulation batch be made into multiple package types?

An adequate stability study should be designed to evaluate the stability of the drug product across batch, strength, and package type. Other information such as previous formulation study results and the variability of the assay method are also helpful in the design of a stability study. Previous formulation study results may provide important information about factors that might affect the stability of drug products of this kind under certain storage conditions. The variability of an assay method, such as that between and within run coefficient of variations (CVs) are useful for sample size determination, which ensures the establishment of a reliable drug shelf-life.

4.1.2 Regulatory Considerations

For a stability study, the FDA requires that pharmaceutical companies describe how the study is to be designed and carried out. An appropriate stability design can help to achieve the objective of a stability study. The 1987 FDA stability guideline provides design considerations for long-term stability studies under ambient conditions. These design considerations include:

- Batch sampling considerations

- Container-closure and drug product sampling

- Sampling-time considerations

These designs considerations focus primarily on how data are to be collected. The purpose of these considerations is to ensure that the data are representative of all future production batches.

For batch sampling considerations, the FDA stability guidelines recommend that at least three batches, and preferably more, be tested to allow for batch-to-batch variability and to test the hypothesis that a single expiration dating period for all batches is justifiable (see also Ahn et al., 1995). For a consideration of container-closure and drug product sampling, the FDA emphasizes that the selection of containers from the batches chosen for inclusion in the study should be carried out so as to ensure that the samples chosen represent the batch as a whole. For sampling time, the FDA stability guidelines suggest that stability testing be done at 3-month intervals during the first year, 6-month intervals during the second year, and annually thereafter. For a drug product that degrades rapidly, more frequent sampling is necessary.

4.1.3   Design Factors

As a drug product may be available in different strengths and different container sizes, a long-term stability study may involve the following design factors: strength, container size, and batch (or lot). As indicated in Chow and Liu (1995) and Nordbrock (1992), the following hypotheses should be examined:

- Do all package-by-strength combinations have the same stability at room temperature?

- Do all packages have the same stability at room temperature?

- Do all batches have the same stability at room temperature?

- Do all strengths have the same stability at room temperature?

- Do all storage conditions have the same stability for all packages?

- Do all storage conditions have the same stability for all batches?

- Do all storage conditions have the same stability for all strengths?

An appropriate stability design can provide useful information to address these questions. For example, we may examine the following hypotheses at room temperature in order to address some of the above questions.

- Are degradation rates among packages consistent across strengths?

- Are degradation rates the same for all packages?

- Are degradation rates the same for all strengths?

- Are degradation rates the same for all batches?

If we fail to reject the hypotheses above, stability data may be pooled to establish a single shelf-life that can reflect the expiration period for all batches manufactured under similar circumstances. The establishment of a single shelf-life applicable to all future batches is the primary objective of a stability study. However, if the hypothesis that degradation rates among packages are consistent across strengths is rejected, the differences in degradation rate between package types are not consistent across strengths. In this case, there is a significant interaction between package type and strength. Thus, the data cannot be pooled and shelf-lives for each combination of package types and strengths should be established. In addition to interaction, an appropriate design should be able to separate any possible confounding effects.

4.1.4 Sampling Time Considerations

The 1987 FDA stability guideline suggests that sampling times be chosen so that any degradation can be adequately characterized (i.e., at a sufficient frequency to determine with reasonable assurance the nature of the degradation curve). Usually, the relationship can be represented adequately by a linear, quadratic, or cubic function on an arithmetic or logarithmic scale of the percent of label claim. As a rule of thumb, more frequent sampling should be taken where a curvature of the degradation curve is expected to occur in order to adequately characterize degradation of the drug product. The 1987 FDA stability guideline also encourages testing an increased number of replicates at later sampling times, particularly the latest sampling time, because this will increase the average sampling time toward the desired expiration dating period.

Assuming that the drug characteristic is expected to decrease with time, for long-term stability studies under ambient conditions such as NDA stability studies, the 1987 FDA stability guideline suggests that stability testing be done at 3-month intervals during the first year, 6-month intervals during the second year, and annually thereafter. However, for drug products predicted to degrade rapidly, more frequent sampling is necessary. For marketing stability studies, less frequent sampling is usually considered from more batches. Note that if a reduced design (e.g., some selected levels of combinations of design factors with fewer sampling time points) is to be applied to a long-term stability study (For example, a 2 year stability study), the FDA suggests that every selected level combination of design factors such as batch, stength, and package type should be tested at 0, 12, and 24 months and at least one additional time point within the first year. As indicated earlier, the purposes of a stability test are to characterize the degradation of the drug product and, consequently, to establish drug shelf-life. For these purposes, the following statistical issues are of concern and must be considered when planing a stability study:

- How can the number and allocation of time points be selected such that the degradation of ingredients of a drug product will be adequately characterized?

- How frequently is sampling necessary to have a desired degree of accuracy and precision for the estimated shelf-life?

- Is it necessary to have replicates at each sampling time point?

- How can the number of assays at each sampling time point be allocated efficiently if a fixed number of assays are to be done?

Mathematically, if the degradation curve is linear, it can be determined uniquely by two time points. One may consider having stability testing at the initial (i.e., the time at which the batch is manufactured) and the latest sampling time points. However, it should be noted that, statistically, there are no degrees of freedom for the error term if only two sampling time points are considered. In practice pharmaceutical companies are usually interested in acquiring stability information regarding the drug product within a short period of time after the drug product is manufactured. If only two time points are considered in a long-term stability study, no information about degradation can be obtained between the two time points. In addition, if the second sampling time point is too close to the initial point, the fitted degradation line may not be reliable for establishing an expiration dating period beyond the time interval under study because we may not observe a significant degradation in a short period of time.

It is therefore of interest to study the impact of the frequency of the sampling on the characterization of the degradation curve and the determination of drug shelf-life. Moreover, the 95% confidence interval for mean degradation at time points such as the initial and final time points, which is further away from the middle of the range of time points, could be very wide. Consequently, the estimated shelf-life may not be reliable. The reliability of an estimated drug shelf-life beyond the time interval under study is then an interesting and important topic in stability analysis, which is worthy of further research.

4.1.5 Sample Size

Suppose the desired shelf-life of a drug product is 4 years. It is easy to determine the number of stability tests for a given combination of design factors according to the FDA's suggestions for sampling intervals. In practice, however, it may be too costly or time consuming to perform stability tests at every time point for each combination of design factors. Therefore, pharmaceutical companies often have an interest in reducing the total number of stability tests by either reducing the combinations of design factors or performing stability tests at the selected time points. This is often done, provided that the reduction of stability tests can reach an acceptable degree of precision for the established shelf-life without losing much information.

Basically, before the sample can be chosen in an appropriate stability design, the following issues should be addressed.

- How many observations are to be taken?

- How large a difference in degradation rate is to be detected between design factors?

- How much variation is present?

For the first issue, if stability tests are to be performed at every sampling interval, the total number of assays can be determined. In some cases, as indicated earlier, pharmaceutical companies may not be able to perform stability tests at every time point owing to limited resources. Instead, stability tests may be done at selected time points. Suppose that a pharmaceutical company is able to handle N assays according to its capacity or budget constraint. These N assays may not be large enough to cover every time point but will cover some selected time points or a few time points twice. In other words, we may test the stability at some time points once or have replicates at a few time points. In this case it is important to determine how many observations are to be taken at each time point, which can have an impact on the estimated shelf-life.

When there are a number of design factors in a stability study, it is often of interest to investigate the impact of these design factors on stability. For example, it is of interest to determine whether degradation rates are the same for all packages. To address this questions, the sample size should be chosen so that there is sufficient statistical power for detection of a meaningful difference in stability loss between packages. Therefore, it is important to determine how large a difference in degradation rate detected between packages is meaningful. The sample size selected should have sufficient power for detection of meaningful differences in degradation rate among the design factors under study.

Sample size determination and justification are usually done based on a prestudy power analysis, which is very sensitive to the variability. Thus, it is important to have some prior knowledge regarding the variation. The variation may include variabilities from different sources, such as location, analyst, and manufacturing process. For example, with large containers, dosage units near the cap of a bottle may have different stability properties than dosage units in other parts of the container. Thus, it may be desirable to sample dosage units from all parts of the container. In this case the location in the container from which they are drawn should be identified and taken into consideration for sample size justification.

4.1.6 Statistical Analysis

For a given stability design, statistical methods used for data analysis should reflect the design to provide a valid statistical inference for the established shelf-life. For example, we may test the hypotheses listed in the previous subsections based on a linear model using PROC from GLM of SAS. Let Y be the potency (expressed as percent of label claim) and T be the time (in months), which is a continuous covariate. Also, let B, S, and P denote the batch, strength, and package, respectively. Note that B, S, and P are class variables. PROC GLM models are as follows (see also, Lin, 1990; Lin and Chow, 1992; Lin and Lin, 1993; Chen et al., 1997):

- Model 1: $Y = B \ S \ B*S \ T \ B*T \ S*T \ P*T \ S*P*T$

- Model 2: $Y = B \ S \ B*S \ T \ B*T \ S*T \ P*T$

Note that the term P is not included in the models above because there is no reason to believe that the initial potency will be different for different package types. Each model has separate intercepts for each batch-by-strength combination. Model 1 has separate slopes for each package-by-strength combination and has separate slopes for each batch. Model 2 has separate slopes for each batch, for each strength, and for each package. Model 1 is used to test the $P*S*T$ term and associated contrasts. If the $P*S*T$ term and associated contrasts are not significant, model 2 is used for further analysis.

In the models above, we assume that the term B is a fixed class variable. However, Chow and Shao (1991) indicate that the term B should reflect the shelf-life for all future batches. The following chapters discuss statistical methods for estimating of drug shelf-life under the assumptions that: (a) the term *batch* is a fixed effect (Chapter 5), (b) the term *batch* is a random effect (Chapter 6), and (c) the term *batch* is a linear mixed effects model (Chapter 7).

4.1.7 Other Considerations

The 1987 FDA guideline indicates that an appropriate stability design should take into consideration the variability of individual dosage units, of containers within a batch, and of batches to ensure that the resulting data for each batch are truly representative of the batch as a whole and to quantify the variability from batch to batch. Other sources of variation may also affect the efficiency of the stability design. These sources of variation may include variabilities from different assay methods, different analysts, different laboratories, and different locations or sites. In addition, expected and unexpected variabilities that may occur during each stage of the manufacturing process can have an impact on the stability design. An appropriate stability design should be able to avoid bias and achieve the minimum variability. As a result, the FDA requires that the analytical method for stability-indicating assay be validated according to some validation performance characteristics before it can be applied to the intended stability study.

It should also be noted that for any drug product that is intended for use as an additive to another drug product, the possibility of incompatibilities may exist. In such a case the FDA guidelines require that a drug product labeled to be administered by addition to another drug product (e.g., parenterals or aerosols) be studied for stability and compatibility in a mixture with the other product. A suggested stability protocol should provide for tests to be conducted at 0-, 6-, to 8-, and 24-hour intervals, or as appropriate over the intended period of use. These tests should include assay of the drug product and additive, pH (especially for unbuffered large-volume parenterals), color, clarity, particulate matter, and interaction with the container.

4.2 Long-Term Stability Designs

In this section we introduce some stability designs that are commonly used when conducting stability studies. These designs include complete (or full) factorial design, fractional factorial design, and some reduced designs proposed by Nordbrock (1992). To illustrate these designs consider the following example.

Suppose a newly developed pharmaceutical compound is to be manufactured in tablets with three different strengths: 15 mg, 30 mg, and 60 mg. To fulfill the needs of various markets, it is suggested that three packaging types be taken into consideration: glass bottle, PVC (polyvinyl chloride) blister, and PE (polyethylene) tubes. In addition, to study the impact of heat and moisture on the product over the desired expiration dating period (e.g., 60 months), we have to store three batches of the finished product to simulating the adverse effects of storage under the conditions to which the product might be subjected during distribution, shipping, handling, and dispensing. The storage conditions of interest are:

- 21° C and 45% relative humidity

- 25° C and 60% relative humidity

- 30° C and 35% relative humidity

- 30° C and 70% relative humidity

In other words, we store three batches of three different strengths kept in three types of packages under four different conditions. In this section we assume that the degradation curve is linear. If there is an exponential decay, it may be linearized by transformation. The drug characteristic of interest is the potency (percent of label claim). Stability testing is to be done at the following times if we consider stability testing up to 4 years:

$$T_1 = \{0, \ 3, \ 6, \ 9, \ 12, \ 18, \ 24, \ 36, \ 48\}$$

In the interest of balance, we assume that sampling times are fixed across all design factors.

4.2.1 Factorial Design

Suppose we are interested in conducting a stability study under ambient conditions (e.g., 25° C room temperature and 60% relative humidity). A full factorial design consists of $3^3 = 27$ combinations (see Table 4.1). If each combination is to be tested at T_1 time points, there are a total of

$$N = 3 \times 3 \times 3 \times 10 = 270$$

assays. In practice, if every batch by strength-by-package combination is tested (i.e., a complete factorial design is used), a substantial expense is involved. Besides, it is in the best interest of the pharmaceutical companies that a longer shelf-life can be claimed by testing fewer batches as to strength-by-package combinations within a short period of time. Therefore, for considerations of time and cost, a fractional factorial design is often used to reduce the total number of tests (or assays). Although a fractional factorial design is preferred in the interest of reducing the number of tests (i.e., cost), it has the following disadvantages. First, if there are interactions such as a strength-by-package interaction, the data cannot be pooled to establish a single shelf-life. In this case it is recommended that individual shelf-lives be established for each combination of strength and package. However, we may not have three batches for each combination of strength and package for a fractional factorial design. Second, we may not have sufficient precision for the estimated drug shelf-life.

As pointed out by Lin and Chen (2003), a good stability study design is the key to a successful stability program. The program should start with a stability proto-col that clearly specifies the study objective, the study design, technical details of

TABLE 4.1: 3^3 Complete Factorial Design

Combination	Batch	Strength	Package
1	1	15	Bottle
2	1	15	Blister
3	1	15	Tube
4	1	30	Bottle
5	1	30	Blister
6	1	30	Tube
7	1	60	Bottle
8	1	60	Blister
9	1	60	Tube
10	2	15	Bottle
11	2	15	Blister
12	2	15	Tube
13	2	30	Bottle
14	2	30	Blister
15	2	30	Tube
16	2	60	Bottle
17	2	60	Blister
18	2	60	Tube
19	3	15	Bottle
20	3	15	Blister
21	3	15	Tube
22	3	30	Bottle
23	3	30	Blister
24	3	30	Tube
25	3	60	Bottle
26	3	60	Blister
27	3	60	Tube

drug substance and excipients, batch and packaging information, specification, time points, storage conditions, sampling plan, and planned statistical analysis. The protocol should be well designed and followed rigorously, and data collection should be complete and in accordance with the protocol. The planned statistical analysis should be described in the protocol to avoid the appearance of choosing an approach to produce the most desirable outcome at the time of data analysis. Any departure from the design makes it difficult to interpret the resulting data. Any changes made to the design or analysis plan without modification to the protocol or after the examination of the data collected should be clearly identified.

A full design can provide not only valid statistical tests for the main effects of the design factors under study, but also estimates for interactions with better precision. Hence, the precision of the estimated drug shelf-life for a full design is better than a reduced design. A reduced design is preferred to a full design for the purpose of reducing the number of test samples and consequently the cost. However, it has the following disadvantages: (a) one may not be able to evaluate some interaction effects for certain designs; (b) if there are interactions between two factors, the data cannot be pooled to establish a single shelf-life; (c) if there are many missing factor combinations, there may not be sufficient precision for the estimated shelf-life.

In practice it is generally impossible to test the assumption that the higher-order terms are negligible. Hence, if the design does not permit the estimation of interactions or main effects, it should be used only when it is reasonable to assume that these interactions are very small. This assumption must be made on the basis of theoretical considerations of the formulation, manufacturing process, chemical and physical characteristics, or data from other studies. Thus, to achieve a better precision of the estimated shelf-life, a design should be chosen to avoid possible confounding and interaction effects. Once the design is chosen, statistical analysis should reflect the nature or the design selected.

4.2.2 Reduced Designs

An appropriate stability design can help achieve the objective of a stability study. Basically, a stability design consists of two parts: the selection of design factors (e.g., batch, strength, and package type) and the choice of sampling intervals (e.g., 3 month during the first year, 6 month during the second year, and yearly thereafter). We will refer to the set of sampling time points or intervals as *time protocol* or *time vector*. For selection of design factors, the stability designs commonly employed are the full factorial design and fractional factorial designs. In the interest of reducing the number of sampling intervals for stability testing, Nordbrock (1992) introduced various choices of subsets of sampling intervals for the intended stability studies for up to four years. Nordbrock (1992) considered the following subsets:

$$T_2 = \{0, 3, 9, 18, 36, 48\},$$
$$T_3 = \{0, 6, 12, 24, 48\},$$
$$T_5 = \{0, 3, 12, 36, 48\},$$
$$T_6 = \{0, 6, 18, 48\},$$
$$T_7 = \{0, 9, 24, 48\},$$
$$T_8 = \{0, 3, 9, 12, 24, 36, 48\},$$
$$T_9 = \{0, 3, 6, 12, 18, 36, 48\},$$
$$T_A = \{0, 6, 9, 18, 24, 48\}.$$

In practice the above subsets can be divided into three groups: (a) T_2 and T_3, (b) T_5, T_6, and T_7, and (c) T_8, T_9, and T_A. For a given stability design, if the first group of subsets is applied, half of the stability tests will be done at 3, 6, 9, 18, 36, and 48 months, and the other half will be done at 6, 12, 24, and 48 months. Half of the stability tests will be done at each time point except at 48 months. A design of this type is usually referred to as *one-half design*. If the first group of subsets is applied to a complete factorial design, we will refer to the design as a complete– one-half design. Similarly, for a given design, if a second group of subsets (i.e., T_5, T_6, and T_7) is applied, one-third of stability tests will be done at 3, 12, 36, and 48 months; one-third will be tested at 6, 18, and 48 months; and one-third will be tested at 9, 24, and 48 months. The design then becomes a *one-third design*. For the third group of subsets (i.e., T_8, T_9, and T_A), the resulting design is a *two-thirds design* because each group consists of two-thirds of the total time points. Note that the idea for applying the three groups of subsets above to a design is to reduce the total number of stability tests by one-half, one-third, and two-thirds, respectively, at each time point except the last. In such a case, the test results obtained at each time point are *balanced*.

Based on the appropriate choice of a subset of T_1, Nordbrock (1992) provided some useful reduced stability designs for long-term stability studies at room temperature:

Design 1 (complete): Every batch-by-strength-by-package combination is chosen. All combinations chosen are tested at every time point.

Design 2 (complete–two-thirds): Every batch-by-strength-by-package combination is chosen. Two-thirds of the combinations chosen are tested at each time point. One-third of the combinations chosen are tested at 3, 9, 12, 24, 36, and 48 months; one-third are tested at 3, 6, 12, 18, 36, and 48 months; and one-third are tested at 6, 9, 18, 24, and 48 months.

Design 3 (complete–one-half): Every batch-by-strength-by-package combination is chosen. Half the combinations chosen are tested at 3, 9, 18, 36, and 48 months, and the other half are tested at 6, 12, 24, and 48 months.

Design 4: (complete–one-third): Every batch-by-strength-by-package combination is chosen. One-third of the combinations chosen are tested at 3, 12, 24, and 48 months; one-third are tested at 6, 18, 36, and 48 months; and one-third are tested at 9, 24, and 48 months.

Design 5 (fractional): Two-thirds of the batch-by-strength-by-package combinations are chosen. All combinations chosen are tested at every time point.

Design 6 (two strengths per batch): Two strengths per batch are selected, and then all packages for this selection are chosen. All combinations chosen are tested at every time point.

Design 7 (two packages per strength): Two packages per strength are selected, and then all batches for this selection are chosen. All combinations chosen are tested at every time point.

Design 8 (fractional–one-half): Two-thirds of the batch-by-strength-by-package combinations are chosen. Half the combinations chosen are tested at 3, 9, 18, 36, and 48 months, and the other half are tested at 6, 12, 24, and 48 months.

Design 9 (two strengths per batch–one-half): Two strengths per batch are selected, and then all packages for this selection are chosen. Half the combinations chosen are tested at 3, 9, 18, 36, and 48 months, and the other half are tested at 6, 12, 24, and 48 months.

Design 10 (two packages per strength–one-half): Two packages per strength are selected, and then all batches for this selection are chosen. Half the combinations chosen are tested at 3, 9, 18, 36, and 48 months, and the other half are tested at 6, 12, 24, and 48 months.

Details of the designs above are given in Table 4.2. For each reduced stability design, the total number of stability tests required and the relative percentage of reducing stability tests as compared to the full design (i.e., the complete factorial design, including every time point) are summarized in Table 4.3. It can be seen from the table that designs 8 to 10 may reduce the total number of stability tests as much as 59.3%. Note that sample sizes in terms of the number of assays for designs 2 and 4 to 10 are all the same except at 48 months. The reduced stability designs described above can easily be modified if the four storage conditions described at the beginning of this section are to be incorporated.

4.2.3 Matrixing and Bracketing Designs

Generally, a reduction of stability tests could be achieved if we applied a different method, such as bracketing and matrixing, which are special cases of fractional factorial designs. There is no universal definition for a matrixing design. For example, the ICH Q1A (R2) (2003) defines a matrixing design as the design of a stability schedule such that a selected subset of the total number of possible samples for all factor combinations is tested at a specified time point. At a subsequent time point, another subset of samples for all factor combinations is tested. The design assumes that the stability of each subset of samples tested represents the stability of all samples at a given time point. The differences in the samples of the same drug product should be identified as, for example, covering different batches, different strengths, different sizes of the same container closure system, and possibly in some cases, different container closure systems. The ICH Q1A (R2) guideline for stability indicates that matrixing can cover reduced testing when more than one design factor is being evaluated. As a result, the design of the matrix will be dictated by the factors needing to be covered and evaluated. For a matrixing design, the ICH suggests that in every case all batches be tested initially and at the end of the long-term testing. As an alternative, Chow (1992) gave a definition of a matrixing design, suggesting that any subset of a complete factorial design be considered as a matrixing design. For example, in Table 4.1, if we only consider two packages per strength and batch or two strengths per package and batch, these two types of designs are considered matrixing designs.

Helboe (1992) discussed some examples of applications of matrixing designs for long-term stability studies. For example, he considered an example concerning a drug product that was manufactured at three dosage strengths from three different batches of granulation. The same granulations A, B, and C, were used in all three dosage strengths. Three types of containers, one blister pack and two sizes of high-density polyethylene (HDPE) bottle, were considered. As a result, the stability study involves three design factors, with three levels in each. In this case if a full factorial design is adopted, a total of 27 combinations need to be tested at each sampling time point. Instead, one may consider a matrixing design to reduce the total number of tests.

TABLE 4.2:   Long-Term Stability Designs

Batch	Strength	Bottle	Blister	Tube	Bottle	Blister	Tube
		Design 1: Complete Factorial Design			Design 2: Complete Two-Thirds Design		
1	15	T_1	T_1	T_1	T_8	T_9	T_A
	30	T_1	T_1	T_1	T_9	T_A	T_8
	60	T_1	T_1	T_1	T_A	T_8	T_9
2	15	T_1	T_1	T_1	T_A	T_8	T_9
	30	T_1	T_1	T_1	T_8	T_9	T_A
	60	T_1	T_1	T_1	T_9	T_A	T_8
3	15	T_1	T_1	T_1	T_9	T_A	T_8
	30	T_1	T_1	T_1	T_A	T_8	T_9
	60	T_1	T_1	T_1	T_8	T_9	T_A
		Design 3: Complete One-Half Design			Design 4: Complete One-Third Design		
1	15	T_2	T_3	T_2	T_5	T_6	T_7
	30	T_3	T_3	T_2	T_6	T_7	T_5
	60	T_3	T_2	T_3	T_7	T_5	T_6
2	15	T_2	T_2	T_3	T_7	T_5	T_6
	30	T_2	T_3	T_3	T_5	T_6	T_7
	60	T_3	T_3	T_2	T_6	T_7	T_5
3	15	T_3	T_2	T_3	T_6	T_7	T_5
	30	T_3	T_2	T_2	T_7	T_5	T_6
	60	T_2	T_3	T_2	T_5	T_6	T_7
		Design 5: Fractional Factorial Design			Design 6: Two Strength Per Batch Design		
1	15	T_1	T_1	—	T_1	T_1	T_1
	30	T_1	—	T_1	T_1	T_1	T_1
	60	—	T_1	T_1	—	—	—
2	15	—	T_1	T_1	—	—	—
	30	T_1	T_1	—	T_1	T_1	T_1
	60	T_1	—	T_1	T_1	T_1	T_1
3	15	T_1	—	T_1	T_1	T_1	T_1
	30	—	T_1	T_1	—	—	—
	60	T_1	T_1	—	T_1	T_1	T_1
		Design 7: Two Packages Per Strength Design			Design 8: Fractional One-Half Design		
1	15	T_1	—	T_1	T_2	T_3	—
	30	T_1	T_1	—	T_3	—	T_2
	60	—	T_1	T_1	—	T_2	T_3
2	15	T_1	—	T_1	—	T_2	T_3
	30	T_1	T_1	—	T_2	T_3	—
	60	—	T_1	T_1	T_3	—	T_2
3	15	T_1	—	T_1	T_3	—	T_3
	30	T_1	T_1	—	—	T_2	T_2
	60	—	T_1	T_1	T_2	T_3	—

(Continued)

TABLE 4.2: Long-Term Stability Designs (Continued)

Batch	Strength	Bottle	Blister	Tube	Bottle	Blister	Tube
		Design 9: Two Strength per Batch One-Half Design			Design 10: Two-Half Packages per Strength One-Half Design		
1	15	T_2	T_3	T_2	T_2	—	T_3
	30	T_3	T_3	T_2	T_3	T_2	—
	60	—	—	—	—	T_3	T_2
2	15	—	—	—	T_2	—	T_3
	30	T_2	T_3	T_3	T_2	T_3	—
	60	T_3	T_3	T_2	—	T_2	T_3
3	15	T_3	T_2	T_3	T_3	—	T_2
	30	—	—	—	T_3	T_2	—
	60	T_3	T_3	T_2	—	T_3	T_2

Table 4.4 presents a two-thirds matrixing design. Helboe (1992) indicated that this design was actually implemented at one major pharmaceutical company in the United States. In Table 4.4, at each time point, the combinations in parentheses were not tested. It can be seen from Table 4.4 that only two-thirds of 27 combinations of the

TABLE 4.3: Summary of Long-Term Stability Designs

Design Description	Sample Time Intervals	Number of Assays[†]
Complete	All tested using T_1	$9kab$
Complete–two-thirds	One-third tested using T_8 One-third tested using T_9 One-third tested using T_A	$20kab/3^*$
Complete–one-half	One-half tested using T_2 One-half tested using T_3	$11kab/2^*$
Complete–one-third	One-third tested using T_5 One-third tested using T_6 One-third tested using T_7	$13kab/3^*$
Fractional	All tested using T_1	$6kab$
Two strengths per batch	All tested using T_1	$18ka$
Two packages per strength	All tested using T_1	$18kb$
Fractional One-half	One-half tested using T_2 One-half tested using T_3	$11kab/3^*$
Two strengths per batch–one-half	One-half tested using T_2 One-half tested using T_3	$11ka$
Two packages per strength–one-half	One-half tested using T_2 One-half tested using T_3	$11kb$

[†] a: number of packages, b: number of strengths, k = the number of batches
[*] Integer part plus 1 if this number is not an integer

TABLE 4.4:   Two-Thirds Matrixing Design

	Dosage Strength/Lot of Granulation[a]								
	50 mg			**75 mg**			**100 mg**		
Package Type	**A**	**B**	**C**	**A**	**B**	**C**	**A**	**B**	**C**
Blister	+	+	(+)	(+)	+	+	+	(+)	+
HDPE1	(+)	+	+	+	(+)	+	+	+	(+)
HDPE2	+	(+)	+	+	+	(+)	(+)	+	+

[a](+) not tested at this time point
Source: Helboe, P. (1992). *Drug Information Journal,* 26, 629–634.

full factorial design were tested. At each sampling time point each container, dosage strength, and granulation were tested six times.

Another design of particular interest is so-called bracketing design. The ICH Q1A (R2) guideline for stability defines a bracketing design as the design of a stability schedule such that only the samples on the extremes of certain design factors, for example, strength and package size, are tested at all time points as in a full design. The design assumes that the stability of any intermediate levels is represented by the stability of the extremes tested. Where a range of strengths is to be tested, bracketing is appliable if the strangths are identical or very closely related in composition (e.g., for a table range made with different compression weights of a similar basic granulation or a capsule range made by filling different plug fill weights of the same basic composition into different-size capsule shells). Bracketing can be applied to different container sizes or different fills in the same container-closure system. In this case, it is believed that testing the highest and lowest strengths will provide sufficient stability information for the drug product. Therefore, there is no need to test the middle strengths. The samples for the middle strengths, which are bracketed, are kept as backup in case there is a significant difference in stability loss between the highest and lowest strengths. The same idea can be applied to packaging. We only test for the largest and smallest package sizes and leave the middle size samples as backup. When we have a stable active ingredient in the form of tablets, if we take the example given earlier, we would test samples kept at 25° C/60% relative humidity and 30° C/70% relative humidity over the expiration dating period. The other samples, which are bracketed, are kept as backup in case one condition is found to be

TABLE 4.5:   Example of Bracketing Design

Storage Condition		Testing Interval[a] (Months)							
Temp. (°C)	**Relative Humidity (%)**	**3**	**6**	**9**	**12**	**18**	**24**	**36**	**48**
21	45	(+)	(+)	(+)	(+)	(+)	(+)	(+)	(+)
25	60	+	+	+	+	+	+	+	+
30	35	(+)	(+)	(+)	(+)	(+)	(+)	(+)	(+)
30	70	+	+	+	+	+	+	+	+

[a] Parentheses indicate that the corresponding stability tests are omitted.
Source: Helboe, P. (1992). *Drug Information Journal,* 26, 629–634.

TABLE 4.6: Example of Bracketing Design

Package Type	Dosage Strength/Raw Material Lot[a]								
	50 mg			**75 mg**			**100 mg**		
	A	B	C	A	B	C	A	B	C
Blister	+	+	+	(+)	(+)	(+)	+	+	+
HDPE/15	+	+	+	(+)	(+)	(+)	+	+	+
HDPE/100	(+)	(+)	(+)	(+)	(+)	(+)	(+)	(+)	(+)
HDPE/500	+	+	+	(+)	(+)	(+)	+	+	+

[a](+): Not tested at this time point.
Source: Helboe, P. (1992). *Drug Information Journal*, 26, 629–634.

too severe. The details of this bracketing design are given in Table 4.5. For strength, we only test the lowest and highest strengths (i.e., 15 mg and 60 mg) and bracket the 30-mg tablets. As a result, the design achieves a 50% reduction in total tests. It should be noted that this design cannot provide stability information for 75-mg strength or for HDPE/100.

Note that the bracketing design given in Table 4.5 does not use the lower extremes of 21° C and relative humidity 35% to bracket temperature and relative humidity factors. Hence, strictly speaking, it is not a bracketing design with respect to temperature and relative humidity. To provide a better understanding of bracketing design, consider the example given in Helboe (1992). This stability study considers three strengths, three batches of granulations, with each granulation used in all three strengths, and one size of blister pack and three sizes of HDPE bottles. If the study is bracketed on strength and bottle size, the resulting bracketing design is as given in Table 4.6. Helboe pointed out that the fundamental assumption for the validity of a bracketing design is that the stability of multiple levels for a design factor can be determined by the stability of the extremes.

4.2.3.1 Uniform Matrix Designs

As discussed above, various subsets of T_1 can be applied to a reduced design with respect to the design factors of batch, strength, package type, and storage condition. As an alternative, Murphy (1996) proposed so-called uniform matrix stability designs. A uniform matrix design is defined as a matrix design in which the same time protocol is used for all combinations of the other design factors. For example, Table 4.7 and Table 4.8 illustrate how a uniform matrix design differs from a typical two-thirds matrix design. For this particular situation, three batches, A, B, and C, together with three package types, 1, 2, and 3, and the three time protocols, T_7, T_8, and T_9, are arranged in a 3×3 Latin square design. Note that other designs, not necessarily Latin squares, would be used for cases where the number of package types is different or where there are other or additional factors. As can be seen in Table 4.8, by contrast, the corresponding uniform matrix design uses the same time protocol U_4 for all three batches and all three package types. It can be easily verified that the uniform time protocol U_4 is derived from the standard full time protocol by eliminating the 6-month and 9-month time points and moving the 18-month time point to 21 months.

TABLE 4.7:   Typical Two-Thirds Matrix Design

Sample Times	Batch A			Batch B			Batch C		
	Pkg 1	Pkg 2	Pkg 3	Pkg 1	Pkg 2	Pkg 3	Pkg 1	Pkg 2	Pkg 3
0	X	X	X	X	X	X	X	X	X
3	X	X		X		X		X	X
6		X	X	X	X		X		X
9	X		X		X	X	X	X	
12	X	X		X		X		X	X
18		X	X	X	X		X		X
24	X		X		X	X	X	X	
36	X	X	X	X	X	X	X	X	X
	T7	T8	T9	T8	T9	T7	T9	T7	T8

Source: Murphy J.R. (1996). *Journal of Biopharmaceutical Statistics*, 6, 477–494.

As indicated by Murphy (1996), the term *uniform matrix design* emphasizes that the same uniform time protocol is used for every batch and every package type in the study.

The motivation for the uniform matrix design is the simple idea that when fitting a regression line, a more reliable estimate of the regression slope results when the data points are concentrated at the beginning and at the end of the study. For the example given in Table 4.8, it is assumed that statistical analysis of the study would be desired at the 12-month time point, at the 24-month time point, and/or at the 36-month time point. For the situation where only 24-month dating is desired, the uniform matrix design could simply eliminate the 36-month time point. For the two-thirds matrix design, a different matrix utilizing different time protocols would be necessary, since the ICH guidelines for stability specify that testing be done at the beginning and at the end of the study.

TABLE 4.8:   Uniform Matrix Design

Sample Times	Batch A			Batch B			Batch C		
	Pkg 1	Pkg 2	Pkg 3	Pkg 1	Pkg 2	Pkg 3	Pkg 1	Pkg 2	Pkg 3
0	X	X	X	X	X	X	X	X	X
3	X	X	X	X	X	X	X	X	X
6									
9									
12	X	X	X	X	X	X	X	X	X
21	X	X	X	X	X	X	X	X	X
24	X	X	X	X	X	X	X	X	X
36	X	X	X	X	X	X	X	X	X
	U4	U4	U4	U4	U4	U4	U4	U4	U4

Sources: Murphy J.R. (1996). *Journal of Biopharmaceutical Statistics*, 6, 477–494.

4.2.3.2   Factors Acceptable for Matrixing Design

Lin (1994, 1999a, 1999b) studied the applicability of matrixing and bracketing to stability study designs. She indicated that a matrixing design may be applicable to strength if there is no change in proportion of active ingredients, container size, and intermediate sampling time points. The application of a matrixing design to situations such as (a) closure systems; (b) orientation of containers during storage; (c) packaging form, such as: glass, plastic, and foil; (d) manufacturing process (e.g., mixing times); and (e) batch size (e.g., proportion of container filled, such as liquid) should be evaluated carefully. Lin also discussed some situations where a matrixing design is not applicable. For example, if there is a significant change in proportions of active ingredients, the matrixing design is not suitable for strength. Lin indicated that a matrixing design should not be applied to sampling times at two endpoints (i.e., the initial and the last) or at any time points beyond the desired expiration date. If the drug product is sensitive to temperature, humidity, and light, the matrixing design should be avoided. Nordbrock (2003) provided a summary of when factors are acceptable to a matrix based on a document prepared by the PhRMA Stability Working Group (Nordbrock and Valvani, 1995):

- It is acceptable to use a matrix at all stages of development for a drug product and also for a drug substance. It is acceptable for NDA studies, investigational new drug application (IND) studies, supplements, and marketed product studies.

- It is acceptable to use a matrix for all types of products, such as solids, semisolids, liquids, and aerosols.

- It is acceptable to use a matrix after bracketing.

- It is acceptable to use a matrix when there are multiple sources of raw materials (e.g., drug products).

- It is acceptable to use a matrix if there are multiple sites of drug manufacture.

- It is acceptable to use a matrix when identical formulations are manufactured into several strengths.

- It is acceptable to use a matrix if formulations are closely related (e.g., difference in colorant or flavoring).

- Matrixing design is applicable to the orientation of containers during storage.

- Matrixing design may be applicable in certain cases when closely related formulations are used for different strengths (e.g., if an inactive is replaced by an active gredient).

- A matrix across container and closure systems may be applicable if justified.

- It is acceptable to use a matrix within a package composition type, for example for, of different sizes if the fill (i.e., head space) is the same, or for the same size but different fills (head space). It may be acceptable to use a matrix if container size and fill size change, if there is adequate explanation. It is not acceptable to use a matrix across package composition types (e.g., blister and HDPE).

- It is not acceptable to use a matrix across storage conditions. However, it is acceptable to do a separate matrix design for each storage condition.

- It is not acceptable to use a matrix across parameters, such as dissolution and potency. However, it is acceptable to do a separate design for each parameter.

- Matrixing design is applicable regardless of method precision. However, when using a matrix design, the resulting shelf-life is generally shorter than when a complete design is used and that when the method precision is larger, the difference between a complete design and a matrixed design will be larger (i.e., a larger penalty to the sponsor, resulting in a shorter shelf-life for the matrix design than the complete design).

- Matrixing design is applicable regardless of the stability of the product. However, comments similar to those in the preceding point apply, and if a product has a poor stability profile (e.g., shelf-life of 1 year), matrixing design will usually result in an even shorter shelf-life.

Nordbrock (2003) also indicated that matrix designs are generally applicable to many situations and can result in significant savings, with the two-thirds matrix on time being readily acceptable for stable products. Larger reductions in testing than those given by the two-thirds matrix on time are sometimes acceptable. However, it is suggested that several general rules be followed when designing stability studies (Nordbrock, 2003):

- Matrix designs should be approximately balanced (i.e., for all one-way and two-way combinations of batch, package, and strength that are ever tested, approximately the same number of tests should be done cumulatively to every time point).

- When every batch-by-strength-by-package combination is not tested, every strength-by-package combination that is ever tested should be tested in at least two batches (i.e., for every package-by-strength combination that is ever tested, there should be at least two batches tested).

- Unless there are manufacturing restrictions such as in the above example, it is probably acceptable to use a matrix on batch-by-strength-by-package combinations only when there are more than three strengths or more than three packages.

TABLE 4.9: Classification of
Reduced Designs

Design	Factors	Sampling Times
Type 1	Complete	Partial
Type 2	Matrixing	All
Type 3	Matrixing	Partial

4.2.4 Classification of Designs

Chow and Liu (1995) classified reduced stability designs into the following categories
(Table 4.9):

- Type 1: Complete design with partial sampling time points

- Type 2: Matrixing design with all sampling time points

- Type 3: Matrixing design with partial sampling time points

Assuming that there are three batches (1, 2, and 3), three strengths (15 mg,
30 mg, and 60 mg), and three package types (bottle, blister, and tube) and the stability
testing will be done over a 4-year period, for each reduced stability design described
in Table 4.2, the total number of stability tests required and the relative percentage
of reducing stability tests as compared to the full design (i.e., the complete factorial
design, including every time point) are summarized in Table 4.10. It can be seen from
Table 4.10 that type 3 designs might reduce the total number of stability tests as much
as 59.3%. Note that type 1, 2, and 3 designs are also referred to as reduced designs.

TABLE 4.10: Number of Stability Tests Required for Various Designs

Type	Design Description	Number of Assays	Percent Reduced[a]
—	Complete	243	—
1	Complete–two-thirds	180	25.9
1	Complete–one-half	142	41.6
1	Complete–one-third	117	51.9
2	Fractional	162	33.3
2	Two strengths per batch	162	33.3
2	Two packages per strength	162	33.3
3	Fractional–one-half	99	59.3
3	Two strengths per batch–one-half	99	59.3
3	Two packages per strength–one-half	99	59.3

[a]Compared to the complete factorial design

4.3 Design Selection

Many criteria for the selection of an appropriate design have been proposed in the literature. For example, Nordbrock (1992) proposed a criterion for choosing a design with the highest power among designs with the same sample size. Murphy (1996) suggested choosing a design under the criteria of moment, D-efficiency, uncertainty, G-efficiency, and statistical power. DeWoody and Raghavarao (1997) proposed a method for choosing the time vectors such that the design is optimal for comparing slope differences in terms of maximum information per unit cost. DeWoody and Raghavarao (1997) and Pong and Raghavarao (2000) considered choosing a design in terms of power for detection of significant difference between slopes. These criteria mainly focus on the power of detecting factor effect, which are briefly described below.

4.3.1 Moment

Statistically, design moment measures the degree to which the time protocols for a given stability design are spread out toward the endpoints. In other words, when fitting least squares regression lines, the greater the design moment, the less uncertain are the estimated slopes of the lines. For a stability study with multiple batches, the design moment is defined as

$$Moment = \sum_{i,j} (x_{ij} - \bar{x}_i)^2$$

where x_{ij} is the jth time point of the ith batch. Note that moment of a design is a function only of the times at which the results are measured and does not depend on the results themselves.

4.3.2 D-Efficiency

D-efficiency is a standard metric for comparing designs and is related to how well regression coefficients of a particular model are estimated. As a result, D-efficiency is defined as

$$D - efficiency = 100 \times \frac{1}{n} |X'X|^{1/p}$$

where $|X'X|$ is the determinant of the design matrix, n is the number of points in the design, and p is the number of parameters.

4.3.3 Uncertainty

As indicated in Murphy (1996), another comparison of interest is the reliability of the shelf-life estimates, which depends on the reliability of the estimated slopes. Another measure of uncertainty is width of a confidence interval on the fitted slope, which, for a fixed amount of residual variation, depends only on the number and spacing of the data points. Murphy (1996) defined an uncertainty index at time x_t as

$$U_{Index}(x_t) = x_t t_{0.95, n-2} \left[\frac{k}{M} \right]^{1/2},$$

where $t_{0.95, n-2}$ is the one-sided t-value for 95% confidence, M is the total moment, and k is the number of fitted lines. Note that the uncertainty index is a 90% confidence interval half width for a typical fitted slope (assuming that the residual variance is equal to 1), multiplied by time. Multiplying by time permits a comparison of uncertainty on a common basis at different times.

4.3.4   G-Efficiency

Uncertainty is closely related to another standard metric for comparing designs called G-efficiency, which is defined as

$$G - efficiency = 100 \frac{\left[\frac{p}{n} \right]^{1/2}}{\sigma_{max}},$$

where σ_{max} is the maximum standard error of prediction over the design, n is the number of points in the design, and p is the number of parameters. Note that for a typical stability design, the maximum standard error of prediction will occur at the last time point, for example, $t = x_T$ and

$$\sigma_{max} = \sigma \left[\frac{1}{n} + \frac{(x_T - \bar{x})^2}{S_{xx}} \right],$$

where S_{xx} is the sum of squares of the x values (time points), n is the number of points in the time protocol, and σ is the residual error.

4.3.5   Statistical Power

Nordbrock (1992) proposed a criterion for design selection based on the power of detection of a significant difference between slopes (stability loss). For a fixed sample size, the design with the highest power of detection of a significant difference between slopes is the best design. For a fixed power, the design with the smallest sample size is the best design. Consider the following model for a single batch:

$$y_j = \alpha + \beta x_j + e_j, \quad j = 1, ..., n,$$

where y_j is the assay result at time x_j. The power for detecting a significant degradation (e.g., $\beta = \Delta$) for the null hypothesis of $\beta = 0$ is given by

$$Power = P \left\{ |\hat{\beta}| > z_{1-\alpha/2} \frac{\sigma}{\sqrt{S_{xx}}} | \beta = \Delta \right\},$$

where $\hat{\beta}$ is the least squares estimator of β, which follows a normal distribution with mean β and variance σ^2 / S_{xx} and

$$S_{xx} = \sum_{j=1}^{n} (x_j - \bar{x})^2.$$

It can be seen from the above that the power increases as S_{xx} increases. When there are two batches, statistical power for detection of a significant difference rate per time unit can be obtained similarly. Let y_{ij} be the assay result of the ith batch at time x_j. Then, y_{ij} can be described by the following linear model:

$$y_{ij} = \alpha_i + \beta_i x_{ij} + e_{ij}, \quad j = 1, ..., n, \quad i = 1, 2,$$

where it is assumed that e_{ij} is an independent and identically distributed normal with mean zero and variance σ^2. The power for detecting a significant difference between β_1 and β_2 (i.e., $\beta_1 - \beta_2 = \Delta$) can be obtained similarly, as follows:

$$Power = 1 - \Phi \left[z_{1-\alpha/2} - \frac{\Delta}{\sigma} \left(\frac{1}{S_1^2} + \frac{1}{S_2^2} \right)^{-1/2} \right]$$
$$+ \Phi \left[-z_{1-\alpha/2} - \frac{\Delta}{\sigma} \left(\frac{1}{S_1^2} + \frac{1}{S_2^2} \right)^{-1/2} \right],$$

where

$$S_i^2 = \sum_{j=1}^{n} (x_{ij} - \bar{x}_i)^2,$$

and

$$\bar{x}_i = \frac{1}{n} \sum_{j=1}^{n} x_{ij}.$$

Note that when

$$S_1^2 = S_2^2 = S_{xx},$$

the above power reduces to

$$Power = 1 - \Phi \left[z_{1-\alpha/2} - \frac{\Delta}{\sqrt{2}\sigma} \sqrt{S_{xx}} \right]$$
$$+ \Phi \left[-z_{1-\alpha/2} - \frac{\Delta}{\sqrt{2}\sigma} \sqrt{S_{xx}} \right].$$

4.3.6 Remarks

The criterion of statistical power for detecting a significant difference in degradation rate, however, may not be appropriate because the primary objective of a stability study is to establish drug shelf-life rather than to examine the effect of strength, package, batch, and storage time. Information regarding the differences in stability losses among packages, across strengths, between packages, between strengths, and between batches is useful for deciding whether the data should be pooled to establish a single shelf-life. As an alternative, Chow (1992) and Ju and Chow (1995) proposed the following critera:

Criterion 1: For a fixed sample size, the design with the best precision for shelf-life estimation is the best design.

Criterion 2: For a fixed desired precision of shelf-life estimation, the design with the smallest sample size is the best design.

As a result, Ju and Chow proposed the following measure of relative efficiency between two candidate designs: design A and design B

$$\lambda = \frac{\bar{x}'(t)(X'_B X_B)^{-1}\bar{x}(t)}{\bar{x}'(t)(X'_A X_A)^{-1}\bar{x}(t)},$$

where $\bar{x}'(t)$ is the average drug characteristic at time t, and X_A and X_B are the design matrices for design A and design B, respectively. λ can be viewed as a relative efficiency index of design B as compared to design A. For example, if $\lambda < 1$, we conclude that design B is at least as efficient as design A. If $\lambda > 1$, design A is superior to design B in terms of its relative efficiency.

4.4 Discussion

For a long-term stability study, it is of interest to adopt a complete factorial design by testing stability at every time point. A complete factorial design can not only provide valid statistical tests for main effects of design factors under study, but can also provide estimates for interactions with the maximum precision. As a result, it improves the precision of the estimated drug shelf-life. A complete factorial design is considered an ideal design if applied to more homogenous drug products. However, a complete factorial design is usually too costly and time consuming to perform. In practice, it is of interest to consider a matrixing or bracketing design. It should be noted that a matrixing or bracketing design may not be able to evaluate interaction effects. For example, for a 2^{4-1} fractional factorial design, two-factor effects are confounded with each other. In this case we might not be able to determine whether the data should be pooled.

In practice it is preferred to have a single shelf-life applicable to all future batches. For this purpose, it is desirable to pool the data across all design factors to achieve a better statistical inference on the estimated shelf-life. It is suggested, however, that some preliminary tests for interactions be conducted to determine whether the data should be pooled. If a significant interaction is observed, the data should not be pooled. On the contrary, individual shelf-lives for each combination of the factors with significant interaction should be carefully evaluated. Thus, to achieve a better statistical inference on the estimated shelf-life, a design should be chosen to avoid possible confounding and interaction effects. Once the design is chosen, statistical analysis should reflect the nature of the design selected.

Since the primary objective of a stability study is to establish an expiration dating period, the selection of an appropriate design should be based on the precision of shelf-life estimation rather than statistical power for detection of a meaningful difference. The relative efficiency index λ proposed by Ju and Chow (1995) can be applied to evaluate the efficiency of a stability design. If two designs have the same efficiencies, the relative merits and disadvantages should be taken into account for design selection. In practice since all the reduced stability designs can be classified into one of the three type of designs (i.e., types 1, 2, and 3) for a given sample size N (e.g., the total number of stability tests or assays), it is of interest to determine an optimal design within each type of design. This needs further investigation.

As mentioned earlier, a complete factorial design for any stability study is fully efficient in the sense that it provides 100% information regarding design factors considered in the stability program. Matrixing and bracketing designs are versions of fractional factorial designs that will not provide full efficiency for all design factors. If previous experience showed that the variability in degradation is small across different strengths that might be manufactured by increasing the size of the tablets but that were made from the same granulation by the same manufacturing process, the strength might be a candidate for matrixing or bracketing. However, the relative efficiency of such fractional factorial design should be evaluated carefully based on the criterion proposed by Ju and Chow (1995).

Following the concept of the criterion proposed by Ju and Chow (1995), Hedayat, et al. (2006) proposed a number of optimality criteria for choosing an appropriate design under different considerations of design factors of batch, strength, package type, and sampling times. More details are provided in the last chapter of this book.

Chapter 5

Stability Analysis with Fixed Batches

As mentioned in previous chapters, the labeled expiration dating period or shelf-life of a drug product is usually established based on the primary stability data obtained from long-term stability studies that are conducted under ambient conditions. For the determination of a labeled shelf-life, the FDA stability guidelines require that at least three batches, and preferably more, be tested to allow for a reliable estimate of batch-to-batch variability and to test the hypothesis that a single expiration dating period for all batches is justifiable. In addition to individual shelf-lives estimated from each batch, it is desirable to establish a single shelf-life for a drug product. As indicated by the FDA stability guidelines, this single labeled expiration dating period should be applicable to all future batches. Before one can combine stability data from all batches, it is required by the FDA stability guidelines to perform preliminary tests for batch similarity. Batch similarity is usually evaluated by testing the equality of intercepts and the equality of slopes of degradation lines among different batches. For testing the hypotheses of the equality of intercepts and the equality of slopes among batches, the FDA stability guidelines suggest the 0.25 level of significance be used. If the hypotheses of equal intercepts and equal slopes are not rejected at the 0.25 level of significance, a single expiration dating period can be estimated using the methods described in Chapter 3 by fitting a single degradation curve based on pooled stability data of all batches under the assumption that batch effects are fixed. If the hypotheses of equal intercepts and equal slopes are rejected at the 0.25 level of significance, the FDA suggests that a single expiration dating period of the drug product be determined based on the minimum of the individual shelf-lives obtained from each batches. This method, however, lacks statistical justification (Chow and Shao, 1991). Under the assumption of fixed batch effects, as an alternative, Ruberg and Hsu (1992) proposed a method for estimating an expiration dating period using multiple comparison techniques for pooling stability data with the worst batches.

Since the method for determining an expiration dating period for a single batch suggested in the FDA stability guidelines has been described in Chapter 3, in the next section, the method of analysis of covariance is applied to derive tests for the hypotheses of equal intercepts and equal slopes for batch similarity. The minimum approach for estimating a single expiration dating period for multiple batches described in the FDA stability guidelines is given in Section 5.2. In Section 5.3 we describe the multiple comparison procedures proposed by Ruberg and Hsu (1992) for pooling stability data with the worst batch. A numerical example is provided in Section 5.4 to illustrate various methods of determining drug shelf-life with fixed batch effects. A brief discussion is given in Section 5.5 including the use of the nonparametric

method described in Chapter 3 and a discussion of the choice between the original
and logarithmic scale of percent of label claim for determining of drug shelf-life.

5.1   Preliminary Test for Batch Similarity

The FDA stability guideline requires that at least three batches, and preferably more,
be tested to allow some estimate of batch-to-batch variability and at the same time to
test the hypothesis that a single expiration dating period for all batches is justifiable.
Justification for a single expiration dating period estimated from the pooled stability
data of all batches can be verified, as suggested in the FDA stability guideline, by
testing batch similarity. The 1987 FDA stability guideline also points out that batch
similarity of the degradation lines can be evaluated in terms of the equality of slopes
and the equality of intercepts obtained from the stability data of individual batches.
If the batch-to-batch variability is small, it would be advantageous to combine the
data from different batches for an overall shelf-life estimation with high precision.
However, combining the data from different batches should be supported by a preliminary test of batch similarity. The 1987 FDA stability guideline recommends a
preliminary test for batch-to-batch variation be performed at the significance level of
0.25 (Bancroft, 1964), though it has been criticized by many researchers (see, e.g.,
Ruberg and Stegeman, 1991; Chow and Liu, 1995). In this section we will introduce
some tests proposed by Chow and Shao (1989) for testing batch-to-batch variability.

Let y_{ij} be the assay results (percent of label claim) of the ith batch for a drug
product at sampling time x_{ij}. If the degradation is to decrease linearly over time for
all batches, Model 3.1 for a single batch can be extended to describe the degradation
for multiple batches as follows:

$$y_{ij} = \alpha_i + \beta_i x_{ij} + e_{ij}, \quad i = 1, \ldots, K, \quad j = 1, \ldots, n_i, \qquad (5.1)$$

where α_i and β_i are the intercept and slope of the degradation line for batch i, and
e_{ij} are assumed to be independent and identically distributed as a normal distribution
with mean zero and variance σ^2. Note that α_i can be viewed as the ith batch effect at
time zero, and β_i is the degradation rate (or stability loss) per time unit.

Tests for the hypotheses of equality of slopes and equality of intercepts are in
essence tests for homogeneity of the degradation lines among batches, which can be
examined by testing the following hypotheses for intercepts and slopes, respectively;

$$H_{0\alpha} : \alpha_i = \alpha_{i'} \text{ for all } i \neq i' \qquad (5.2)$$

$$H_{0\beta} : \beta_i = \beta_{i'} \text{ for all } i \neq i'. \qquad (5.3)$$

The typical analysis of covariance (ANCOVA) for a completely randomized design
can be applied to obtain test statistics for both hypotheses 5.2 and 5.3 (see, e.g.,

Snedecor and Cochran [1980] and Wang and Chow [1994]). Let $S_{xx}(i)$, $S_{yy}(i)$, and $S_{xy}(i)$ be the sum of squares of time points, the percent of label claim, and the sum of cross products between time points and percent of label claims for batch i individually and

$$S_{xx}(W) = \sum_{i=1}^{K} S_{xx}(i),$$

$$S_{yy}(W) = \sum_{i=1}^{K} S_{yy}(i),$$

$$S_{xy}(W) = \sum_{i=1}^{K} S_{xy}(i),$$

where $N = \sum n_i$. The least squares estimate of the slope and the intercept for batch i are given, respectively, by

$$\hat{\beta}_i = \frac{S_{xy}(i)}{S_{xx}(i)},$$

and

$$\hat{\alpha}_i = \bar{y}_{i.} - \hat{\beta}_i \bar{x}_{i.},$$

where

$$\bar{x}_{i.} = \frac{1}{n_i} \sum_{j=1}^{n_i} x_{ij} \quad \text{and} \quad \bar{y}_{i.} = \frac{1}{n_i} \sum_{j=1}^{n_i} y_{ij}.$$

The residual sum of squares for batch i is given as

$$SSE(i) = S_{yy}(i) - \frac{\left[S_{xy}(i)\right]^2}{S_{xx}(i)}.$$

The combined residual sum of squares is the sum of the individual residual sum of squares over all batches, that is,

$$SSE = \sum_{i=1}^{K} SSE(i),$$

which has $N - 2K$ degrees of freedom. The residual sum of squares computed from $S_{xx}(W)$, $S_{xy}(W)$, and $S_{yy}(W)$ is given by

$$SSE(W) = S_{yy}(W) - \frac{\left[S_{xy}(W)\right]^2}{S_{xx}(W)}, \tag{5.4}$$

which has $N - K - 1$ degrees of freedom. It follows that the sum of squares for the difference in slopes is given by

$$SS(\beta) = SSE(W) - SSE.$$

Therefore, the null hypothesis of equality of slopes is rejected at the α level of significance if

$$F_\beta = \frac{MS(\beta)}{MSE} > F_{\alpha, K-1, N-2K},$$

where

$$MS(\beta) = \frac{SS(\beta)}{K - 1},$$

and

$$MSE = \frac{SSE}{N - 2K},$$

and $F_{\alpha, K-1, N-2K}$ is the αth upper quantile of a central F distribution with $K - 1$ and $N - 2K$ degrees of freedom.

Let $w_i = 1/S_{xx}(i)$ and $\bar{\beta}$ be the weighted mean of the least squares estimates of slopes, that is,

$$\bar{\beta} = \frac{\sum\limits_{i=1}^{K} w_i \hat{\beta}_i}{\sum\limits_{i=1}^{K} w_i}.$$

The sum of squares for the differences of slopes can be obtained as

$$SS(\beta) = \sum_{i=1}^{K} w_i (\hat{\beta}_i - \bar{\beta})^2. \tag{5.5}$$

If the null hypothesis of equal slopes is not rejected at the α level of significance, β_i in Model 5.1 can be replaced by a common slope β, and Model 5.1 reduces to

$$y_{ij} = \alpha_i + \beta x_{ij} + e_{ij}, \quad i = 1, \ldots, K, \quad j = 1, \ldots, n_i. \tag{5.6}$$

The null hypothesis of equal intercepts can be treated as the null hypothesis of equal batch effects after adjustment for the linear regression between time and percent of label claim. This can be obtained by a direct application of the analysis of covariance.

Now, define

$$S_{xx}(T) = \sum_{i=1}^{K} \sum_{j=1}^{n_i} (x_{ij} - \bar{x}_{..})^2,$$

$$S_{yy}(T) = \sum_{i=1}^{K} \sum_{j=1}^{n_i} (y_{ij} - \bar{y}_{..})^2,$$

$$S_{xy}(T) = \sum_{i=1}^{K} \sum_{j=1}^{n_i} (x_{ij} - \bar{x}_{..})(y_{ij} - \bar{y}_{..}),$$

$$S_{xx}(B) = \sum_{i=1}^{K} n_i (\bar{x}_{i.} - \bar{x}_{..})^2,$$

$$S_{yy}(B) = \sum_{i=1}^{K} n_i (\bar{y}_{i.} - \bar{y}_{..})^2,$$

$$S_{xy}(B) = \sum_{i=1}^{K} n_i (\bar{x}_{i.} - \bar{x}_{..})(\bar{y}_{i.} - \bar{y}_{..}). \tag{5.7}$$

where

$$\bar{x}_{..} = \frac{1}{N} \sum_{i=1}^{K} \sum_{j=1}^{n_i} x_{ij},$$

$$\bar{y}_{..} = \frac{1}{N} \sum_{i=1}^{K} \sum_{j=1}^{n_i} y_{ij}.$$

The total sum of deviations from regression is given as

$$SST = S_{yy}(T) - \frac{[S_{xy}(T)]^2}{S_{xx}(T)}.$$

Similarly, the within-batch sum of deviations from regression is $SSE(W)$ given in Equation 5.4. The between-batch sum of deviations from regression can be obtained by subtraction as

$$SSB = SST - SSE(W).$$

The degrees of freedom associated with SST, SSW, and SSB are $N - 2$, $N - K - 1$, and $K - 1$, respectively. Hence, the corresponding between- and within-batch mean squares after adjustment for regression are given by

$$MSB = \frac{SSB}{K - 1},$$

and

$$MSW = \frac{SSE(W)}{N - K - 1},$$

respectively. The null hypothesis of equality of intercepts is rejected if

$$F_\alpha' = \frac{MSB}{MSW} > F_{\alpha,K-1,N-K-1},$$

where $F_{\alpha,K-1,N-K-1}$ is the αth upper quantile of a central F distribution with $K-1$ and $N-K-1$ degrees of freedom. Note that the least squares estimate of the common slope in Model 5.6 is given by

$$\hat{\beta} = \frac{S_{xy}(W)}{S_{xx}(W)}.$$

The inference about β in Model 5.6 can be obtained based on $\hat{\beta}$ and

$$SE(\hat{\beta}) = \frac{MSW}{S_{xy}(W)}.$$

It should also be noted that test statistic F_α' is derived under the assumption that there is a common slope as described in Model 5.6. Therefore, F_α' is the test statistic for the following hypotheses:

$$H_0 : \alpha_i = \alpha_{i'}; \quad \beta_i = \beta_{i'} \quad \text{for all } i \neq i'$$

versus

$$H_a : \alpha_i \neq \alpha_{i'} \quad \text{for some } i \neq i' \quad \text{and} \quad \beta_i \neq \beta_{i'} \quad \text{for some } i \neq i'.$$

The test statistic for the unrestricted null hypothese of hypotheses 5.2 can be obtained as

$$F_\alpha = \frac{MSB}{MSE}.$$

We then reject the unrestricted null hypothesis of equal intercepts if

$$F_\alpha > F_{\alpha,K-1,N-2K}.$$

These results are summarized in the analysis of covariance table (see Table 5.1). Note that the sum of squares due to the common slope is given by

$$SSS = \frac{\left[S_{xy}(W)\right]^2}{S_{xx}(W)}.$$

It can easily be verified that the sum of squares for the sum of the differences of slopes in Equation 5.5 is also the sum of squares due to the interaction between time and batch.

The 1987 FDA stability guideline indicates that the preliminary tests for the equality of slopes and the equality of intercepts should be performed at the 0.25 level of significance as suggested by Bancroft (1964). Bancroft suggested that one preliminary test be performed for the two commonly used models: the two-stage nested model and

TABLE 5.1: ANCOVA Table for Model (5.1)

Source Variation	df	Sum of Squares	Mean Square	F Statistic
Intercept (batch)	$K-1$	SSB	MSB	$F_\alpha = $ MSB/MSE
Time (common slope)	1	SSS[a]	MSS	$F_s = $ MSS/MSE
Different in slope (batch-by-time)	$K-1$	SS(β)	MS(β)	$F_\beta = $ MS(β)/MSE
Error	$N-2K$	SSE	MSE	
Total	$N-1$	SST		

[a] $\text{SSS} = [S_{xy}(W)]^2 / S_{xx}(W)$.

Source: Chow, S.C. and Liu, J.P. (1995). *Statistical Design and Analysis in Pharmaceutical Science*, Marcel Dekker, New York.

the two-way cross-classification mixed model. However, as discussed above, testing for batch similarity on degradation lines involves two preliminary tests: one for the equality of slopes and the other one for the equality of intercepts. The 1987 FDA stability guideline does not state clearly whether the 0.25 level of significance should be applied to each of the two preliminary tests separately or should be used as the overall significance level for the following joint hypotheses:

$$H_0 : \alpha_i = \alpha_{i'} \quad \text{and} \quad \beta_i = \beta_{i'} \quad \text{for all } i \neq i'$$

versus

$$H_a : \alpha_i \neq \alpha_{i'} \quad \text{or} \quad \beta_i \neq \beta_{i'} \quad \text{for some } i \neq i'.$$

5.2 Minimum Approach for Multiple Batches

If the preliminary tests for the equality of slopes and the equality of intercepts are not rejected at the 0.25 level of significance for the null hypothesis of batch similarity of the degradation lines among batches, all batches are considered from the same population of production batches with a common degradation pattern. As a result, Model 5.1 reduces to

$$y_{ij} = \alpha + \beta x_{ij} + e_{ij}, \quad i = 1, \ldots, K, \quad j = 1, \ldots, n_i, \tag{5.8}$$

where α and β are the common intercept and common slope for Model 5.8 and the same normality assumption is posed for e_{ij}. The procedure for estimating an expiration dating period of a single batch described in Section 3.2 can be applied directly to the pooled stability data from all batches. The least squares estimates of β, α and σ^2

are then given by

$$\hat{\beta}_C = \frac{S_{xy}(T)}{S_{xx}(T)}$$

$$\hat{\alpha}_C = \bar{y}_{..} - \hat{\beta}_C \bar{x}_{..},$$

$$s_C^2 = \frac{S_{yy}(T) - \hat{\beta}_C S_{xy}(T)}{N - 2}$$

where $S_{xx}(T)$, $S_{yy}(T)$, $S_{xy}(T)$, $\bar{x}_{..}$, $\bar{y}_{..}$, and N are as defined in Equation 5.7. The least squares estimate of the mean degradation line at time point $t = x$ is then given as

$$y(x) = \hat{\alpha}_C + \hat{\beta}_C x$$

with its least squares estimate of the variance

$$\hat{V}[y(x)] = s_C^2 \left[\frac{1}{N} + \frac{(x - \bar{x}_{..})^2}{S_{xx}(T)} \right].$$

Therefore, the 95% lower confidence limit for the mean degradation line is given as

$$L_C(x) = \hat{\alpha}_C + \hat{\beta}_C x - t_{0.05, N-2} SE(x),$$

where

$$SE(x) = (\hat{V}[y(x)])^{1/2},$$

and $t_{0.05, N-2}$ is the 5% upper quantile of a central t distribution with $N - 2$ degrees of freedom. Hence, the overall expiration dating period can be estimated as the small root $x_L(C)$ of the following quadratic equation:

$$\left[\eta - (\hat{\alpha}_C + \hat{\beta}_C x) \right]^2 = t_{0.05, N-2}^2 s_C^2 \left[\frac{1}{N} + \frac{(x - \bar{x}_{..})^2}{S_{xx}(T)} \right]. \tag{5.9}$$

The conditions for the existence of the root $x_L(C)$ for the quadratic equation above are the same as those given in Equations 3.5 and 3.6 for a single batch. However, the standard error of the slope and intercept estimated from the pooled stability data should be used to evaluate the two conditions. Under the assumption that all batches come from a population for which the degradation pattern can be described accurately by Model 5.8 with the common slope and intercept, the pooled stability data from all batches provide more precise least squares estimates of slope, intercept, and variability. Hence, the 95% confidence limit for the mean degradation line becomes much narrower because it is based on the pooled stability data with $N - 2$ degrees of freedom. Therefore, if Model 5.8 is adequate, the expiration dating period estimated from the pooled stability data is longer than those estimated from individual batches.

If preliminary tests based on the F statistics described in the previous section are rejected at the 0.25 level of significance, the degradation lines of individual batches cannot be considered the same because of different slopes or different intercepts.

TABLE 5.2: ANCOVA Table for Data Set 1

Source of Variation	df	Sum of Squares	*F*-Statistic	*P*-Value
Intercepts	5	0.60	2.85	0.038
Time	1	21.52	503.24	<0.001
Difference in slopes	5	0.35	1.64	0.186
Error	25	1.07		

Source: Ruberg S.J. and Stegeman J.W. (1991). *Biometrics*, 47, 1059–1069.

In this case, according to the 1987 FDA stability guideline, the overall expiration dating period may depend on the minimum time that a batch is expected to remain within acceptable limits. Therefore, let $x_L(i)$ be the estimated shelf-life for batch i, $i = 1, \ldots, K$. An intuitive estimate of the overall expiration dating period, which meets the FDA requirements, may be obtained as follows:

$$x_L(\min) = \min\{x_L(1), \ldots, x_L(K)\}. \tag{5.10}$$

Since $x_L(\min)$ is the shortest shelf-life among all batches, this estimate will provide a 95% confidence that the strength of the drug product will remain above the acceptable lower specification limit η until $x_L(\min)$ for all batches. However, $x_L(\min)$ is a conservative estimate of the overall expiration dating period because it provides more than 95% confidence for all batches except the batch from which it is estimated. The approach above is usually referred to as the minimum approach.

The minimum approach for estimating of the overall shelf-life of a drug product is used when the preliminary tests for batch similarity are rejected (i.e., there are different intercepts and different slopes) at the 0.25 level of significance. However, the procedure for testing the batch similarity of degradation lines and the minimum approach for determining shelf-life as the estimated overall expiration dating period have received considerable criticism because of their drawbacks. For example, Chow and Shao (1991) indicated that the minimum approach lacks statistical justification. Ruberg and Stegeman (1991) and Ruberg and Hsu (1992) illustrated the disadvantages of using the minimum approach through two numerical data sets of six batches. Ruberg and Stegeman analysis of covariance tables for the two stability data sets are reported in Tables 5.2 and Table 5.3, respectively. The results of fitting the least squares to individual batches and to the pooled stability data with the estimated shelf-lives for both data sets are summarized in Tables 5.4 and 5.5, respectively. Figures 5.1 and 5.2 give the stability data and individual regression lines. Note that the overall shelf-life

TABLE 5.3: ANCOVA Table for Data Set 2

Source of Variation	df	Sum of Squares	*F*-Statistic	*P*-Value
Intercepts	5	15.93	4.01	0.009
Time	1	4.14	5.21	0.032
Difference in slopes	5	4.60	1.16	0.359
Error	25	18.28		

Source: Ruberg S.J. and Stegeman J.W. (1991). *Biometrics*, 47, 1059–1069.

TABLE 5.4: Results of Least Squares Regression for Stability Data Set 1

Batch	n	Intercept	Slope	s^2	S_{xx}	Shelf-Life
1	9	100.49	−1.515	0.019	14.63	6.7
2	7	103.66	−1.449	0.043	7.83	6.8
3	7	101.25	−1.682	0.062	5.14	5.5
4	6	102.45	−1.393	0.035	1.38	6.2
5	4	102.45	−1.999	0.011	0.61	4.4
6	4	99.98	−1.701	0.124	0.61	3.5
Pooled		99.90	−1.534	0.047		6.2

Source: Ruberg, S.J. and Stegeman, J.W. (1991). *Biometrics*, 47, 1059–1069.

presented in Tables 5.4 and 5.5 were obtained under Equation 5.6 with a common slope and different intercepts in Ruberg and Stegeman.

From Table 5.2, the F statistic for overall difference in slopes obtained from stability data set 1 is 1.64 with a p-value of 0.186. Hence, the null hypothesis of equal slopes is rejected at the 0.25 level of significance. According to the FDA stability guidelines, the stability data of six batches in data set 1 cannot be pooled. Hence, the overall expiration dating period was determined using the minimum approach. Since batch 6 gives the minimum shelf-life of the six batches, the estimated shelf-life of 3.5 years from batch 6 is used as the overall expiration dating period. For data set 2, the null hypothesis of equal slopes is not rejected at the 0.25 level of significance, because the p-value for the difference in slopes is 0.359, as shown in Table 5.3. Ruberg and Stegeman (1991) pooled the stability data over six batches to estimate the overall expiration dating period, which gives a shelf-life of 18.2 years under Model 5.6.

It is evident from Figures 5.1 and 5.2 that the variability of stability data set 1 is much smaller than that of data set 2. However, following the FDA stability guideline for the use of the significance level of 0.25 for the null hypothesis of the equality of slopes, stability data set 2, with a much larger variability, can be pooled, but stability data set 1, with a smaller variability, cannot be pooled over batches. Therefore, a paradox occurs. As indicated by Ruberg and Stegeman (1991), well-designed and carefully executed stability studies generate reliable and less variable data to provide least squares estimates of the mean degradation lines and their variance with high precision and efficiency. However, the accuracy, precision, and efficiency of these estimates cannot be utilized for estimation of the overall expiration dating period simply because

TABLE 5.5: Results of Least Squares Regression for Stability Data Set 2

Batch	n	Intercept	Slope	s^2	S_{xx}	Shelf-Life
1	9	100.48	−0.109	0.343	14.63	32.3
2	6	103.63	−0.449	0.703	2.96	9.6
3	7	101.24	−0.778	1.189	5.14	7.0
4	5	102.21	0.194	0.221	1.26	16.2
5	4	102.21	−2.218	0.256	0.61	3.2
6	4	100.07	−1.045	2.973	0.61	1.7
Pooled		100.41	−0.330	0.628		18.2

Source: Ruberg S.J. and Stegeman J.W. (1991). *Biometrics*, 47, 1059–1069.

the less variable data are able to detect smaller differences of no practical importance in slopes as being of statistical significance because of an arbitrary choice of the significance level of 0.25. Therefore, good stability studies are penalized for their small variability by recommendation of the choice of the significance level in the FDA stability guidelines.

Another possible explanation for the paradox is that the null hypotheses of equal slopes and equal intercepts are wrong for batch similarity of degradation lines among batches. Failure to reject the null hypotheses of equal intercepts and equal slopes does not prove that the slopes and intercepts are the same across all batches. As an alternative, the following hypotheses may be more reasonable for batch similarity:

$$H_{0\beta} : \max |\beta_i - \beta_{i'}| \geq \Delta_\beta \quad \text{for some } i \neq i' \qquad (5.11)$$
$$\text{vs.} \quad H_{a\beta} : \max |\beta_i - \beta_{i'}| < \Delta_\beta \quad \text{for all } i \neq i'$$

$$H_{0\alpha} : \max |\alpha_i - \alpha_{i'}| \geq \Delta_\alpha \quad \text{for some } i \neq i' \qquad (5.12)$$
$$\text{vs.} \quad H_{a\alpha} : \max |\alpha_i - \alpha_{i'}| < \Delta_\alpha \quad \text{for all } i \neq i'$$

where Δ_β and Δ_α are prespecified equivalence limits for the allowable differences between batches for slopes and intercepts, respectively.

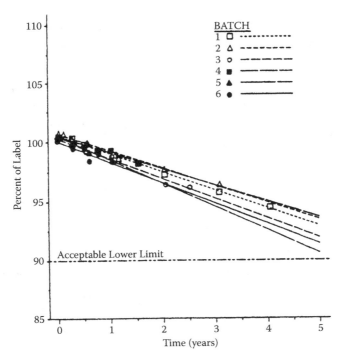

Figure 5.1: Stability data and individual regression lines for data set 1. [From Ruberg and Stegeman (1991).]

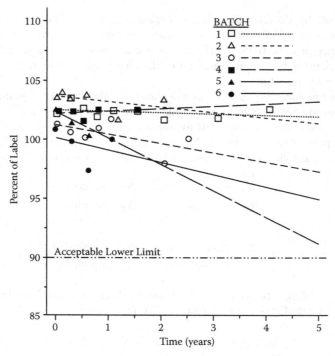

Figure 5.2: Stability data and individual regression lines for data set 2. [From Ruberg and Stegeman (1991).]

Since each batch represents an independent sample from the population of production batches, the least squares estimates of slopes and intercepts of different batches are independent of each other. Therefore, an estimated variance of the difference in least squares estimates of slopes between any pair of batches is given by

$$\hat{V}(\hat{\beta}_i - \hat{\beta}_{i'}) = MSE(w_i + w_{i'}), \quad 1 \le i \ne i' \le K$$

where $w_i = 1/S_{xx}(i)$. Then

$$T_{\hat{\beta}}(i, i') = \frac{\hat{\beta}_i - \hat{\beta}_{i'} - (\beta_i - \beta_{i'})}{SE(\hat{\beta}_i - \hat{\beta}_{i'})}, \quad 1 \le i \ne i' \le K$$

follows a central t distribution with $N - 2K$ degrees of freedom, where

$$SE(\hat{\beta}_i - \hat{\beta}_{i'}) = \left[\hat{V}(\hat{\beta}_i - \hat{\beta}_{i'})\right]^{1/2}. \qquad (5.13)$$

Let

$$T_{\max}(\beta) = \max_{1 \le i < i' \le K} \left|T_{\hat{\beta}}(i, i')\right|.$$

Then, $\sqrt{2}T_{max}(\beta)$ is distributed as the range of K independent standard normal random variables divided by the square root of a central chi-square random variable with $N - 2K$ degrees of freedom divided by $N - 2K$. Since

$$\max|\beta_i - \beta_{i'}| = \beta_{max} - \beta_{min}, \quad 1 \le i \ne i' \le K,$$

the hypotheses of the similarity for slopes in Equation 5.11 can be reformulated as

$$H_{0\beta} : \beta_{max} - \beta_{min} \ge \Delta_\beta \quad \text{for some } i \ne i' \tag{5.14}$$
$$\text{vs.} \quad H_{a\beta} : \beta_{max} - \beta_{min} < \Delta_\beta \quad \text{for all } i \ne i',$$

where

$$\beta_{max} = \max\{\beta_1, \dots, \beta_K\} \tag{5.15}$$
$$\beta_{min} = \min\{\beta_1, \dots, \beta_K\}.$$

Then, the $(1 - \alpha) \times 100\%$ upper confidence interval for the maximum of all pair-wise differences between β_i and $\beta_{i'}$ for $i \ne i'$ or $\beta_{max} - \beta_{min}$ is given as $U_{max}(\beta)$, where

$$U_{max}(\beta) = \hat{\beta}_{max} - \hat{\beta}_{min} + (0.5)^{1/2} Q_{\alpha,K,N-2K} SE(\hat{\beta}_{max} - \hat{\beta}_{min}) \tag{5.16}$$

where $\hat{\beta}_{max}$, $\hat{\beta}_{min}$, and $SE(\hat{\beta}_{max} - \hat{\beta}_{min})$ are defined similarly to Equations 5.15 and 5.13, respectively, and $Q_{\alpha,K,N-2K}$ is the αth upper quantile of studentized range distribution with K and $N - 2K$ degrees of freedom. A procedure for testing the interval hypothesis of Equation 5.11 of the similarity of slopes among batches is to reject $H_{0\beta}$ if the $(1 - \alpha) \times 100\%$ upper confidence interval $U_{max}(\beta)$ is smaller than Δ_β. Let

$$T_U(\beta) = \frac{\hat{\beta}_{max} - \hat{\beta}_{min} - \Delta_\beta}{(0.5)^{1/2} SE(\hat{\beta}_{max} - \hat{\beta}_{min})} \tag{5.17}$$

Then, $H_{0\beta}$ in Equation 5.14 is rejected at the α level of significance if

$$T_U(\beta) < -Q_{\alpha,K,N-2K}$$

Note that the confidence limit approach given in Equation 5.16 is operationally equivalent to the hypothesis testing procedure described in Equation 5.17. Similarly, the $(1 - \alpha) \times 100\%$ upper confidence interval and the testing procedure can be applied to Equation 5.12 for the similarity of intercepts among batches. Note that the estimated variance of the difference in least squares estimates of intercepts between two batches is given as

$$\hat{V}(\hat{\alpha}_i - \hat{\alpha}_{i'}) = MSE \left[\frac{US_{xx}(i)}{n_i S_{xx}(i)} + \frac{US_{xx}(i)}{n_i S_{xx}(i)} \right],$$

where $1 \le i \ne i' \le K$, and $US_{xx}(i)$ is the uncorrected sum of squares of x for batch i.

5.3   Multiple Comparison Procedure for Pooling Batches

As discussed in the previous section, the current FDA recommendation for pooling stability data over batches has some disadvantages. First, the choice of the significance level of 0.25 for testing batch similarity may penalize good stability studies with small variabilities. Second, the minimum approach ignores the information from other batches for estimation of an overall expiration dating period. Finally, the use of wrong hypotheses of equality to test batch similarity in terms of intercepts and slopes of degradation lines among batches may alter the poolability of stability data.

Under the assumption of fixed batch effects for Model 5.1, Ruberg and Hsu (1992) proposed an approach using the concept of multiple comparison to derive criteria for pooling batches with the worst batches. Instead of testing the null hypothesis of the equality of slopes, they suggested investigating a simultaneous confidence interval for

$$\theta_i = \beta_i - \min_{i \neq i'} \beta_{i'} \quad \text{for } i = 1, \ldots, K. \tag{5.18}$$

From Equation 5.18, the worst batch is defined as the batch with the largest degradation rate or minimum slope. If Δ_β is some equivalence limit for the allowable difference between batches, Ruberg and Hsu's (1992) idea for combining stability data from a certain number of batches is to pool the batches:

$$\theta_i = \beta_i - \min_{i \neq i'} \beta_{i'} < \Delta_\beta.$$

In other words, Ruberg and Hsu's procedure is to pool the batches that have slopes similar to the worst degradation rate with respect to the equivalence limit Δ_β. To illustrate this procedure, let us start with an arbitrary batch, batch i, as the reference batch. We first calculate all possible lower confidence limits:

$$l_{ii'} = (\hat{\beta}_i - \hat{\beta}_{i'}) - d_i(\alpha)SE(\hat{\beta}_i - \hat{\beta}_{i'}) \quad \text{for } 1 \leq i \neq i' \leq K,$$

where $SE(\hat{\beta}_i - \hat{\beta}_{i'})$ is as defined before, and $d_i(\alpha)$ is the $(1 - \alpha) \times 100\%$ critical values for the confidence interval of $\beta_i - \min_{i \neq i'} \beta_{i'}$, which depends on the degrees of freedom for the combined residual sum of squares and the $(K - 1) \times (K - 1)$ correlation matrix of

$$\hat{\beta}_i - \hat{\beta}_{i'}, \quad i = 1, .., K.$$
$$\scriptstyle i \neq i'$$

The lower limit of the $(1 - \alpha) \times 100\%$ confidence interval for θ_i is then given by

$$L_i = \min(l_i, 0), \quad i = 1, \ldots, K, \tag{5.19}$$

where

$$l_i = \max_{i \neq i'}(l_{ii'}), \quad i = 1, \ldots, K.$$

We then repeat the procedure above with each batch as the reference batch to compute the lower limit $L_i, i = 1, \ldots, K$. Note that if $l_{ii'} > 0$, according to Equation 5.19, the lower limit for the $(1 - \alpha) \times 100\%$ confidence interval is zero. Therefore, the degradation rate is statistically significantly different from the true minimum slope, and the ith batch is not the worst batch. Let G be the set of all batches with the smallest slope:

$$G = \{i, \, l_i < 0\}.$$

Hence, G is the set that contains all possible worst batches. The computation of the upper limit of the $(1 - \alpha) \times 100\%$ confidence interval for θ_i depends on the number of batches in G. If G contains only a single batch g, batch g has the largest degradation rate, and the upper limit of the $(1 - \alpha) \times 100\%$ confidence interval for θ_i is then given by

$$U_i = \begin{cases} 0 & \text{if } i = g \\ (\hat{\beta}_i - \hat{\beta}_g) + d_g(\alpha)SE(\hat{\beta}_i - \hat{\beta}_g) & \text{if } i \neq g. \end{cases}$$

If G contains more than one batch, calculate all possible upper confidence limits using the batches in G as the candidate batches for the worst batches as follows:

$$u_{ig} = (\hat{\beta}_i - \hat{\beta}_g) + d_g(\alpha)SE(\hat{\beta}_i - \hat{\beta}_g)$$

for all $g \in G$ and $i \neq g$. The upper limit of the $(1 - \alpha) \times 100\%$ confidence interval for θ_i is then given by

$$U_i = \max(u_i, 0),$$

where

$$u_i = \max_{i \neq g}(u_{ig}).$$

Note that calculation of critical values $d_i(\alpha)$ depends on the correlation matrix of

$$\hat{\beta}_i - \hat{\beta}_{i'}, \quad i = 1, .., K. \tag{5.20}$$
$$\scriptstyle i \neq i'$$

Let

$$\lambda_{i'} = \left[1 + \frac{S_{xx}(i)}{S_{xx}(i')}\right]^{-1/2}.$$

Then, the off-diagonal elements of the correlation matrix of Equation 5.20 are given as

$$r_{i'i''} = \lambda_{i'}\lambda_{i''}, \quad i' \neq i''.$$

If $S_{xx}(i) = S_{xx}(i')$ for $1 \leq i \neq i'$, then $d_i(\alpha)$ is Dunnett's one-sided αth upper quantile with $K - 1$ and $N - 2K$ degrees of freedom. However, if $S_{xx}(i) \neq S_{xx}(i')$ for some $i \neq i'$, $d_i(\alpha)$ is the solution of integration of the multivariate t distribution, as shown in Ruberg and Hsu (1992). In practice, numerical integration is required to obtain $d_i(\alpha)$. Since numerical integration of the multivariate t distribution is sometimes quite tedious and computer routines for numerical integration may not be available, Ruberg and Hsu (1992) suggested the use of the Tukey-Kramer procedure for all pair-wise comparisons of degradation rates. The critical values used in the Tukey-Kramer procedure are the upper quantile of the studentized range distribution $Q_{\alpha, K, N-2K}$ given in Equation 5.16. The Tukey-Kramer $(1 - \alpha) \times 100\%$ simultaneous confidence interval for all pair-wise comparisons between β_i and $\beta_{i'}$ is given by

$$u_{ii'}(l_{ii'}) = \hat{\beta}_i - \hat{\beta}_{i'} \pm (0.5)^{1/2} Q_{\alpha, K, N-2K} SE(\hat{\beta}_i - \hat{\beta}_{i'})$$

where $1 \leq i \neq i' \leq K$. Consequently, the lower and upper limits for the $(1 - \alpha) \times 100\%$ confidence interval for

$$\theta_i = \beta_i - \min_{i \neq i'} \beta_{i'}$$

are given, respectively, as

$$L_i = \max_{i \neq i'}(l_{ii'}),$$

and

$$U_i = \max_{i \neq i'}(u_{ii'}).$$

Ruberg and Hsu (1992) also suggested two sets of decision rules for pooling stability data over batches. The first uses an FDA-like approach (FDA, 1987), and the second is a bioequivalence-like approach (Chow and Liu, 2000). In the FDA-like approach, 75% simultaneous confidence intervals are calculated. The decision rules for pooling batches are as follows:

- If $L_i = 0$, batch i is not a candidate for the worst batch.

- Since the event $U_i = 0$ can only occur for at most one batch, if $U_i = 0$, batch i is the only candidate for the worst batch.

- If $L_i < 0$ and $U_i > 0$, pool all such batches for slope estimation because they are candidates for the worst batch.

Since this approach is the version of confidence intervals for the null hypothesis of the equality of slopes recommended in the FDA stability guidelines, it suffers the same drawbacks. Stability studies with a poor design and large variability will produce a wide simultaneous confidence interval for θ_i and allow too many batches to be pooled. Fewer batches from good stability studies cannot be pooled for an accurate, precise, and efficient estimation because of narrower confidence intervals for θ_i due to small variability. For the bioequivalence-like approach, we first compute the 95% upper confidence limits U_i for all θ_i, $i = 1, \ldots, K$. Then we pool batches with $U_i > \Delta_\beta$, where Δ_β is a prespecified upper allowable specification limit. If all $U_i > \Delta_\beta$, no batch can be pooled. In this case shelf-lives are computed separately for each batch, and the minimum among all batches is used to estimate the overall expiration dating period. This is a much more reasonable approach because it is based on the concept of similarity, as discussed in Section 5.2. As a result, good studies with tight upper confidence limits will not be penalized any more by using the bioequivalence-like approach. The most recent development on the application of the equivalence-like approach for batch pooling can be found in Yoshioka et al. (1997) and Tsong et al. (2003b).

5.3.1 Remarks

For the method of Ruberg and Stegeman's equivalence testing of slopes, Lin and Tsong (1991) pointed out that given the same Δ_β, when pooling slopes if $\max |\beta_i - \beta_{i'}| < \Delta_\beta$ leads to a different impact when $\beta_i \approx 0$ than β_i is much smaller than 0 (see also Tsong et al., 2003b). For Ruberg and Hsu's (1992) method of pooling slopes that is equivalent to the worse slope, it should be noted that the worst slope is not necessarily the batch of worst shelf-life. This approach may lead to pooling the direction opposite to the expected direction. For a given equivalence limit, the impact on shelf-life can be large. In addition, with the regulatory limitation of extrapolation, a complicated pooling approach could lead to impractical differences.

5.4 An Example

To illustrate the statistical methods described in the previous sections, consider the following example. A stability study was conducted on 300-mg tablets of a drug product to establish an overall expiration dating period. Tablets from five batches were stored at room temperature in two types of containers (i.e., high-density polyethylene bottle and blister package). The tablets were tested for potency at 0, 3, 6, 9, 12, and 18 months. The assay results (expressed as percent of label claim) were reported in Shao and Chow (1994). These results are given in Table 5.6. Note that this data set contains sampling time intervals only up to 18 months, which does not cover the entire range of sampling time points up to 60 months or more. In addition, assays of all batches were performed at the same time points. Therefore, this data set represents the simplest structure of a stability study. For simplicity, assay results from bottles are selected to illustrate statistical concepts and computational procedures discussed in this chapter.

Figure 5.3 displays the scatter plot of stability data (for bottle) given in Table 5.6 with the acceptable lower specification limit equal to 90%. The stability data of batch 1 using the bottle in Table 5.6 are chosen to illustrate the computation of the shelf-life

TABLE 5.6:   Assay Results in Percent of Label Claim

Package	Batch	Sampling Time (months)					
		0	**3**	**6**	**9**	**12**	**18**
Bottle	1	104.8	102.5	101.5	102.4	99.4	96.5
	2	103.9	101.9	103.2	99.6	100.2	98.8
	3	103.5	102.1	101.9	100.3	99.2	101.0
	4	101.5	100.3	101.1	100.6	100.7	98.4
	5	106.1	104.3	101.5	101.1	99.4	98.2
Blister	1	102.0	101.6	100.9	101.1	101.7	97.1
	2	104.7	101.3	103.8	99.8	98.9	97.1
	3	102.5	102.3	100.0	101.7	99.0	100.9
	4	100.1	101.8	101.4	99.9	99.2	97.4
	5	105.2	104.1	102.4	100.2	99.6	97.5

Source: Shaw & Chow (1944). *Biometrics*, 50, 753–763.

for a single batch. It can be easily verified that

$$\bar{x} = 8,$$
$$\bar{y} = 101.183,$$
$$S_{xx} = 210,$$
$$S_{xy} = -88.9,$$

and

$$S_{yy} = 41.508.$$

Therefore, the least squares estimates of slope, intercept, and error variance are given, respectively, as

$$\hat{\beta} = \frac{S_{xy}}{S_{xx}} = -\frac{88.9}{210} = -0.423,$$
$$\hat{\alpha} = \bar{y} - \hat{\beta}\bar{x} = 101.183 - (-0.423)(8) = 104.57,$$
$$s^2 = \frac{S_{yy} - \hat{\beta}S_{xy}}{n-2} = \frac{41.508 - (-0.423)(-88.9)}{6-2} = 0.969.$$

Hence, the least squares estimate of the mean degradation line at time $t = x$ is given as

$$y(x) = 104.57 - 0.423x.$$

The corresponding ANOVA table is provided in Table 5.7, and the standard error of the least squares estimates for slope and intercept are given by

$$SE(\hat{\beta}) = \left[\frac{s^2}{S_{xx}}\right]^{1/2} = \left[\frac{0.969}{210}\right]^{1/2} = 0.0679,$$

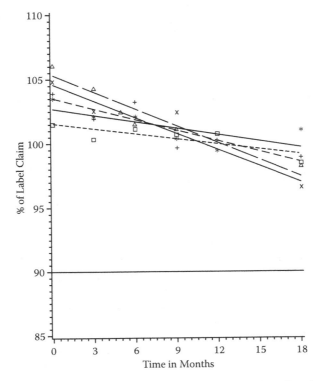

Figure 5.3: Scatter plot of stability data (bottle) in Table 5.6 ×, Batch 1; + batch 2; *. batch 3; ▢, batch 4, Δ, batch 5.

and

$$SE(\hat{\alpha}) = \left[s^2 \left(\frac{1}{n\,S_{xx}} \sum_{j=1}^{n} x_j^2 \right) \right]^{1/2}$$

$$= \left[0.969 \left(\frac{594}{6 \times 210} \right) \right]^{1/2}$$

$$= 0.676,$$

TABLE 5.7: ANOVA Table for Batch 1 of Bottle Container Stability Data Set in Table 5.6

Source Variation	df	Sum of Squares	Mean Squares	*F*-Value	*P*-Value
Regression	1	37.634	37.634	38.858	0.0034
Residual	4	3.874	0.969		
Total	5	41.508			

respectively. Before we apply Equation 3.3 to calculate the shelf-life, we need to check Equations 3.5 and 3.6. For $\eta = 90$,

$$T_{\hat{\alpha}} = \frac{\hat{\alpha} - \eta}{SE(\hat{\alpha})} = \frac{104.57 - 90}{0.676} = 21.55,$$

with a p-value less than 0.001 and

$$T_{\hat{\beta}} = \frac{\hat{\beta}}{SE(\hat{\beta})} = \frac{-0.423}{0.0679} = -6.234$$

with a p-value of 0.0017. Both conditions are met. Thus, the shelf-life is estimated as the smaller root of the following equation:

$$\{90 - [104.57 + (-0.423)x]\}^2 = (2.132)^2(0.969)\left[\frac{1}{6} + \frac{(x - 8)^2}{210}\right].$$

Therefore, the estimated expiration dating period for batch 1 with a bottle is 27.5 months. The stability data with estimated regression line, the 95% lower confidence limit for the mean degradation line, and the acceptable lower specification limit of 90% are plotted in Figure 5.4.

A summary of the results of estimating slopes and intercepts by least squares regression is provided in Table 5.8 along with the corresponding standard errors. It can be verified that Equations 3.5 and 3.6 are both satisfied for all batches except for batch 3. The slope of batch 3 is estimated as -0.168 with a standard error of 0.08. Therefore, $T_{\hat{\beta}}$ is -2.1, with a p-value of 0.053. Thus, the slope of batch 3 is not statistically significantly smaller than zero at the 5% level of significance. Consequently, the shelf-life cannot be estimated by the method described in Section 3.2. Table 5.9 provides the ANCOVA table for the data set given in Table 5.6. The F-value for the difference in slopes is 4.36, with a p-value of 0.0107. Thus, according to the 1987 FDA stability guideline, the stability data presented in Table 5.6 cannot be pooled for estimation of a common slope for all batches.

Suppose from previous experience that the specification for the minimum allowable degradation rate is 0.35% of the label claim per month. From Table 5.7, $\hat{\beta}_{\max} = -0.135$ and $\hat{\beta}_{\min} = -0.441$. Thus,

$$T_U(\beta) = \frac{-0.135 - (-0.441) - 0.35}{0.0676} = -0.651$$

TABLE 5.8: Summary Results of Least Squares Estimation of Slopes and Intercepts

Batch	Intercept	SE(α)	Slope	SE(β)	s^2
1	104.57	0.676	-0.423	0.068	0.969
2	103.50	0.742	-0.280	0.075	1.166
3	102.67	0.800	-0.168	0.080	1.358
4	101.51	0.488	-0.135	0.049	0.505
5	105.29	0.614	-0.441	0.062	0.800
Pooled	103.51	0.374	-0.289	0.038	

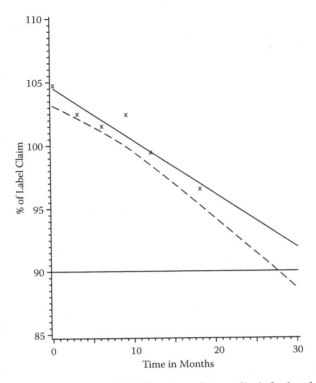

Figure 5.4: Degradation line and 90% lower confidence limit for batch 1.

which is larger than $-Q_{0.05,5,20} = -4.232$. Hence, the null hypothesis of hypotheses 5.4 is not rejected at the 5% level of significance. Thus, the stability data given in Table 5.5 cannot be pooled with respect to a Δ_β of 0.35% per month. It can be verified that if Δ_β was chosen to be 0.6%, the null hypothesis would have been rejected. Note that $\Delta_\beta = 0.35\%$ (or 0.6%) per month is equivalent to 4.2% (or 7.2%) per year, which are considered too big for Δ_β in practice. Therefore, knowledge and experience with the maximum allowable degradation rate of a drug product is very important in making the decision about pooling stability data.

TABLE 5.9: ANCOVA Table for the Stability Data Set in Table 5.6

Source Variation	df	Sum of Squares	Mean Squares	F-Value	p-Value
Intercept	4	19.09	4.77	4.97	0.0060
Time	1	87.84	87.84	91.53	<0.0001
Difference in slopes	4	16.75	4.19	4.36	0.0107
Error	20	19.19	0.96		
Total	29	129.37			

TABLE 5.10: 95% Confidence Interval of $\theta_I = \beta_i - \min\beta_i^{\dagger}$

Method*	Batch	95% Confidence Interval
MCW	1	$(-0.21, 0.24)$
	2	$(-0.07, 0.39)$
	3	$(0.00, 0.50)$
	4	$(0.00, 0.53)$
	5	$(-0.24, 0.21)$
TK	1	$(-0.27, 0.30)$
	2	$(-0.12, 0.45)$
	3	$(-0.01, 0.56)$
	4	$(0.02, 0.59)$
	5	$(-0.30, 0.27)$

$^{\dagger}d_i(\alpha) = 2.39$, the Dunnett's upper one-sided critical value with parameter 5 and 20 degrees of freedom; $Q(0.05,5,20) = 4.232$.

*MCW: Multiple comparison with the worst; TW: Tukey–Kramer simultaneous confidence intervals.

Since $S_{xx}(i) = S_{xx}(i') = 210$ for all i, the Dunnett's upper 5% one-sided critical value with parameter $K - 1 = 5$ and degrees of freedom $N - 2K = 20$ is used for illustration of Ruberg and Hsu's multiple comparison procedure with the worst batch. The results are summarized in Table 5.10 with those computed from the Tukey-Kramer simultaneous confidence intervals. Since $l_1 > 0$ for batches 3 and 4, the set of all possible worst batches includes batches 1, 2, and 5. Therefore, the upper limits of the 95% confidence interval for θ_i were computed from all possible confidence bounds using batches 1, 2, and 5 as candidates for the worst batches by the multiple comparison procedure.

From Table 5.11, the 95% confidence interval of $\beta_i - \min_{i \neq i'} \beta_i$ by a multiple comparison procedure are narrower than those obtained from the Tukey-Kramer method. However, if $\Delta_{\beta} = 0.35\%$ per month is still selected as the upper allowable

TABLE 5.11: Summary of Estimation of Shelf-Lives for Data Set in Table 5.6

Method	Batch	Shelf-Life (months)
Individual	1	27.5
	2	33.5
	3	—
	4	51.4
	5	28.6
Pooled	All	39.5
FDA minimum		27.5
MCW/TK[a]	1,5	30.3
	2,3,4	49.2

†MCW-TK, multiple comparison with the worst and Tukey–Kramer simultaneous confidence interval.

degradation rate, by the bioequivalence-like approach, batches 1 and 5 can be pooled for estimation of a common slope and the overall expiration dating period. Table 5.11 lists the shelf-lives by individual batches. Batch 4 has the longest shelf-life, while batch 1 has the shortest shelf-life. The shelf-life estimation from the data by pooling all five batches is 39.5 months. The pooled data set from batches 1 and 5 generates a shelf-life of 30.3 months compared to the minimum approach of 27.5 months for batch 1. The combined data set from the other three batches (2, 3, and 4) gives a shelf-life of 49.2 months. Examination of slopes in Table 5.8 and Figure 5.3 reveals that those five batches can basically be classified into two groups: one group consists of batches 1 and 5, which have a degradation rate more than 0.4% per month, and the other group contains batches 2, 3, and 4, which have a degradation rate less than 0.3% per month. Therefore, the overall expiration dating periods estimated either by batch 1 alone or by batches 1 and 5 combined using the multiple comparison procedure are still quite conservative because the other three batches have much slower degradation rates.

5.5 Discussion

In pharmaceutical development, the data from stability tests are required to be submitted to the regulatory authority to establish the labeled shelf-life of the drug product at the stage of new drug application (NDA). For this reason, data from stability tests are usually limited to a few batches (e.g., three batches) and less than a two-year test. As a result, stability data are often evaluated by statistical reviewers at the regulatory authority using analysis methods with fixed batches.

As shown in Table 5.11 and Figure 5.3, the estimated shelf-lives for all batches for the stability data considered in the example above are beyond the range of observed time intervals. These estimated shelf-lives are obtained based on extrapolation from the estimated regression line over the range of the time intervals observed. However, it is not known whether an empirical linear relationship between the strength and time still holds from the last observed time interval to the estimated shelf-life. If the true relationship is not linear between these two time points, the shelf-life estimated by extrapolation beyond the range of observed time points is seriously biased. It should be noted that current U.S. regulatory practice is to approve labeled shelf-life with an increment of six months when it is longer than one year. For example, when the estimated shelf-life is 27 months based on data of 18 months of testing, the regulatory shelf-life (or labeled shelf-life) is approved at 24 months rather than 27 months.

Comparison of regression lines includes not only examination of the similarity of slopes and intercepts, but also the equivalence of the within-batch variability. However, the similarity of within-batch variability is often ignored during the decision making for pooling stability data over batches. Tables 5.4 and 5.5 indicate that the ratio of the maximum within-batch variability to the minimum is 11.3 and 13.45, respectively, for stability data given in Ruberg and Stegeman (1991), while the same ratio is 2.69 for the data set of the bottle given in Table 5.3. Hence, large differences in within-batch variability exist for both data sets of Ruberg and Stegeman (1991).

More interestingly, the extremes of within-batch variability occur in the batches with the fewest data points in their data sets. Either Bartlett's chi-square test or Hartley's F_{max} test (Gill, 1978) can be applied to examine the null hypothesis of equality of within-batch variability. Bartlett's test is quite sensitive to the departure of normality, although it can be generalized to the unequal number of observations with each batch. However, Hartley's F_{max} test is rather robust against nonnormality, but it can be applied only when each batch has the same number of assays. Since the upper 5% quantile of the distribution for F_{max} with five batches and 4 residual degrees of freedom is 25.2 and the observed F_{max} is 2.69 for the stability data of the bottle for the data set given in Table 5.6, the null hypothesis of equal within-batch variability is not rejected at the 5% level of significance. It should be noted that these procedures are for the equality; further research is required for testing the similarity of within-batch variability.

An ideal situation for examining the similarity of the degradation lines among batches is for all batches to have the same range of time points as shown by the data set given in Table 5.6. However, in practice, during the early stage of drug development, the duration of time points may be quite different from batch to batch. For example, in both data sets presented by Ruberg and Stegeman (1991), the duration of time points was more than 4 years for one batch, between 2 to 3 years for two batches, 1.5 years for one batch, and 1 year for two batches. Since the entire degradation pattern is not fully understood for these batches with a short duration, one probably should not test the similarity of degradation lines unless batches have reached similar ranges of duration.

In Chapter 3 we introduced a nonparametric approach based on rank regression for determining drug shelf-life for a single batch. This method can be similarly applied to multiple batches. Chen et al. (2003) considered the following three models assuming that $n_i = n$ for all i:

$$y_{ij} = \alpha_i + \beta x_{ij} + e_{ij}, \quad i = 1, \ldots, K, \quad j = 1, \ldots, n_i, \quad (5.21)$$

$$y_{ij} = \alpha + \beta_i x_{ij} + e_{ij}, \quad i = 1, \ldots, K, \quad j = 1, \ldots, n_i, \quad (5.22)$$

$$y_{ij} = \alpha_i + \beta_i x_{ij} + e_{ij}, \quad i = 1, \ldots, K, \quad j = 1, \ldots, n_i. \quad (5.23)$$

Models 5.21 to 5.23 describe situations where: (a) there are different intercepts but common slope, (b) there are different slopes but common intercept, and (c) there are different intercepts and different slopes, respectively.

Under Model 5.21, when the batch-to-batch variation is present, consider

$$U(\beta) = \sum_{i=1}^{K} U_i,$$

where

$$U_i = \sum_{j=1}^{n_i} (x_{ij} - \bar{x}_i) R(e_{ij}),$$

in which

$$\bar{x}_i = \frac{1}{n_i} \sum_{j=1}^{n_i} x_{ij},$$

and

$$e_{ij} = y_{ij} - (\alpha_i + \beta x_{ij}).$$

Similar to the case for a single batch, $U(\beta)$ is not able to estimate α_i's. However, the medians of $y_{ij} - \beta x_{ij}$, $j = 1, \ldots, n_i$ can be used to estimate α_i, $i = 1, \ldots, K$, individually. Under Model 5.22, when there is batch-to-batch variation, to estimate individual β_i, consider solving the following equation for $\hat{\beta}_i$ for every individual i

$$U_i(\hat{\beta}_i) = 0.$$

Thus, the medians of $y_{ij} - \beta x_{ij}$, $j = 1, \ldots, n_i$, can be used to estimate α. For the case where there are different intercepts and different slopes, similarly, under Model 5.23, we can estimate individual $\hat{\beta}_i$ by solving $U_i(\hat{\beta}_i) = 0$ and use the medians of $y_{ij} - \beta x_{ij}$, $j = 1, \ldots, n_i$ to estimate α_i, $i = 1, \ldots, K$ individually. Based on the set of $(\hat{\alpha}_i, \hat{\beta}_i)$, the shelf-life can be obtained based on either confidence bound or prediction bounds of the average values of α_i and β_i following the approaches by Chow and Shao (1991), Shao and Chow (1994), and Shao and Chen (1997).

Chapter 6

Stability Analysis with Random Batches

As indicated earlier, to establish an expiration dating period, the FDA stability guidelines require that at least three batches, and preferably more, be tested in stability analysis to account for batch-to-batch variation so that a *single* shelf-life is applicable to all *future* batches manufactured under similar circumstances. Under the assumption that the drug characteristic decreases linearly over time, the FDA stability guidelines indicate that if there is no documented evidence for batch-to-batch variation (i.e., all the batches have the same shelf-life), the single shelf-life can be determined, based on the ordinary least squares method, as the time point at which the 95% lower confidence bound for the mean degradation curve of the drug characteristic intersects the approved lower specification limit. Several methods for determination of drug shelf-life have been proposed, as discussed in detail in the previous chapter. As indicated in the 1987 FDA stability guideline, the batches used in long-term stability studies for establishment of drug shelf-life should constitute a random sample from the population of future production batches. In addition, the guidelines require that all estimated expiration dating periods be applicable to all future batches. In this case the statistical methods discussed in the previous chapter, which are derived under a fixed effects model, may not be appropriate. This is because statistical inference about the expiration dating period obtained from a fixed effects model can only be made to the batches under study and cannot be applied to future batches. Since the ultimate goal of a stability study is to apply the established expiration dating period to the population of all future production batches, statistical methods based on a *random* effects model seem more appropriate. In the past two decades several statistical methods for determining drug shelf-life with random batches have been proposed. See, for example, Chow and Shao (1989, 1991), Murphy and Weisman (1990), Chow (1992), Ho, Liu, and Chow (1993), Shao and Chow (1994), and Shao and Chen (1997).

The difference between a random effects model and a fixed effects model is that the batches used in a random effects model for stability analysis are considered a random sample drawn from the population of all future production batches. As a result, the intercepts and slopes, which are often used to characterize the degradation of a drug product, are no longer fixed unknown parameters but random variables. The objectives of this chapter are twofold: first, to derive some statistical tests for batch-to-batch variability and some related fixed effects, such as the effects due to packages, strength, and nonlinearity over time. Second, the chapter will introduce statistical methods for the determination of drug shelf-life under a random effects model.

The rest of this chapter is organized as follows. In the next section we introduce a linear regression model with random coefficients. Statistical tests for random batch effects and some related fixed effects are discussed in Section 6.2. In Section 6.3 we introduce several methods for estimating drug shelf-life with random batches. These methods include methods proposed by Chow and Shao (1991), Shao and Chow (1994), and Ho, Liu, and Chow (1993). These methods are compared with the FDA's minimum approach in Section 6.4 through a simulation study. Also included in this section is an example of a long-term stability study that is used to illustrate these methods. A statistical method for determining drug shelf-life based on the concept of lower prediction bound proposed by Shao and Chen (1997) is described in Section 6.5. Some concluding remarks are given in the last section of this chapter.

6.1   Linear Regression with Random Coefficients

Consider modifying Model 5.1 as follows:

$$y_{ij} = X'_{ij}\beta_i + e_{ij}, \quad i = 1, \dots, K, \quad j = 1, \dots, n, \quad (6.1)$$

where i is the index for the batch; X_{ij} is a $p \times 1$ vector of nonrandom covariates of the form $(1, t_{ij}, w_{ij})'$, $(1, t_{ij}, t_{ij} w_{ij})'$, or $(1, t_{ij}, w_{ij}, t_{ij} w_{ij})'$; t_{ij} is the jth time point for the ith batch; w_{ij} is the jth value of a vector of nonrandom covariates (e.g., package type and strength); β_i is an unknown $p \times 1$ vector of parameters; and e_{ij}'s are independent random errors in observing y_{ij}'s. Note that $X'_{ij}\beta_i$ is the mean drug characteristic for the ith batch at X_{ij} (conditional on β_i). The primary assumptions for Model 6.1 are summarized below.

- β_i, $i = 1, \dots, K$ are independent and identically distributed (i.i.d.) as $N_p(\beta, \Sigma_\beta)$, where β is an unknown $p \times 1$ vector and Σ_β is an unknown $p \times p$ nonnegative definite matrix.

- $e_{ij}, i = 1, \dots, K, j = 1, \dots, n_i$ are i.i.d. as $N(0, \sigma_e^2)$, and $\{e_{ij}\}$, and $\{\beta_i\}$ are independent.

- $n_i > p$ and $X_i = (X_{i1}, \dots, X_{in_i})'$ is of full rank for all i.

Note that the K batches constitute a random sample from the population of all future batches of the drug product manufactured under similar circumstances. Hence, Σ_β reflects batch-to-batch variation. If $\Sigma_\beta = 0$ (i.e., $\beta_i = \beta$ for all i), there is no batch-to-batch variation. In such a case Model 6.1 reduces to Model 5.1, which is a fixed effects linear regression model. When $\Sigma_\beta \neq 0$, there is batch-to-batch variation and β_i is a $p \times 1$ vector of unobserved random effects. Consequently, Model 6.1 becomes a mixed effects linear regression model. Note that the fixed effects linear regression model, i.e., Model 5.1, is a special case of the general mixed effects model.

Model 6.1 is known as a linear regression model with random coefficients. If K is large, Hildreth and Houck (1968) recommended that a weighted least squares (WLS) approach be used to estimate the mean of β_i. In practice, however, K is usually small except for postapproval (or marketing) stability studies. The current FDA stability guidelines require that only three batches (i.e., $K = 3$) be tested to establish an expiration dating period for the drug product. Thus, the method suggested by Hildreth and Houck is not appropriate. It is interesting to note that Model 6.1 is a special case of a regression model with random coefficients based on the fact that the values of the regressor X_i are the same for each batch i.

6.2 Random Batch Effect and Other Fixed Effects

Under Model 6.1, we are able to test for the random batch effect and other fixed effects, such as the package effect, the strength effect, and the effect due to nonlinearity over time. Let

$$X_i = (X_{i1}, \ldots, X_{in_i})',$$
$$y_i = (y_{i1}, \ldots, y_{in_i})',$$
$$\epsilon_i = (e_{i1}, \ldots, e_{in_i})'.$$

Then, Model 6.1 can be rewritten as follows:

$$y_i = X_i \beta_i + \epsilon_i, \quad i = 1, \ldots, K. \tag{6.2}$$

Under the assumptions of Model 6.1 described in the previous section, y_i, $i = 1, \ldots, K$ are independently distributed as

$$N_{n_i}(X_i \beta, \ D_i),$$

where

$$D_i = X_i \Sigma_\beta X_i' + \sigma_e^2 I_{n_i},$$

and I_{n_i} is the identity matrix of order n_i. If $\Sigma_\beta = 0$, then $\beta_i = \beta$ is nonrandom and Model 6.2 reduces to

$$y_i = X_i \beta + \epsilon_i, \quad i = 1, \ldots, K, \tag{6.3}$$

which is an ordinary regression model. If $n_i = n$ and $X_i = X$ for all i, Models 6.2 and 6.3 are called balanced models. The maximum likelihood estimator (MLE) of β in the balanced case is the same as the ordinary least squares estimator, which is given by

$$\bar{\beta} = (X'X)^{-1} X' \bar{y},$$

where

$$\bar{y} = \frac{1}{K} \sum_{i=1}^{K} y_i.$$

Since the y_i are normal, we have

$$\bar{\beta} \sim N_p \left(\beta, \ \frac{1}{K} [\Sigma_\beta + (X'X)^{-1}] \right).$$

The $l'\bar{\beta}$ is the uniformly minimum variance unbiased estimator (UMVUE) of $l'\beta$, where l is any fixed $p \times 1$ vector. When $\Sigma_\beta \neq 0$, $\bar{\beta}$ is the ordinary least squares estimator of β and $l'\bar{\beta}$ is still the UMVUE of $l'\beta$ for any fixed $p \times 1$ vector l, since $\bar{\beta}$ is, in fact, obtained from

$$\bar{\beta} = (X'D^{-1}X)^{-1}X'D^{-1}\bar{y},$$

which is well known in the literature. See, for example, Section 4.1 in Laird et al. (1987).

6.2.1   Testing for Batch-to-Batch Variation

For illustration and simplicity, we first consider the special case where $X_{ij} = (1, x_j)'$ and $\beta_i = (\alpha_i, \beta_i)'$. In this case, Model 6.2 reduces to the following model considered by Chow and Shao (1989):

$$y_{ij} = \alpha_i + \beta_i x_j + e_{ij}, \quad i = 1, \ldots, K, \quad j = 0, 1, \ldots, n-1, \tag{6.4}$$

where α_i and β_i are random variables with distributions $N(\alpha, \sigma_\alpha^2)$ and $N(\beta, \sigma_\beta^2)$, respectively, and α_i, β_i, and e_{ij} are mutually independent. If $\sigma_\alpha^2 = 0$ and $\sigma_\beta^2 = 0$, Model 6.4 becomes

$$y_{ij} = \alpha + \beta x_j + e_{ij}, \quad i = 1, \ldots, K, \quad j = 0, 1, \ldots, n-1.$$

If $\sigma_\alpha^2 > 0$ but $\sigma_\beta^2 = 0$ (i.e., different intercepts but common slope), the ordinary least squares method for the estimation of drug shelf-life described in the previous chapters is still valid since

$$y_{ij} = \alpha + \beta x_j + u_{ij}$$

with

$$u_{ij} = \alpha_i - \alpha + e_{ij}$$

being independent $N(0, \sigma_\alpha^2 + \sigma_e^2)$. If, however, $\sigma_\beta^2 > 0$, the ordinary least squares method is not appropriate. A commonly accepted test procedure for the hypothesis that $\sigma_\beta^2 = 0$ is the likelihood ratio test. Let

$$L = L\left(\alpha, \ \beta, \ \sigma^2, \ \sigma_\beta^2 \right)$$

denote the log-likelihood function (given the data y_{ij}), where $\sigma^2 = \sigma_\alpha^2 + \sigma_e^2$. The likelihood ratio test rejects the null hypothesis that $\sigma_\beta^2 = 0$ if

$$A > \chi_{\alpha,1}^2,$$

where

$$A = \sup_{\alpha,\beta,\sigma^2} L(\alpha,\ \beta,\ \sigma^2,\ 0) - \sup_{\alpha,\beta,\sigma^2,\sigma_\beta^2} L(\alpha,\ \beta,\ \sigma^2,\ \sigma_\beta^2),$$

and $\chi_{\alpha,1}^2$ is the αth upper quantile of the chi-square distribution with 1 degree of freedom. A straightforward calculation shows that

$$A = K \sum_{j=1}^{n-1} \ln\left(\hat\sigma_\beta^2 x_j^2 + \hat\sigma^2\right) - Kn\ln(\tilde\sigma^2),$$

where

$$\tilde\sigma^2 = \frac{1}{nK-2} \sum_{i=1}^{K}\sum_{j=1}^{n-1}(y_{ij} - \hat\alpha - \hat\beta x_j)^2,$$

$\hat\alpha$ and $\hat\beta$ are the ordinary least squares estimates of α and β under Model 6.4, and $\hat\sigma_\beta^2$ and $\hat\sigma^2$ are solutions of the following system:

$$\sum_{i=1}^{K}\sum_{j=0}^{n-1} \frac{y_{ij} - \alpha - \beta x_j}{\sigma_\beta^2 x_j^2 + \sigma^2} = 0,$$

$$\sum_{i=1}^{K}\sum_{j=0}^{n-1} x_j \frac{y_{ij} - \alpha - \beta x_j}{\sigma_\beta^2 x_j^2 + \sigma^2} = 0,$$

$$\sum_{i=1}^{K}\sum_{j=0}^{n-1} \frac{(y_{ij} - \alpha - \beta x_j)^2}{\sigma_\beta^2 x_j^2 + \sigma^2} = Kn,$$

$$\sum_{i=1}^{K}\sum_{j=0}^{n-1} \left(\frac{y_{ij} - \alpha - \beta x_j}{\sigma_\beta^2 x_j^2 + \sigma^2}\right)^2 = \sum_{j=0}^{n-1} \frac{1}{\sigma_\beta^2 x_j^2 + \sigma^2}.$$

In addition to the likelihood ratio test, under the assumption that $\sigma_\alpha^2 = 0$, Chow and Shao (1989) proposed the following three statistics for testing the null hypothesis that $H_0 : \sigma_\beta^2 = 0$. The case where $\sigma_\alpha^2 \neq 0$ was examined in Chow and Shao (1991) and will be discussed later in this section. The first test procedure, which was referred to as test procedure I in Chow and Shao (1989), is described below.

Since $x_0 = 0$, y_{i0}, $i = 1, \ldots, K$ are independently distributed as $N(\alpha, \sigma^2)$, where $\sigma^2 = \sigma_\alpha^2 + \sigma_e^2$,

$$s_0^2 = \frac{1}{K-1} \sum_{i=1}^{K}(y_{i0} - \bar y_0)^2,$$

where

$$\bar{y}_0 = \frac{1}{K} \sum_{i=1}^{K} y_{i0}$$

is an unbiased and consistent (as $K \to \infty$) estimator of σ^2, and

$$\frac{(K-1)s_0^2}{\sigma^2}$$

is distributed as a chi-square variable with $K - 1$ degrees of freedom. Under the hypothesis that $\sigma_\beta^2 = 0$,

$$y_{ij} = \alpha + \beta x_j + u_{ij}, \quad i = 1, \dots, K, \quad j = 1, \dots, n-1$$

is a simple linear regression model with errors u_{ij} independently distributed as $N(0, \sigma^2)$. Let

$$\bar{x}_{(0)} = \frac{1}{n-1} \sum_{j=1}^{n-1} x_j,$$

$$\bar{y}_{(0)} = \frac{1}{K(n-1)} \sum_{i=1}^{K} \sum_{j=1}^{n-1} y_{ij},$$

$$s_1 = K \sum_{j=1}^{n-1} (x_j - \bar{x}_{(0)})^2,$$

$$s_2 = \sum_{i=1}^{K} \sum_{j=1}^{n-1} (x_j - \bar{x}_{(0)})(y_{ij} - \bar{y}_{(0)}).$$

Then, $\hat{\beta} = s_2/s_1$ and $\hat{\alpha} = \bar{y}_{(0)} - \hat{\beta}\bar{x}_{(0)}$ are the ordinary least squares estimators of β and α under Model 6.1. Let

$$r_{ij} = y_{ij} - \hat{\alpha} - \hat{\beta}x_j, \quad i = 1, \dots, K, \quad j = 1, \dots, n-1.$$

Under the null hypothesis that $\sigma_\beta^2 = 0$, we have

$$s_r^2 = \frac{1}{m-2} \sum_{i=1}^{K} \sum_{j=1}^{n-1} r_{ij}^2,$$

where $m = K(n-1)$ is an unbiased and consistent estimator of σ^2, and

$$\frac{(m-2)s_r^2}{\sigma^2}$$

is distributed as a chi-square variable with $m - 2$ degrees of freedom. If $\sigma_\beta^2 > 0$,

$$E\left(s_r^2\right) = \frac{1}{m-2} \sum_{i=1}^{K} \sum_{j=1}^{n-1} E\left(r_{ij}^2\right)$$

$$= \sigma^2 + \left[\frac{K}{m-2} \sum_{j=1}^{n-1}(1-w_j)x_j^2\right]\sigma_\beta^2,$$

where

$$w_j = \frac{\sum_{l=1}^{n-1} x_l^2 - 2x_j \sum_{l=1}^{n-1} x_l + (n-1)x_j^2}{K\left[(n-1)\sum_{l=1}^{n-1} x_l^2 - \left(\sum_{l=1}^{n-1} x_l\right)^2\right]}.$$

Since $0 < w_j < 1$ and $x_j^2 > 0$ for $j = 1, \ldots, n-1$, $E(s_r^2)$ is larger than σ^2 if $\sigma_\beta^2 > 0$. Hence,

$$\frac{s_r^2}{s_0^2} >> 1$$

indicates that $\sigma_\beta^2 > 0$. Since s_0^2 and s_r^2 are independent under the null hypothesis that $\sigma_\beta^2 = 0$, s_r^2/s_0^2 is distributed as an F distribution with $m - 2$ and $K - 1$ degrees of freedom. Hence, we reject the null hypothesis that $\sigma_\beta^2 = 0$ at the α level of significance if

$$\frac{s_r^2}{s_0^2} > F_{\alpha, m-2, K-1},$$

where $F_{\alpha, m-2, K-1}$ is the αth upper quantile of the F distribution with $m - 2$ and $K - 1$ degrees of freedom.

For the second procedure (i.e., test procedure II), Chow and Shao (1989) considered

$$s_j^2 = \frac{1}{K-1} \sum_{i=1}^{K}(y_{ij} - \bar{y}_j)^2,$$

where

$$\bar{y}_j = \frac{1}{K} \sum_{i=1}^{K} y_{ij}.$$

It can be verified that

$$E\left(s_j^2\right) = \sigma_\beta^2 x_j^2 + \sigma^2, \quad j = 1, \ldots, n-1.$$

The s_j^2 are independent and identically distributed if $\sigma_\beta^2 = 0$. Note that

$$E\left(s_0^2\right) < E\left(s_1^2\right) < \cdots < E\left(s_{n-1}^2\right)$$

provided that $\sigma_\beta^2 > 0$. Chow and Shao (1989) suggested using the following test procedure, which rejects the null hypothesis that $\sigma_\beta^2 = 0$ if

$$s_0^2 < s_1^2 < \cdots < s_{n-1}^2.$$

Under the hypothesis that $\sigma_\beta^2 = 0$

$$P\left\{s_0^2 < s_1^2 < \cdots < s_{n-1}^2\right\} = \frac{1}{n!}.$$

Hence, this test has level $1/n!$. Chow and Shao (1989) indicated that test procedure II is a robust procedure because it does not require the normality assumptions for e_{ij}, α_i, and β_i. However, this procedure has some disadvantages. First, one cannot choose a desired test level, and the p-value is unknown. Second, if $n \geq 6$, the test level $1/n!$ is either too small or too large. In this case Chow and Shao suggested that the test procedure be modified. Finally, test procedure II requires that there be exactly the same number of observations at each x_j. Thus, unlike test procedure I, one cannot use test procedure II when there are missing observations.

When K is small (e.g., $K = 3$) and n is larger than 5, test procedures I and II may not be appropriate. Chow and Shao (1989) suggested the following alternative (test procedure III). Let $h = n/2$ if n is even and $h = (n-1)/2$ if n is odd. Denote the least squares estimators of α and β under the model

$$y_{ij} = \alpha + \beta x_j + u_{ij}, \quad i = 1, \ldots, K, \quad j = 0, 1, \ldots, h - 1$$

by $\hat{\alpha}_1$ and $\hat{\beta}_1$, respectively, and under the model

$$y_{ij} = \alpha + \beta x_j + u_{ij}, \quad i = 1, \ldots, K, \quad j = h, \ldots, n - 1$$

by $\hat{\alpha}_2$ and $\hat{\beta}_2$, respectively. Also, let

$$s_{r1}^2 = \frac{1}{Kh - 2} \sum_{i=1}^{K} \sum_{j=1}^{h-1} (y_{ij} - \hat{\alpha}_1 + \hat{\beta}_1 x_j)^2,$$

$$s_{r2}^2 = \frac{1}{K(n-h) - 2} \sum_{i=1}^{K} \sum_{j=h}^{n-1} (y_{ij} - \hat{\alpha}_2 + \hat{\beta}_2 x_j)^2.$$

Under the hypothesis that $\sigma_\beta^2 = 0$,

$$\frac{(Kh - 2)s_{r1}^2}{\sigma^2} \quad \text{and} \quad \frac{[K(n-h) - 2]s_{r2}^2}{\sigma^2}$$

are distributed as chi-square variables with $Kh - 2$ and $K(n-h) - 2$ degrees of freedom, respectively. Since s_{r1}^2 and s_{r2}^2 are independent, s_{r2}^2/s_{r1}^2 is distributed as an F with $K(n-h) - 2$ and $Kh - 2$ degrees of freedom. Thus, we reject the null hypothesis that $\sigma_\beta^2 = 0$ if

$$\frac{s_{r2}^2}{s_{r1}^2} > F_{\alpha, K(n-h)-2, Kh-2}.$$

The above test procedure is of level α. It's p-value can be calculated as follows:

$$P\left\{F > \frac{s_{r2}^2}{s_{r1}^2}\right\}.$$

For the use of the three test procedures described above, Chow and Shao (1989) have recommended the following:

- When K is large or moderate, the first test procedure is preferred.

- When K is small, but n is larger than 5, the third test procedure is recommended.

- When both K and n are small (e.g., $n = 4$ or $n = 5$), it is suggested that the second test procedure be used.

As indicated by Chow and Shao (1989), test procedure III provides a quick examination of batch-to-batch variation and is robust against non-normality. However, the restriction on n limits its utility. Test procedures I through III are valid only under the assumption that $\sigma_\alpha^2 = 0$. In practice, σ_α^2 may not be zero, and there exists variability between batches with respect to other class variables, such as package type or strength. In this case the following test procedures for general case are useful.

Consider the general case and the following hypotheses:

$$H_0 : \Sigma_\beta = 0 \quad \text{vs.} \quad K_0 : \Sigma_\beta \neq 0 \tag{6.5}$$

If K_0 is concluded, there is batch-to-batch variation. The sum of squared ordinary least squares residuals can be decomposed as follows:

$$SSR = tr(S) + SE,$$

where

$$SE = K\bar{y}'[I_n - X(X'X)^{-1}X']\bar{y}, \tag{6.6}$$

and $tr(S)$ is the trace of the matrix

$$S = \sum_{i=1}^{K}(y_i - \bar{y})(y_i - \bar{y})'. \tag{6.7}$$

Thus, we have the following results.

Theorem 6.1 *Under the assumptions of Model 6.1, we have the following:*

(a) *SE/σ_e^2 is distributed as the χ_{n-p}^2, the chi-square random variable with $n - p$ degrees of freedom.*

(b) *S in Equation 6.7 has a Wishart distribution $W(K - 1, X\Sigma_\beta X' + \sigma_e^2 I_n)$ and*

$$E[tr(S)] = (K - 1)\left[tr(X\Sigma_\beta X') + n\sigma_e^2\right]$$

(c) *If H_0 in Hypotheses 6.5 holds, $tr(S)/\sigma_e^2$ is distributed as $\chi^2_{n(K-1)}$.*

(d) *Statistics $\bar{\beta}$, SE, and S are independent.*

Proof (a) follows directly from Equation 6.6. Since y_i are i.i.d. normal, S has a Wishart distribution $W(K-1, D)$, where

$$D = X\Sigma_\beta X' + \sigma_e^2 I_n.$$

Hence,

$$\begin{aligned}
E[tr(S)] &= (K-1)tr(D) \\
&= (K-1)\left[tr(X\Sigma_\beta X') + n\sigma_e^2\right].
\end{aligned}$$

This proves (b). If H_0 in Hypotheses 6.5 holds $D = \sigma_e^2 I_n$, and therefore the diagonal elements of S are independent. Thus, (c) follows since each diagonal element of S is distributed as $\sigma_e^2 \chi_{K-1}^2$. Since y_i are i.i.d. normal, $\bar{y}$ and S are independent. Hence, S and $(\hat{\beta}, SE)$ are independent. Since

$$(I_n - H)X = 0,$$

where

$$H = X(X'X)^{-1}X'$$
$$E[(I_n - H)\bar{y}] = (I_n - H)X\beta = 0,$$
$$\begin{aligned}
Cov[(I_n - H)\bar{y}, X'\bar{y}] &= E[(I_n - H)\bar{y}\bar{y}'X] \\
&= (I_n - H)[D/K + E(\bar{y})E(\bar{y}')]X \\
&= (I_n - H)[(X\Sigma_\beta X' + \sigma_e^2 I_n)/K + X\beta\beta'X']X \\
&= 0.
\end{aligned}$$

As a result, $(I_n - H)\bar{y}$ and $X'\bar{y}$ are independent. Therefore,

$$\begin{aligned}
SE &= K\bar{y}'(I_n - H)\bar{y} \\
&= K[(I_n - H)\bar{y}]'(I_n - H)\bar{y},
\end{aligned}$$

and

$$\bar{\beta} = (X'X)^{-1}X'\bar{y}$$

are independent. This proves (d).

From the above theorem, $SE/(n - p)$ estimates σ_e^2. Under H_0,

$$\frac{tr(S)}{n(K - 1)}$$

also estimates σ_e^2. Therefore, the statistic

$$T = \frac{(n - p)tr(S)}{n(K - 1)SE} \tag{6.8}$$

should be around 1. If $\Sigma_\beta \neq 0$, there is another positive component,

$$\frac{1}{n}tr(X\Sigma_\beta X')$$

in the expectation of

$$\frac{tr(S)}{n(K - 1)},$$

and T in Equation 6.8 would be large. Hence, we may use T to test H_0 in Hypotheses 6.5. Under H_0, T follows an F distribution with $n(K - 1)$ and $n - p$ degrees of freedom. At the α level of significance, we would reject H_0 if

$$T > F_{\alpha, n(K-1), n-p},$$

where $F_{\alpha, n(K-1), n-p}$ is the αth upper quantile of an F distribution with $n(K - 1)$ and $n - p$ degrees of freedom. If H_0 is rejected, we conclude that there is significant batch-to-batch variation.

6.2.2 Estimation of Batch Variation

For the estimation of σ_α^2 and σ_β^2 under Model 6.4 for the special case, Chow and Shao (1989) considered the following consistent estimators. Let d_{jl} be the (i, l)th element of $\hat{D}$, where

$$\hat{D} = \frac{1}{K - 1} \sum_{i=1}^{K} (y_i - \bar{y})(y_i - \bar{y})'.$$

Then, d_{jl} is a consistent estimator of σ_{jl}, where

$$\sigma_{jl} = \begin{cases} \sigma_\alpha^2 + \sigma_\beta^2 x_j^2 + \sigma_e^2 & \text{if } j = l \\ \sigma_\alpha^2 + \sigma_\beta^2 x_j x_l & \text{if } j \neq l \end{cases}.$$

Since σ_{jl} is a linear function of σ_α^2, σ_β^2, and σ_e^2, consistent estimators of σ_α^2 and σ_β^2 can be obtained by solving the following linear equations:

$$\hat{\sigma}_\alpha^2 = \frac{1}{n - 1} \sum_{l=2}^{n} d_{1l},$$

$$\hat{\sigma}_\beta^2 = \frac{\sum_{j=2}^{n} \sum_{l=j+1}^{n} d_{jl} - (n - 2)/2 \sum_{l=1}^{n} d_{1l}}{\sum_{j=2}^{n} \sum_{l=j+1}^{n} x_j x_l}.$$

As an alternative, Chow and Wang (1994) proposed two unbiased estimators for batch-to-batch variation based on a transformed model under certain conditions. The idea can be applied to some general estimation procedures such as a restricted maximum likelihood estimator.

6.2.3   Tests for Fixed Effects

In stability analysis, we may need to test the effects of some covariates, such as package type and dosage strength. In practice, we may want to determine the following:

- Is the degradation rate the same from package type to package type and from strength to strength?

- Are the differences in degradation rate between strengths the same across package types?

- Is the degradation rate linear or quadratic in t?

Each of the above questions can be formulated and tested as follows. Let m be a fixed integer; $1 \le m < p$; and $i_1, i_2, \ldots, i_m$ be given integers between 1 and p. For any $p \times 1$ vector l, let $l_{(m)}$ be the subvector of l containing the (i_1)th, (i_2)th, $\ldots$, and (i_m)th components of l. Similarly, for any $p \times p$ matrix A, let $A_{(m)}$ be the $m \times m$ submatrix of A containing elements that are in the (i_1)th, (i_2)th, $\ldots$, and (i_m)th rows and columns of A. Since β_i is random, the (i_1)th, (i_2)th, $\ldots$, and (i_m)th terms in Model 6.1 have no effect if and only if both $\beta_{(m)} = 0$ and $\Sigma_{\beta_{(m)}} = 0$. Hence, we consider the following hypotheses:

$$H_m : \beta_{(m)} = 0 \text{ and } \Sigma_{\beta_{(m)}} = 0$$
$$vs. \quad K_m : H_m \text{ does not hold.}$$

Under H_m, we have $(\hat{\beta}_{(m)} \sim N_m(0, K^{-1}\sigma_e^2 A_{(m)},)$ where $\hat{\beta}_{(m)}$ is the corresponding subvector of $\hat{\beta}$,

$$A_{(m)} = [(X'X)^{-1}]_{(m)},$$
$$K(\hat{\beta}_{(m)})' A_{(m)} \hat{\beta}_{(m)} \sim \sigma_e^2 \chi_{(m)}^2.$$

Let ζ be the $p \times 1$ vector whose (i_1)th, (i_2)th, $\ldots$, and (i_m)th components are 1, and the other components are 0. Under H_m, $\zeta' \Sigma_\beta \zeta = 0$. Hence, by (b) of the theorem described earlier,

$$S_m = \frac{\zeta'(X'X)^{-1} X' S X (X'X)^{-1} \zeta}{\zeta'(X'X)^{-1}\zeta}$$
$$\sim \sigma_e^2 \chi_{K-1}^2.$$

Furthermore, from (d) of the theorem,

$$SE + S_m \sim \sigma_e^2 \chi_{n+K-p-1}^2$$

and is independent of $(\hat{\beta}_{(m)})' A_{(m)} \hat{\beta}_{(m)}$. Therefore,

$$T_m = \frac{n + K - p - 1}{m} \frac{K(\hat{\beta}_{(m)})' A_{(m)} \hat{\beta}_{(m)}}{SE + S_m}$$

has an F distribution with m and $n + K - p - 1$ degrees of freedom. At the α level of significance, we would reject H_m if

$$T_m > F_{\alpha, m, n+K-p-1}.$$

6.3 Shelf-Life Estimation with Random Batches

For simplicity, the following notations are used in this chapter. We let t_{label} denote the established expiration dating period that appears on the container label of a drug product. Also, t_{true} is the true shelf-life of a particular batch of the drug product. Since t_{true} is unknown, it is reasonable to assume that t_{label} will not be granted by the regulatory agency unless $t_{true} \geq t_{label}$ is statistically justified. According to the FDA stability guidelines, under a fixed effects model, it can be shown that t_{label} is a confidence lower bound for t_{true}, and therefore, if t_0 is chosen to be less than or equal to t_{label}, $t_{true} \geq t_0$ provides strong statistical evidence. As indicated earlier, in long-term stability studies X_{ij} is usually chosen to be x_j for all i, where x_j is a $p \times 1$ vector of nonrandom covariates, which could be of the form $(1, t_j, w_j)'$, $(1, t_j, t_j w_j)'$, or $(1, t_j, w_j, t_j w_j)'$, where t_j is the jth time point and w_j is the jth value of a $q \times 1$ vector of nonrandom covariates (e.g., package type and strength). For convenience, let $x_j = x(t_j, w_j)$, where $x(t, w)$ is a known function of t and w. In practice, most of the time we need a labeled shelf-life for a fixed value of covariate (e.g., a fixed package type and a fixed dosage strength). Thus, for simplicity, when w is fixed, we denote $x(t, w)$ by $x(t)$.

In the simple case where $\Sigma_\beta = 0$ (i.e., there is no batch-to-batch variation), the average drug characteristic at time t is $x(t)'\beta$, and the true shelf-life is equal to

$$\bar{t}_{true} = \inf \left\{ t : x(t)'\beta \leq \eta \right\},$$

which is an unknown but nonrandom quantity, where η is the approved lower specification limit for the drug characteristic. In this case, as indicated in the previous chapter, the labeled shelf-life can be determined as follows.

Suppose $x(t)'\beta$ is continuous and decreasing in t. Under Model 6.3, for any given t, an $(1 - \alpha) \times 100\%$ lower confidence bound for $x(t)'\beta$ is given by

$$L(t) = x(t)'\bar{\beta} - t_{\alpha,nK-p} \left[\frac{x(t)'(X'X)^{-1}x(t)}{K(nK - p)} SSR \right]^{1/2},$$

where SSR is the usual sum of squared residuals from the ordinary least squares regression under Model 6.3, and $t_{\alpha,u}$ is the αth upper quantile of the t distribution with u degrees of freedom. Define

$$\hat{t} = \inf \left\{ t : L(t) \leq \eta \right\}. \tag{6.9}$$

The 1987 FDA stability guideline suggests that $\hat{t}$ be used as the labeled shelf-life (i.e., $t_{label} = \hat{t}$). This is based on the fact that $\hat{t}$ is an $(1 - \alpha) \times 100\%$ lower confidence bound for $\hat{t}_{true}$, that is,

$$\begin{aligned}
P_Y \left\{ \hat{t} < \bar{t}_{true} \right\} &= P_Y \left\{ L(\bar{t}_{true}) \leq \eta \right\} \tag{6.10} \\
&= P_Y \{ L(\bar{t}_{true}) \leq x(\bar{t}_{true})'\hat{\beta} \} \\
&= 1 - \alpha,
\end{aligned}$$

where P_Y is the probability with respect to $y_1, \ldots, y_K$. The last quantity of Equation 6.10 follows from the fact that $\bar{t}_{true}$ is nonrandom when $\Sigma_\beta = 0$ and $L(\bar{t}_{true})$ is an $(1 - \alpha) \times 100\%$ lower confidence bound for $x(t)'\beta$.

6.3.1   Chow and Shao's Approach

Under Model 6.2, for a given batch i, $x(t)'\beta_i$ is the average characteristic of the drug product at time t, where β_i is random and distributed as $N_p(\beta, \Sigma_\beta)$. The true shelf-life for this batch is then given by

$$t_{true} = \inf\{t : x(t)'\beta_i \le \eta\}.$$

When there is batch-to-batch variation (i.e., $\Sigma_\beta \ne 0$), t_{true} is random since β_i is random. In this case $x(t)'\beta_i$ follows a normal distribution with mean $x(t)'\beta$ and variance

$$\frac{1}{K} x(t)' \big[\Sigma_\beta + \sigma_e^2 (X'X)^{-1}\big] x(t).$$

Consequently, the procedure described above is not appropriate even if Equation 6.10 holds. It should be noted that

$$P_Y \{t_{label} \le t_{true}\}$$

might be much smaller than

$$P_Y \{t_{label} \le \bar{t}_{true}\}$$

since $\bar{t}_{true}$ is the median of t_{true}. Also, if $t_{label} \le \bar{t}_{true}$,

$$P_{\beta_i} \{t_{label} \le t_{true}\}$$

could be quite high, where P_{β_i} is the probability with respect to β_i. Define

$$
\begin{aligned}
\Psi(t) &= P_{\beta_i} \{t_{true} \le t\} \\
&= P_{\beta_i} \{x(t)'\beta_i \le \eta\} \\
&= \Phi\left(\frac{\eta - x(t)'\beta}{\sigma(t)}\right),
\end{aligned}
\tag{6.11}
$$

where Φ is the standard normal distribution function and

$$\sigma(t) = [x(t)'\Sigma_\beta x(t)]^{1/2}$$

is the standard deviation of $x(t)'\beta_i$.

Chow and Shao (1991) and Shao and Chow (1994) proposed an $(1 - \alpha) \times 100\%$ lower confidence bound of the ϵth quantile of t_{true} as the labeled shelf-life, where ϵ is a given small positive constant. We will refer to this method as Chow and Shao's approach. That is,

$$P_Y \{t_{label} \le t_\epsilon\} \ge 1 - \alpha,
\tag{6.12}$$

where t_ϵ satisfies

$$P_{\beta_i}\{t_{true} \le t_\epsilon\} = \epsilon.$$

It follows from Equation 6.11 that

$$t_\epsilon = \inf\{t : x(t)'\beta - \eta = z_\epsilon \sigma(t)\}, \tag{6.13}$$

where $z_\epsilon = \Phi^{-1}(1 - \epsilon)$. The use of a labeled shelf-life satisfying Equation 6.12 can be justified from another point of view. Note that $t_{label} \le t$ if and only if

$$P_{\beta_i}\{t_{true} \le t_{label}\} \le \epsilon.$$

Hence, Equation 6.12 is equivalent to

$$1 - \alpha \le P_Y\left\{P_{\beta_i}\{t_{true} \le t_{label}\} \le \epsilon\right\}$$
$$= P_Y\{\Psi(t_{label}) \le \epsilon\},$$

where $\Psi(t)$ is as defined in Equation 6.11, which can be viewed as the future failure rate (i.e., the percentage of future batches that fail to meet the specification) at time t. Thus Equation 6.12 ensures, with $(1 - \alpha) \times 100\%$ assurance, that the future failure rate at time t_{label} is no more than ϵ. For small K, Shao and Chow (1994) suggested the following improved procedure for both balanced and imbalanced cases. We first consider the balanced case.

6.3.1.1 Balanced Case

Consider the following balanced model (i.e., $n_i = n$ and $X_i = X$ for all i). When Model 6.2 is balanced, it can be written as

$$y_i = X\beta + \epsilon_i^*, \quad i = 1, \ldots, K, \tag{6.14}$$

where

$$\epsilon_i^* = X(\beta_i - \beta) + \epsilon_i$$

are independently distributed as $N_n(0, D)$ with $D = X\Sigma_\beta X' + \sigma_e^2 I_n$. Under Model 6.14, the ordinary least squares estimator $\bar{\beta}$ of β is given by

$$\bar{\beta} = (X'X)^{-1}X'\bar{y}.$$

Since the covariance matrix D has a special structure, $x(t)'\bar{\beta}$ is the best linear estimator of $x(t)'\beta$ under Model 6.14 (see, e.g., Rao [1973], p. 312). Note that $\hat{t}$ in Equation 6.9 is equal to

$$\hat{t} = \inf\left\{t : x(t)'\bar{\beta} \le \hat{\eta}(t)\right\},$$

where

$$\hat{\eta}(t) = \eta + t_{\alpha, nK-p}\left[\frac{x(t)'(X'X)^{-1}x(t)SSR}{K(nK - p)}\right]^{1/2}$$

can be viewed as an adjusted lower specification limit. In the case of $\Sigma_\beta \neq 0$, we may apply the same idea to obtain a valid t_{label}. Let

$$v(t) = \frac{1}{K-1} x(t)'(X'X)^{-1} X' S X (X'X)^{-1} x(t),$$

where S is as defined in Equation 6.7. From Theorem 6.1, $(K-1)v(t)$ is distributed as $a(t)\chi^2_{K-1}$, where

$$a(t) = x(t)' \left[\Sigma_\beta + \sigma_e^2 (X'X)^{-1} \right] x(t) \tag{6.15}$$
$$\geq x(t)' \left[\Sigma_\beta + \sigma_e^2 (X'X)^{-1} \right] x(t)$$
$$= [\sigma(t)]^2.$$

Define

$$\tilde{\eta}(t) = \eta + c_K(\epsilon, \alpha) z_\epsilon \sqrt{v(t)},$$
$$\tilde{t} = \inf\{t : x(t)'\bar{\beta} \leq \tilde{\eta}(t)\}, \tag{6.16}$$

where for given ϵ, α, and K,

$$c_K(\epsilon, \alpha) = \frac{1}{\sqrt{K} z_\epsilon} t_{\alpha, K-1, \sqrt{K} z_\epsilon}, \tag{6.17}$$

and $t_{\alpha, K-1, \sqrt{K} z_\epsilon}$ is the αth upper quantile of the noncentral t distribution with $K-1$ degrees of freedom and noncentrality parameter $\sqrt{K} z_\epsilon$. Similar to $\hat{\eta}(t)$, $\tilde{\eta}(t)$ is an adjusted lower specification limit. The values of $c_K(\epsilon, \alpha)$ are given in Table 6.1. Shao and Chow (1994) proposed using $\tilde{t}$ in Equation 6.16 as the labeled shelf-life. This is justified by the following result.

Theorem 6.2 *For any given ϵ, α, and K, Equation 6.9 holds for $t_{label} = \tilde{t}$. If we let σ_e tend to zero or $X'X$ tend to infinity, the quantity in Equation 6.9 holds.*

Proof Let $T(K, \lambda)$ denote a noncentral t random variable with K degrees of freedom and noncentrality parameter λ. By the Theorem 6.1, we have

$$\frac{\sqrt{K}[x(t_\epsilon)'\hat{\beta} - \eta]}{\sqrt{v(t_\epsilon)}} = T(K-1, \lambda_\epsilon),$$

TABLE 6.1: Values of c_K (ε, 0.05) Defined by Equation 6.17

				K				
ε	3	4	5	6	7	8	9	10
0.01	4.536	3.027	2.468	2.176	1.995	1.872	1.781	1.711
0.02	4.570	3.056	2.493	2.199	2.016	1.891	1.799	1.729
0.03	4.599	3.081	2.515	2.218	2.034	1.907	1.815	1.743
0.04	4.627	3.104	2.535	2.236	2.051	1.923	1.829	1.757
0.05	4.654	3.127	2.555	2.254	2.067	1.938	1.843	1.770
0.10	4.803	3.248	2.658	2.346	2.150	2.015	1.915	1.837
0.15	4.993	3.396	2.784	2.457	2.250	2.107	2.001	1.918

where

$$\lambda_\epsilon = \frac{\sqrt{K}[x(t_\epsilon)'\beta - \eta]}{\sqrt{a(t_\epsilon)}}$$

t_ϵ and $a(t)$ are given in Equations 6.13 and 6.15, respectively. Thus,

$$
\begin{aligned}
P_Y\{t_\epsilon < \tilde{t}\} &= P_Y\{x(t_\epsilon)'\hat{\beta} > \tilde{\eta}(t_\epsilon)\} \\
&= P_Y\left[\frac{x(t_\epsilon)'\hat{\beta} - \eta}{\sqrt{v(t_\epsilon)}} > c_K(\epsilon, \alpha)\right] \\
&= P_Y\{T(K-1, \lambda_\epsilon) > \sqrt{K}c_K(\epsilon, \alpha)z_\epsilon\} \\
&\le P_Y\{T(K-1, \sqrt{K}z_\epsilon) > \sqrt{K}c_K(\epsilon, \alpha)z_\epsilon\} \\
&= \alpha
\end{aligned}
$$

where the first equality follows from the definition of $\tilde{t}$, while the second and third equalities follow from the definitions of $\tilde{\eta}(t)$ and $T(K-1, \lambda_\epsilon)$. The inequality follows from

$$\lambda_\epsilon \le \sqrt{K}z_\epsilon$$

under Equations 6.13 and 6.15 and the last equality follows from Equation 6.17. The last assertion follows from the fact that

$$a(t) \to [\sigma(t)]^2 \quad \text{as} \quad \sigma_e \to 0 \quad \text{or} \quad X'X \to \infty.$$

This completes the proof.

To provide a better understanding of the above procedure, Figures 6.1 and 6.2 give graphical presentation of the relationship among mean degradation curve, lower specification limit, and the labeled shelf-life for $\hat{t}$ and $\tilde{t}$, respectively.

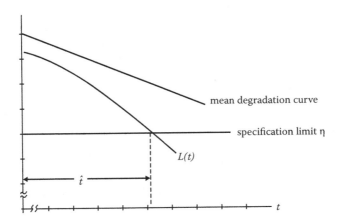

Figure 6.1: Labeled shelf-life $\hat{t}$.

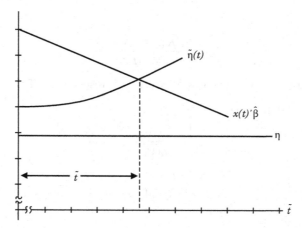

Figure 6.2:   Labeled shelf-life $\tilde{t}$.

6.3.1.2   Unbalanced Case

In the unbalanced case where the n_i are not equal or the X_i are not identical, it is difficult to obtain exact fixed sample (small K) results without any further assumptions. In some cases we may have a large number of observations for each tested batch, and therefore we may assume that

$$n_i \to \infty \quad \text{and} \quad X_i'X_i \to \infty, \quad i = 1,.., K. \tag{6.18}$$

Apparently, Equation 6.18 cannot always be fulfilled. We can, however, adopt an alternative approach, which we will refer to as the small-error asymptotic approach, by assuming that

$$\sigma_e^2 \to 0. \tag{6.19}$$

Equation 6.19 simply means that the assay measurement errors are small to ensure that their variance σ_e^2 is small. This can be done for stability data obtained under well-controlled conditions. With either Equation 6.18 or Equation 6.19, we can extend the above result to the unbalanced case as follows. Let

$$\hat{\beta}_i = (X_i'X_i)^{-1}X_i'y_i,$$

where X_i and y_i are as defined before. Under the assumptions of Model 6.1, the $\hat{\beta}_i$ are independent, and

$$\hat{\beta}_i \sim N_p\left(\beta, \Sigma_\beta + \sigma_e^2(X_i'X_i)^{-1}\right). \tag{6.20}$$

Under either Equation 6.18 or Equation 6.19, it follows from Equation 6.20 that, approximately,

$$\hat{\beta}_i \sim N_p(\beta, \Sigma_\beta).$$

Hence, approximately,

$$\tilde{\beta} = \bar{\beta} = \frac{1}{K} \sum_{i=1}^{K} \hat{\beta}_i \sim N_p(\beta, \Sigma_{\hat{\beta}}/K),$$

where

$$\Sigma_{\hat{\beta}} = \Sigma_\beta + \frac{1}{K} \sigma_e^2 \sum_{i=1}^{K} (X_i' X_i)^{-1}.$$

Define

$$\tilde{v}(t) = \frac{1}{K-1} \sum_{i=1}^{K} x'(t)(\hat{\beta}_i - \bar{\beta})(\hat{\beta}_i - \bar{\beta})' x(t). \tag{6.21}$$

Let $\tilde{\eta}(t)$ be defined as in Equation 6.16 with $v(t)$ replaced by $\tilde{v}(t)$ in Equation 6.21, and let $\tilde{t}$ be defined as in Equation 6.16 with $\bar{\beta}$ replaced by $\tilde{\beta}$. Then, Equation 6.12 still holds approximately for $\tilde{t}$. Therefore, $\tilde{t}$ can be used as the labeled shelf-life.

In the balanced case where $n_i = n$ and $X_i = X$ for all i, it follows that

$$\tilde{\beta} = \bar{\beta} = (X'X)^{-1} X' \bar{y}. \tag{6.22}$$

Similarly, from Equations 6.21 and 6.22, we have

$$\tilde{v}(t) = \frac{1}{K-1} \sum_{i=1}^{K} x'(t)(X'X)^{-1} X'(y_i - \bar{y})(y_i - \bar{y})' X(X'X)^{-1} x(t),$$

which is the same as $v(t)$. Thus, the results described in the previous section (Section 6.2) are a special case of that derived in this section, except that the result in Section 6.2 is exact.

In the pharmaceutical industry the imbalance of the model is often caused by missing values. That is, in the original design, $n_i = n$ and $X_i = X$ for all i, but the actual data set, $n_i \leq n$ and X_i, is a submatrix of X, $i = 1, \ldots, K$. In such a case, even if neither Equation 6.18 nor Condition 6.19 is satisfied, the use of $\tilde{t}$ as the labeled shelf-life is still approximately valid as long as

$$(X_i' X_i)^{-1} (X'X) - I_p$$

is nearly equal to the zero matrix for all i.

6.3.1.3 Sampling Distribution of the Estimated Shelf-Life

Sun et al. (1999) examined sampling distribution of the estimated shelf-life proposed by Shao and Chow (1994). Consider Model 6.2. Let $S_{\hat{\alpha}}^2$ and $S_{\hat{\beta}}^2$ be the sample variances of $\hat{\alpha}_i$ and $\hat{\beta}_i$, respectively, and $S_{\hat{\alpha}\hat{\beta}}$ be the sample covariance of $\hat{\alpha}_i$ and $\hat{\beta}_i$. Then

$$v^2(t) = S_{\hat{\alpha}}^2 + 2 S_{\hat{\alpha}\hat{\beta}} t + S_{\hat{\beta}}^2 t^2.$$

The covariance matrix of $(\hat{\alpha}_i, \hat{\beta}_i)$ is given by

$$W = \begin{pmatrix} S_\alpha^2 & S_{\alpha\beta} \\ S_{\alpha\beta} & S_\beta^2 \end{pmatrix}$$

$$= \begin{pmatrix} \sigma_\alpha^2 + \sigma_e^2 \dfrac{\sum\limits_{j=1}^{n} x_j^2}{n \sum\limits_{j=1}^{n}(x_j-\bar{x})^2} & \sigma_{\alpha\beta} - \sigma_e^2 \dfrac{\bar{x}}{\sum\limits_{j=1}^{n}(x_j-\bar{x})^2} \\ \sigma_{\alpha\beta} - \sigma_e^2 \dfrac{\bar{x}}{\sum\limits_{j=1}^{n}(x_j-\bar{x})^2} & \sigma_\alpha^2 + \sigma_e^2 \dfrac{1}{\sum\limits_{j=1}^{n}(x_j-\bar{x})^2} \end{pmatrix}. \qquad (6.23)$$

Assume $\beta < 0$, $z_\epsilon S_\alpha < \alpha - \eta$, and $\beta^2 \neq z_\epsilon^2 S_\beta^2$. Let

$$\rho_K = c_K(\epsilon, \alpha) z_\epsilon.$$

First, we show that the equation

$$\hat{\alpha} + \hat{\beta} t = \eta + \rho_K v(t) \qquad (6.24)$$

has a unique positive solution $\tilde{t}$ as the estimated drug shelf-life. A simple calculation shows that

$$|W| = S_\alpha^2 S_\beta^2 - S_{\alpha\beta}^2 > 0.$$

Hence, we have $S_{\hat{\alpha}}^2 S_{\hat{\beta}}^2 - S_{\hat{\alpha}\hat{\beta}}^2 > 0$ almost surely for large K. This implies that $v^2(t) > 0$ for all t and

$$v'(t) = \frac{S_{\hat{\beta}}^2 t + S_{\hat{\alpha}\hat{\beta}}}{v(t)},$$

and

$$v''(t) = \frac{S_{\hat{\alpha}}^2 S_{\hat{\beta}}^2 - S_{\hat{\alpha}\hat{\beta}}^2}{v^3(t)} > 0.$$

Hence, $v(t)$ is a positive convex function. Under the conditions that $\beta < 0$, $z_\epsilon S_\alpha < \alpha - \eta$, and $\beta^2 \neq z_\epsilon^2 S_\beta^2$, we have $\hat{\beta} < 0$, $v(0) = \rho_K S_{\hat{\alpha}} < \hat{\alpha} - \eta$, and $\hat{\beta}^2 \neq \rho_K^2 S_{\hat{\beta}}^2$ for large K. Therefore, Equation 6.24 has unique positive solution $\tilde{t}$. Squaring both sides of Equation 6.24 and solving the equation gives the solutions

$$\tilde{t} = \frac{-[\hat{\beta}(\hat{\alpha} - \eta) - \rho_K^2 S_{\hat{\alpha}\hat{\beta}}] \pm \hat{\theta}}{\hat{\varphi}}, \qquad (6.25)$$

where

$$\hat{\varphi} = \hat{\beta}^2 - \rho_K^2 S_{\hat{\beta}}^2,$$

and

$$\hat{\theta} = \left[\left(\rho_K^2 S_{\hat{\alpha}\hat{\beta}} - \hat{\beta}(\hat{\alpha} - \eta) \right)^2 \right.$$
$$\left. - \left((\hat{\alpha} - \eta)^2 - \rho_K^2 S_{\hat{\alpha}}^2 \right) \left(\hat{\beta}^2 - \rho_K^2 S_{\hat{\beta}}^2 \right) \right]^{1/2}.$$

If $-z_\epsilon S_\beta > \beta$, then for large K, we have

$$-\rho_K v'(\infty) = -\rho_K S_{\hat\beta} > \hat\beta,$$

which implies that both solutions are positive and $\tilde{t}$ is the smaller one. The estimated shelf-life $\tilde{t}$ is the one found by using the negative sign in Equation 6.25 since $\hat\varphi$ is positive. If $-z_\epsilon S_\beta < \beta$, then for large K, we have

$$\rho_K v'(-\infty) = -\rho_K S_{\hat\beta} < \hat\beta,$$

which implies that one solution is positive and the other one is negative, and $\tilde{t}$ is the positive one found by using the negative sign in Equation 6.25 since in this case $\hat\varphi$ is negative. The estimated shelf-life is given by

$$\tilde{t} = f\left(\hat\alpha - \eta,\ \hat\beta,\ S_{\hat\alpha}^2,\ S_{\hat\beta}^2,\ S_{\hat\alpha\hat\beta},\ \rho_K^2\right)$$

$$\approx \frac{-\left[\hat\beta(\hat\alpha - \eta) - \rho_K^2 S_{\hat\alpha\hat\beta}\right] - \hat\theta}{\hat\phi}. \tag{6.26}$$

By the law of large numbers, we know that $\hat\alpha$, $\hat\beta$, $S_{\hat\alpha}^2$, $S_{\hat\beta}^2$, and $S_{\hat\alpha\hat\beta}$ have respective limits α, β, S_α^2, S_β^2, and $S_{\alpha\beta}$. Let φ and θ be the limits of $\hat\varphi$ and $\hat\theta$, respectively. To assess the sampling distribution of $\tilde{t}$, it is helpful to introduce the following notations:

$$f_0 = \left(-\left[\beta(\alpha - \eta) - \rho_K^2 S_{\alpha\beta}\right] - \theta\right)/\varphi, \tag{6.27}$$

$$f_\alpha' = \left(-\beta - \rho_K^2\left[(\alpha - \eta)S_\beta^2 - \beta S_{\alpha\beta}\right]/\theta\right)/\varphi,$$

$$f_\beta' = \left(\left[-\alpha + \eta - \rho_K^2\left(\beta S_\alpha^2 - (\alpha - \eta)S_{\alpha\beta}\right)/\theta\right]\phi - 2\beta\varphi f_0\right)/\varphi^2,$$

$$f_{S_\alpha^2}' = \left(-\left[-\rho_K^4 S_\beta^2 + \rho_K^2\beta^2\right]\right)/(2\varphi\theta),$$

$$f_{S_{\alpha\beta}}' = \left(\rho_K^2\left[1 - \left(\rho_K^2 S_{\alpha\beta} - \beta(\alpha - \eta)\right)/\theta\right]\right)/\varphi,$$

$$f_{S_\beta^2}' = \left(-\left[-\rho_K^4 S_\alpha^2 + \rho_K^2(\alpha - \eta)^2\right]\right)/(2\varphi\theta) + \rho_K^2 f_0/\varphi.$$

Here, f_α', f_β', $f_{S_\alpha^2}'$, $f_{S_{\alpha\beta}}'$, and $f_{S_\beta^2}'$ are the respective partial derivatives of the function $f(\alpha - \eta, \beta, S_\alpha^2, S_{\alpha\beta}, S_\beta^2, \rho_K^2)$. Using the standard delta method and well-known results on the asymptotic distribution of $(\hat\alpha, \hat\beta, S_{\hat\alpha}^2, S_{\hat\alpha\hat\beta}, S_{\hat\beta}^2)$ (see, e.g., Muirhead, 1982, p. 43), we obtain the asymptotic distribution of $\tilde{t}$ as $K \to \infty$ as

$$\sqrt{K}\left[\tilde{t} - f(\alpha - \eta, \beta, S_\alpha^2, S_{\alpha\beta}, S_\beta^2, \rho_K^2)\right] \xrightarrow{d} N(0,\ \sigma^2), \tag{6.28}$$

where

$$\sigma^2 = \left(f_\alpha',\ f_\beta'\right)W\left(f_\alpha',\ f_\beta'\right)^T$$

$$+ \left(f_{S_\alpha^2}',\ f_{S_{\alpha\beta}}',\ f_{S_\beta^2}'\right)U\left(f_{S_\alpha^2}',\ f_{S_{\alpha\beta}}',\ f_{S_\beta^2}'\right)^T,$$

and ρ_K in σ^2 is replaced by z_ϵ since $\rho_K \to z_\epsilon$ as $K \to \infty$. Hence, W is as given before and

$$U = \begin{pmatrix} 2S_\alpha^4 & 2S_{\alpha\beta}S_\alpha^2 & 2S_{\alpha\beta}^2 \\ 2S_{\alpha\beta}S_\alpha^2 & S_\alpha^2 S_\beta^2 + S_{\alpha\beta}^2 & 2S_\beta^2 S_{\alpha\beta} \\ 2S_{\alpha\beta}^2 & 2S_\beta^2 S_{\alpha\beta} & 2S_\beta^4 \end{pmatrix}.$$

Thus, W/K is the asymptotic covariance of $(\hat{\alpha}, \hat{\beta})$ and U/K is the asymptotic covariance of $(S_{\hat{\alpha}}^2, S_{\hat{\alpha}\hat{\beta}}, S_{\hat{\beta}}^2)$. When there is no batch-to-batch variation (i.e., $\sigma_\alpha^2 = \sigma_\beta^2 = 0$), the true shelf-life t_{true} is a constant. The estimated shelf-life $\tilde{t}$ may be considered as the $(1 - \alpha)$ level lower confidence bound of the $\epsilon = 0.5$th quantile of the true shelf-life. In what follows we obtain the following expression for the case:

$$\rho_K^2 = t_{\alpha, K-1}^2 / K,$$

$$S_\alpha^2 = \sigma_e^2 \frac{\sum\limits_{j=1}^n x_j^2}{n \sum\limits_{j=1}^n (x_j - \bar{x})^2},$$

$$S_\beta^2 = \frac{\sigma_e^2}{\sum\limits_{j=1}^n (x_j - \bar{x})^2},$$

$$S_{\alpha\beta} = -\sigma_e^2 \frac{\bar{x}}{\sum\limits_{j=1}^n (x_j - \bar{x})^2},$$

$$\theta = \frac{t_{\alpha, K-1} \sigma_e}{\sqrt{K}} \left[\frac{\beta^2 \sum\limits_{j=1}^n x_j^2 + 2\beta(\alpha - \eta)n\bar{x} + (\alpha - \eta)^2 n}{n \sum\limits_{j=1}^n (x_j - \bar{x})^2} \right],$$

and

$$f\left(\alpha - \eta, \beta, S_\alpha^2, S_{\alpha\beta}, S_\beta^2, \rho_K^2\right) \tag{6.29}$$

$$= -\frac{\alpha - \eta}{\beta} - \frac{\theta}{\beta^2} + o(1) \quad \text{as} \quad K \to \infty,$$

where $t_{\alpha, K-1}$ is the $(1 - \alpha)$th quantile of the t distribution with $K - 1$ degrees of freedom. To obtain the asymptotic variance σ^2, by letting $\rho_K = z_{0.5} = 0$, we get

$$f_\alpha' = -1/\beta,$$
$$f_\beta' = \frac{\alpha - \eta}{\beta^2},$$

and

$$f_{S_\alpha^2}' = f_{S_{\alpha\beta}}' = f_{S_\beta^2}' = 0.$$

Thus,

$$\sigma^2 = \left(\frac{\bar{x}}{\beta} - \frac{\eta - \alpha}{\beta^2}\right)^2 \frac{\sigma_e^2}{\sum_{j=1}^n (x_j - \bar{x})^2} + \frac{\sigma_e^2}{n\beta^2}. \tag{6.30}$$

Inserting Equations 6.29 and 6.30 into Equation 6.28, we obtain the following limit distribution of $\tilde{t}$ as $K \to \infty$ for the case when there is no batch-to-batch variation:

$$\sqrt{K}\left(\tilde{t} - \frac{\eta - \alpha}{\beta}\right) \xrightarrow{d} N\left(-\frac{t_{\alpha,K-1}}{\beta^2}\left[\frac{\beta^2}{n} + \frac{(\beta\bar{x} + \alpha - \eta)^2}{\sum_{j=1}^{n}(x_j - \bar{x})^2}\right]^{1/2}, \sigma^2\right),$$

where σ^2 is given in Equation 6.30.

6.3.2   The HLC Method

Chow and Shao's approach described above is based on the concept that the true shelf-life is the minimum of the time point at which any observed total stability loss for any future batch randomly chosen from the population of the production batches is equal to or greater than the lower specification limit η with a high degree of confidence. One approach suggested by the FDA stability guidelines for establishing drug shelf-life is to estimate the expiration dating period as the time point at which the 95% one-sided lower confidence limit for mean degradation curve intersects the acceptable lower specification limit. Therefore, the statistical concept and procedure for estimating shelf-lives described in the FDA stability guidelines is based on the 95% lower confidence limit for the mean degradation curve so that the mean strength of the drug product will remain within specifications until the labeled expiration date for 95% of the future batches (FDA, 1987, 1998).

For an estimation of drug shelf-life with multiple batches, there are two major sources of variations: within-batch variation and between-batch variation. It is suggested that these two sources be taken into account in stability analysis. It should be noted that the statistical methods described in Chapter 5 are derived under a fixed effects model, which considers only the within-batch variability. The consequence of the methods derived from a fixed effects model is that statistical inference can only be made to the batches under study. This is because the statistical inference obtained does not account for the between-batch variability. To account for this problem, Ho, Liu, and Chow (1993) suggested an alternative statistical procedure for estimating the expiration dating period. The method not only uses the 95% lower confidence limit approach, but also takes into account the between-batch variability. In this chapter we refer to their method as the HLC method.

To introduce the HLC method, we first consider a simple linear regression model without covariates for the balanced case where the time points are the same for all batches. Extension to the unbalanced case with different time points will be reviewed later. For a given batch i, consider the following simple linear regression model that was given in Model 5.1:

$$y_{ij} = \alpha_i + \beta_i x_j + e_{ij}, \quad i = 1, \ldots, K, \quad j = 1, \ldots, n,$$

where α_i, β_i, and e_{ij} are defined in Model 5.1. Let B_i denote batch $i, i = 1, \ldots, K$. Then, $\{B_i, i = 1, \ldots, K\}$ represents a random sample from the population of the production batches. Let

$$\beta_i = (\alpha_i, \beta_i)', \quad i = 1, \ldots, K$$

Then, β_i is a random vector that describes the degradation pattern of drug product characteristic. Under the assumptions described under Model 6.1, β_i, $i = 1, \ldots, K$ are i.i.d. as $N_2(\beta, \Sigma_\beta)$, where

$$\beta = (\alpha, \beta)',$$

and Σ_β is a 2×2 nonnegative definite covariance matrix. Therefore, Model 5.1 becomes a regression model with a random coefficient (Gumpertz and Pantula, 1989; Carter and Yang, 1986; Vonesh and Carter, 1987). Let

$$\hat{\beta}_i = (\hat{\alpha}_i, \hat{\beta}_i)', \quad i = 1, \ldots, K$$

be the least squares estimates of β_i for batch i, where $\hat{\alpha}_i$ and $\hat{\beta}_i$ are as defined in Section 5.1. Under the assumptions of Model 6.1, the maximum likelihood estimator (MLE) of β is given by

$$\bar{\beta} = (X'X)^{-1}X'\bar{y}$$
$$= \frac{1}{K} \sum_{i=1}^{K} (X'X)^{-1}X'y_i$$
$$= \frac{1}{K} \sum_{i=1}^{K} \hat{\beta}_i$$
$$= (\bar{\alpha}, \bar{\beta})',$$

where

$$\bar{\alpha} = \frac{1}{K} \sum_{i=1}^{K} \hat{\alpha}_i,$$

and

$$\bar{\beta} = \frac{1}{K} \sum_{i=1}^{K} \hat{\beta}_i.$$

Hence, the MLE of β is simply the average of the ordinary least squares estimates over the K batches. As indicated in Section 6.2, we have

$$\bar{\beta} \sim N(\beta, \Sigma_{\bar{\beta}}),$$

where

$$\Sigma_{\bar{\beta}} = \frac{1}{K}\left[\Sigma_\beta + \sigma_e^2(X'X)^{-1}\right].$$

Note that $\bar{\beta}$ is also an unbiased estimator of β under the balanced case. In addition, the covariance matrix of $\bar{\beta}$ consists of the within-batch variability, that is, $\sigma_e^2(X'X)^{-1}$, and the between-batch variability, that is, Σ_β. Therefore, under the assumptions of Model 6.1, a random effects model can account for both the within-batch and between-batch variabilities for estimation of shelf-life. The unbiased estimators of Σ_β, σ_e^2, and Σ_β are given, respectively, as

$$S_{\hat\beta} = \frac{1}{K-1} \sum_{i=1}^{K} (\hat\beta_i - \bar\beta)(\hat\beta_i - \bar\beta)',$$

$$\hat\sigma_e^2 = MSE = \frac{1}{N-2K} SSE = \frac{1}{N-2K} \sum_{i=1}^{K} SSE(i),$$

$$\hat\Sigma_\beta = S_{\hat\beta} - \hat\sigma_e^2 (X'X)^{-1},$$

where *MSE*, *SSE*, and *SSE(i)* are defined in Section 5.1. $\hat\Sigma_\beta$ is obtained as the difference between $S_{\hat\beta}$ and $\hat\sigma_e^2 (X'X)^{-1}$. It should be noted, however, that $\hat\Sigma_\beta$ might not be positive definite because of the probability that $|\hat\Sigma_\beta| < 0$ may be greater than zero. In this case the estimator suggested by Carter and Yang (1986) is recommended to estimate Σ_β. Note that $\hat\Sigma_\beta$ is not required for estimation of drug shelf-life in a balanced case.

For any arbitrary 2×1 vector $x = (1, x)'$, the minimum variance unbiased estimator is $x'\bar\beta$, which follows a univariate normal distribution with mean $x'\beta$ and variance $\sigma_x^2 = x'\Sigma_\beta x$. An unbiased estimator for σ_x^2 is given as follows:

$$\hat\sigma_x^2 = x' S_{\hat\beta} x,$$

which is distributed as $\sigma_x^2 \chi_{K-1}^2$. Since $x'\bar\beta$ and $\hat\sigma_x^2$ are independent of each other, we have

$$T = \frac{x'(\bar\beta - \beta)}{\sqrt{\hat\sigma_x^2 / K}}, \tag{6.31}$$

which follows a central t distribution with $K - 1$ degrees of freedom. Consequently, the 95% lower confidence limit for the mean degradation line $x'\beta$ at x is given by

$$L(x) = x'\bar\beta - t_{0.05, K-1} \sqrt{\frac{\hat\sigma_x^2}{K}}. \tag{6.32}$$

Note that

$$x'\bar\beta = \bar\alpha + \bar\beta x$$

$$\hat\sigma_x^2 = \frac{1}{K-1} \sum_{i=1}^{K} \left[(\hat\alpha_i - \bar\alpha)^2 + 2x(\hat\alpha_i - \bar\alpha)(\hat\beta_i - \bar\beta) + x^2(\hat\beta_i - \bar\beta)^2 \right].$$

The 95% lower confidence limit for the mean degradation curve $\alpha + \beta x$ is constructed from both within- and between-batch variability. Hence, statistical inference about the expiration dating period based on $L(x)$ in Equation 6.32 can be made to all future production batches.

Similar to the method described in Section 3.2, the time points at which Equation 6.32 intersects the acceptable lower specification limit η (if it exists) are the two roots of the following quadratic equation:

$$[\eta - (\bar{\alpha} + \bar{\beta}x)]^2 = t_{0.05,K-1}^2 \frac{1}{K(K-1)} \sum_{i=1}^{K} [(\hat{\alpha}_i - \bar{\alpha})^2$$
$$+ 2x(\hat{\alpha}_i - \bar{\alpha})(\hat{\beta}_i - \bar{\beta})$$
$$+ x^2(\hat{\beta}_i - \bar{\beta})^2].$$

These two roots, denoted by x_L and x_U, constitute the lower and upper limits of the 90% confidence interval for $(\eta - \alpha)/\beta$. Let

$$SE(\bar{\alpha}) = \left[\frac{1}{K(K-1)} \sum_{i=1}^{K} [(\hat{\alpha}_i - \bar{\alpha})^2 \right]^{1/2},$$

$$SE(\bar{\beta}) = \left[\frac{1}{K(K-1)} \sum_{i=1}^{K} [(\hat{\beta}_i - \bar{\beta})^2 \right]^{1/2},$$

$$T_{\bar{\alpha}} = \frac{\bar{\alpha} - \eta}{SE(\bar{\alpha})},$$

$$T_{\bar{\beta}} = \frac{\bar{\beta}}{SE(\bar{\beta})}.$$

If the slope is statistically smaller than zero and the intercept is statistically greater than η at the 5% level of significance, that is

$$\text{(i) } T_{\bar{\beta}} < -t_{0.05,K-1}, \tag{6.33}$$
$$\text{(ii) } T_{\bar{\alpha}} > t_{0.05,K-1},$$

the 90% confidence interval for $(\eta - \alpha)/\beta$ is a close interval $[x_L, x_U]$. Hence, x_L is defined as the estimated shelf-life for all future production batches. In other cases, however, the 90% confidence interval for $(\eta - \alpha)/\beta$ is either the entire real line or two disjoint open intervals. Consequently, the estimated expiration dating period is not defined.

When the number of time points is the same for all batches but the time points are different from batch to batch, Model 5.1 can be expressed in the following matrix form:

$$y_i = X_i \beta_i + \epsilon_i, \quad i = 1, \ldots, K, \tag{6.34}$$

where

$$y_i = (y_{i1}, \ldots, y_{in}),$$
$$X_i = \begin{bmatrix} 1 & x_{i1} \\ \vdots & \vdots \\ 1 & x_{in} \end{bmatrix},$$

β_i and ϵ_i are as defined in Model 6.2. Under Model 6.34 and the assumptions of Model 6.1, $\bar{\beta}$ is also distributed as a bivariate normal vector with mean vector β and covariance matrix

$$\Sigma_{\bar{\beta}} = \Sigma_{\beta} + \frac{1}{K}\sigma^2 \sum_{i=1}^{K}(X_i'X_i)^{-1}.$$

Hence, $\bar{\beta}$ is still unbiased for β and $S_{\bar{\beta}}$ is an unbiased estimator of $\Sigma_{\bar{\beta}}$. However, for any arbitrary 2×1 vector x,

$$\hat{\sigma}_x^2 = x'S_{\bar{\beta}}x$$

is no longer distributed as a chi-square random variable. As a result, the T statistic defined in Equation 6.31 does not have a central t distribution. Gumpertz and Pantula (1989) showed that when the product of the number of batches and number of time points becomes large, T follows approximately a central t distribution with v degrees of freedom, where

$$v = \frac{v_N}{v_D},$$

$$v_N = \left\{ x'\Sigma_{\beta}x + \frac{1}{K}\sigma_e^2 \left[\sum_{i=1}^{K} x'(X_i'X_i)^{-1}x\right] \right\}^2,$$

$$v_D = \left(\frac{x'\Sigma_{\beta}x}{K-1}\right)^2 + \left\{ \frac{\sigma_e^2}{K^2(n-2)}\left[\sum_{i=1}^{K} x'(X_i'X_i)^{-1}x\right] \right\}^2. \tag{6.35}$$

If the condition in Equation 6.33 is satisfied, the expiration dating period can be estimated by the smaller root x_L of the quadratic equation:

$$[\eta - (\bar{\alpha} + \bar{\beta}x)]^2 = t_{0.05,v}^2 \hat{\sigma}_x^2.$$

Application of Equation 6.35 to estimate the v degrees of freedom involves the unknown time point at which the mean degradation line intersects the lower specification limit η, the unknown between-batch covariate matrix Σ_{β}, and the unknown within-batch variability σ_e^2. However, σ_e^2 and Σ_{β} can be estimated by their unbiased estimators, which are given, respectively, as

$$\hat{\sigma}_e^2 = MSE = \frac{1}{N-2K} \sum_{i=1}^{K} SSE(i),$$

$$\hat{\Sigma}_{\beta} = S_{\bar{\beta}} \sim \hat{\sigma}_e^2 \frac{1}{K} \sum_{i=1}^{K}(X_i'X_i)^{-1}.$$

It is suggested that the unknown time point be estimated by its maximum likelihood estimate

$$\hat{x}_0 = \frac{\eta - \bar{\alpha}}{\bar{\beta}}.$$

For Σ_β, as indicated earlier, $\hat{\Sigma}_\beta$ might not be positive definite. In this case the estimator suggested by Carter and Yang (1986) can be used. Note that when $n_i = n$ and $X_i = X$ for $i = 1, \ldots, K$, the exact inference of the shelf-life is based on the number of batches. However, the approximation by a central t distribution with v degrees of freedom can still be applied as long as nK is large. When the number of time points are different from batch to batch, the estimated generalized least squares estimator for β suggested by Carter and Yang (1986) and Vonesh and Carter (1987) might be useful for construction of the 95% lower confidence limit for the mean degradation line $\alpha + \beta x$. However, the statistical inference of their methods is based on the asymptotic results as either the number of batches becomes large or the minimum number of time points becomes large.

6.4   Comparison of Methods for Multiple Batches

Under the assumptions for the random effects model, two methods for estimating the expiration dating period were presented in the previous sections. Chow and Shao's method was derived from the probability statement for a given future batch selected randomly from the population of the production batches. The HLC method incorporates the between-batch variability into the construction of the 95% lower confidence limit for the mean degradation curve. Under the simple linear regression model without covariates, for the balanced case where all time points are the same, the estimated shelf-life by Chow and Shao's method is the solution to the following equation:

$$\bar{\alpha} + \bar{\beta}x = \eta + c_K(\epsilon, \alpha)z_\epsilon \sqrt{v(t)}.$$

In other words, the smaller root of the following quadratic equation, if it exists, will be the estimated shelf-life

$$[\eta - (\bar{\alpha} + \bar{\beta}x)]^2 = [c_K(\epsilon, \alpha)z_\epsilon]^2 \, v(t). \qquad (6.36)$$

Note that

$$[c_K(\epsilon, \alpha)z_\epsilon]^2 = \left[\frac{t_{\alpha, K-1, \sqrt{K}z_\epsilon}}{\sqrt{K}z_\epsilon} z_\epsilon \right]^2$$

$$= \frac{1}{K} \left[t_{\alpha, K-1, \sqrt{K}z_\epsilon} \right]^2.$$

In addition, since

$$S = \sum_{i=1}^{K} (y_i - \bar{y})(y_i - \bar{y})',$$

it follows that

$$(X'X)^{-1}X'SX(X'X)^{-1}$$
$$= \sum_{i=1}^{K}(X'X)^{-1}X'(y_i - \bar{y})(y_i - \bar{y})'X(X'X)^{-1}$$
$$= \sum_{i=1}^{K}(\hat{\beta}_i - \bar{\beta})(\hat{\beta}_i - \bar{\beta})'$$
$$= (K-1)S_{\hat{\beta}}.$$

As a result, for any arbitrary 2×1 fixed vector x, $v(t)$ gives

$$v(t) = \frac{1}{K-1}x'(X'X)^{-1}X'SX(X'X)^{-1}x$$
$$= \frac{1}{K-1}x'S_{\hat{\beta}}x$$
$$= \hat{\sigma}_x^2.$$

Hence, Equation 6.36 reduces to

$$[\eta - (\bar{\alpha} + \bar{\beta}x)]^2 = \frac{1}{K}\hat{\sigma}_x^2[t_{\alpha, K-1, \sqrt{K}z_\epsilon}]^2. \tag{6.37}$$

Recall that the estimated shelf-life determined by the HLC method is the smaller root, if it exists, of the following equation:

$$[\eta - (\bar{\alpha} + \bar{\beta}x)]^2 = \frac{1}{K}\hat{\sigma}_x^2[t_{\alpha, K-1}]^2. \tag{6.38}$$

If $\epsilon = 0.5$ and $z_\epsilon = 0$, Equation 6.37 becomes Equation 6.38. In addition,

$$\tilde{L}(x) = (\bar{\alpha} + \bar{\beta}x) - \left(\frac{1}{K}\hat{\sigma}_x^2\right)^{1/2}t_{\alpha, K-1, \sqrt{K}z_\epsilon},$$

which is the $(1 - \alpha) \times 100\%$ lower ϵ-content tolerance limit. Since for $\epsilon < 0.5$, a $(1-\alpha) \times 100\%$ lower ϵ-content tolerance limit is the $(1-\alpha) \times 100\%$ lower confidence limit for the ϵth quantile of the distribution under study (Hahn and Meeker, 1991), the estimated expiration dating period by Chow and Shao's method is the time point at which the $(1-\alpha) \times 100\%$ lower confidence limit for the ϵth quantile of the distribution of the random degradation line $\alpha_i + \beta_i x$ intersects the lower specification limit η. When $\epsilon = 0.5$, the 50% quantile of a normal distribution is the mean. Therefore, the HLC method is a special case of Chow and Shao's method. However, if $\epsilon < 0.5$,

$$t_{\alpha, K-1} \leq t_{\alpha, K-1, \sqrt{K}z_\epsilon}.$$

Hence, the estimated shelf-life determined by Chow and Shao's method is always shorter than the shelf-life given by the HLC method. It should be noted that the interpretation of the estimates are different. The estimate by the HLC method provides a 95% confidence that the average characteristic of the dosage units in future batches will remain within specification up to the end of the estimated shelf-life. Chow and Shao's method suggests that use of an estimate that gives a 95% confidence that at least a 1-ϵ proportion of the distribution of the drug characteristic for future batches will be within specifications until the end of the estimated shelf-life.

6.4.1 An Example

To illustrate Chow and Shao's approach and the HLC method for determining drug shelf-life, consider the stability data given in Table 5.6. The lower specification limit for the drug product is $\eta = 90\%$. For simplicity, consider the following random effects model:

$$y_{ij} = x_j'\beta_i + e_{ij}$$

where $i = 1, \ldots, 5$; $j = 1, \ldots, 12$; and $x_j = (1, t_j, w_j, t_j w_j)'$, in which $w_j = 0$ for a bottle and $w_j = 1$ for a blister package, and t_j are sampling times, which are as follows:

j	1	2	3	4	5	6	7	8	9	10	11	12
t_j	0	3	6	9	12	18	0	3	6	9	12	18
w_j	0	0	0	0	0	0	1	1	1	1	1	1

We first use T statistic in Equation 6.8 to examine the random effect. the data set given in Table 5.6 gives $tr(S) = 88.396$ and $SE = 4.376$. Since $n = 12$, $p = 4$, and $K = 5$, we have

$$T = \frac{(n - p)tr(S)}{n(K - 1)SE} = \frac{(12 - 8)(88.396)}{(12)(5 - 1)(4.376)} = 3.367$$

which is greater than $F_{0.05,48,9} = 2.8$. Hence, we conclude that there is a batch-to-batch variation at the 5% level of significance.

Although all of the assay results at 18 months are clearly higher than the lower specification limit $\eta = 90\%$, this does not imply that $t_{label} \geq 18$ months can be used as the labeled shelf-life, since the assay results are from five batches, which should be considered as a random sample from the population of all future batches. A labeled shelf-life should be determined by statistically analyzing the assay results by using the procedures described in the previous sections. For Chow and Shao's method, since the unit of the shelf-life is a month, we need to evaluate $x(t)'\hat{\beta}$ and $\tilde{\eta}(t)$ for integer values of t in a reasonable range in order to calculate $\tilde{t}$. The results of the calculation for both package types are shown in Table 6.2. It shows that if ϵ is selected to be 0.05, we may use 22 months (with $\alpha = 5\%$) as the labeled shelf-life for the blister package. If ϵ is chosen to be 0.01, then 19 months can be used as the labeled shelf-life for both bottle and blister package. Note that in this example the trade-off for reducing ϵ from 0.05 to 0.01 is 2 to 3 months of shelf-life.

We now compare the minimum approach with Chow and Shao's method using the assay results for bottle in this example. The difference between the two methods is

TABLE 6.2: Results from Stability Analysis

Values of Statistics[†]

	Bottle		Blister Package	
t(months)	$\mathbf{X}(t)'\hat{\beta}$	$\sqrt{v(t)}$	$\mathbf{X}(t)'\hat{\beta}$	$\sqrt{v(t)}$
18	98.304	1.140	98.057	1.220
19	98.015	1.275	97.779	1.350
20	97.725	1.411	97.501	1.482
21	97.437	1.548	97.222	1.615
22	97.147	1.686	96.944	1.749
23	96.858	1.825	96.666	1.885
24	96.569	1.963	96.387	2.020
25	96.280	2.102	96.109	2.157
26	95.990	2.242	95.381	2.294

Labeled Shelf-Life t (months: $\alpha = 0.05$)

ε	Bottle	Blister Package
0.01	19	19
0.02	20	19
0.03	21	20
0.04	21	21
0.05	22	21
0.10	23	23
0.15	25	24

Source: Chow, S.C. and Liu, J.P. (1995). *Statistical Design and Analysis in Pharmaceutical Science.* Marcel Dekker, New York.

significant, since $\hat{t}_{min} = 27.5$ months. For $\epsilon = 0.05$, $\hat{t}_{min} - \bar{t}$ is 5.5 months, and for $\epsilon = 0.01$, $\hat{t}_{min} - \bar{t}$ is 8.5 months. However, the minimum approach is not justifiable, since it only ensures that the shelf-lives of the five batches, rather than the shelf-lives of future batches, are longer than the lower specification limit with certainty. $\hat{t} = 27.5$ if ϵ is chosen to be greater than 0.2. This indicates that $t = 27.5$ months is a valid labeled shelf-life if we allow the future failure rate at the indicated date of expiration to be as large as 20%. A risk of 20% future failure rate is usually too high for a pharmaceutical company.

From Table 5.8, it can be verified that

$$\bar{\beta} = (\bar{\alpha}, \bar{\beta})' = (103.51, -0.289),$$

$$S_{\hat{\beta}} = \begin{bmatrix} 2.250 & -0.2058 \\ -0.2058 & 0.01994 \end{bmatrix},$$

$$\hat{\sigma}_e^2 = 0.9597,$$

$$\hat{\sigma}_e^2 (X'X)^{-1} = \begin{bmatrix} 0.452 & -0.0366 \\ -0.0366 & 0.00457 \end{bmatrix},$$

$$\hat{\Sigma}_{\hat{\beta}} = S_{\hat{\beta}} - \hat{\sigma}_e^2 (X'X)^{-1}$$

$$= \begin{bmatrix} 1.798 & -0.1692 \\ -0.1692 & 0.0154 \end{bmatrix}.$$

Hence,

$$SE(\tilde{\alpha}) = 0.671 \quad \text{and} \quad SE(\tilde{\beta}) = 0.0632.$$

Since

$$T_{\tilde{\alpha}} = \frac{103.51 - 90}{0.671} = 20.141 > t_{0.05,4} = 2.132,$$

$$T_{\tilde{\beta}} = \frac{-0.289}{0.0632} = -4.58 < -t_{0.05,4} = -2.132,$$

both conditions in Equation 6.25 are satisfied, and the estimate of the shelf-life by the HLC method is the smaller root of the following equation:

$$[90 - (103.51 - 0.289x)]^2 = (2.132)\left(\frac{1}{5}\right)[2.250 - 0.4116x + (0.01994x)^2],$$

which is 35.1 months. If the asymptotic procedure of the HLC method is applied, then

$$\hat{x}_0 = \frac{90 - 103.51}{-0.289} = 46.71,$$

$$\hat{v}_N = 703.42,$$

$$\hat{v}_D = 97.657.$$

Thus, the estimated degrees of freedom is given by

$$\hat{v} = \frac{\hat{v}_N}{\hat{v}_D} = \frac{703.42}{97.657} = 7.2.$$

The estimated shelf-life is 36.1 months.

6.4.2   Comparison of Methods

Ho, Liu, and Chow (1993) conducted a simulation study to compare four methods of estimating drug shelf-life with multiple batches. These four methods include the FDA's minimum approach, Ruberg and Hsu's method under the fixed effects model discussed in the previous chapter, Chow and Shao's method, and the HLC procedure under the random effects model described in this chapter. Model 5.1 was employed to generate random samples. The intercept and slope were chosen to provide true shelf-life $t_{true} = (90 - \alpha)/\beta$ to be 4, 6.67, and 20 months. Three sets of sampling time points were selected:

- 1. 0, 3, 6, 9, and 12 months

- 2. 0, 3, 6, 9, 12, 18, and 24 months

- 3. 0, 3, 6, 9, 12, 18, 24, 36, and 48 months

The within-batch variabilities σ_e^2 were selected to be 0.25, 0.75, and 1.25. The following three between-batch covariance matrices were considered in the simulation:

$$m_1 = \begin{bmatrix} 0 & 0 \\ 0 & 0 \end{bmatrix}, \quad m_2 = \begin{bmatrix} 1.00 & 0.03 \\ 0.03 & 0.01 \end{bmatrix}, \quad m_3 = \begin{bmatrix} 1.00 & 0.03 \\ 0.03 & 0.02 \end{bmatrix}.$$

In addition, the impact of sample size was also studied by examining three batches: 3, 6, and 9. For each of the 81 combinations, 1000 random samples were generated to estimate the shelf-life by the four methods. For Ruberg and Hsu's method, the bioequivalence-like approach was used with the upper allowable specification limit Δ_β chosen as recommended by Ruberg and Stegeman (1991). The coverage probability, average bias, and mean squared error of the corresponding estimates with respect to $(90 - \alpha)/\beta$ were computed. In general, if the relationship between drug characteristic and time points is still linear beyond the observed range of time points as assumed in the simulation, the results are consistent even when an extrapolation is required. When both variabilities are large and the number of batches is small, the coverage probabilities of all four methods are relatively lower than those obtained from other combinations that have smaller variabilities and large numbers of batches.

Since Chow and Shao's method is not used for the mean drug characteristic but rather for the ϵ quantile of the distribution, where $\epsilon < 0.5$, not surprisingly, their method has the highest, but excessive, coverage probability uniformly greater than the nominal level of 0.95. As a consequence, this method has the largest average bias and mean squared error. This indicates that Chow and Shao's method often underestimates the shelf-life and is the most conservative of the four methods. The FDA's minimum approach exhibits a pattern similar to Chow and Shao's method. It provides a coverage probability greater than the 0.95 nominal level for a large number of batches and small within- and between-batch variabilities. However, the coverage probability can drop below 0.95 in the presence of a large variability with a small number of batches. If the observed range is 12 months, the coverage probability can be below 0.90. For example, with an observed range of 48 months and a true shelf-life of 20 months, the coverage probability for three batches decreases to about 0.90. Although the FDA's minimum approach is conservative, it may not provide enough coverage probability for some cases with three batches.

In summary, the FDA's minimum approach results in large average bias and mean squared errors. Ruberg and Hsu's method yields an adequate coverage probability near the nominal level of 0.95 in the absence of between-batch variability. However, the coverage probability provided by Ruberg and Hsu's method is the lowest of the four methods if there is a batch-to-batch variation. Although both the FDA's minimum approach and Ruberg and Hsu's method were derived from the fixed effects model, the FDA's minimum approach is more robust with respect to between-batch variability than Ruberg and Hsu's procedure. It should be noted that the performance of Ruberg and Hsu's method depends on the choice of Δ_β.

Since the sampling distribution (exact or asymptotic) of the HLC method is based on the number of batches rather than the number of total assays, the coverage probability decreases to 0.87 in some cases when the number of batches is small (e.g., $K = 3$). However, when the number of batches becomes large (e.g., $K = 6$ or 9), the HLC method provides adequate coverage probability around 0.95, which is within the

95% confidence interval $(0.9365, 0.9635)$ obtained from the 1000 random samples with respect to a nominal level of 0.95. The average bias and mean squared error is relatively smaller than Chow and Shao's method and the FDA's minimum procedure. More details regarding the simulation results can be found in Ho, Liu, and Chow (1993) and Chow and Liu (1995).

6.5 Determining Shelf-Life Based on the Lower Prediction Bound

Shao and Chen (1997) considered the following model

$$y_{ij} = x(t_{ij})'\beta_i + e_{ij}, \quad i = 1, \ldots, K, \quad j = 1, \ldots, n_i \qquad (6.39)$$

where i is the index for the batch and the sampled time intervals may differ in different batches. β_i may differ and, therefore, Model 6.39 contains K different regression curves. Under Model 6.39, assume that

$$\beta_i \sim N_p(b, \Sigma),$$
$$e_{ij} \sim N(0, \sigma_e^2),$$

and that β_i and e_{ij} are independent. Let

$$y_i = (y_{i1}, \ldots, y_{in_i})',$$
$$X_i = (x(t_{i1}), \ldots, x(t_{in_i}))',$$

and

$$e_i = (e_{i1}, \ldots, e_{in_i})',$$

where $i = 1, \ldots, K$. Then, we can rewrite Model 6.39 as

$$y_i = X_i\beta_i + e_i, \quad i = 1, \ldots, K. \qquad (6.40)$$

Assume that X_i are of full rank. Then, under the ith model in (6.40), the least squares estimator of β_i is given by

$$\hat{\beta}_i = (X_i'X_i)^{-1}X_i'y_i.$$

Conditional on β_i, we have

$$\hat{\beta}_i \sim N_p\big(\beta_i, \sigma_e^2(X_i'X_i)^{-1}\big).$$

Unconditionally, we have

$$\hat{\beta}_i \sim N_p\big(b, \Sigma + \sigma_e^2(X_i'X_i)^{-1}\big).$$

As a result, an approximate $1 - \alpha$ lower prediction bound for t_{true} can be obtained as

$$\hat{t} = \inf\{t : L(t) \le \eta\}, \qquad (6.41)$$

where η is a given lower specification limit,

$$L(t) = x(t)'\hat{b} - \rho(k, \alpha)\sqrt{v(t)/K} \qquad (6.42)$$

is a lower confidence bound for $x(t)'b$, where

$$\hat{b} = \frac{1}{k}\sum_{i=1}^{K}\beta_i,$$

and

$$v(t) = x(t)'\left[\frac{1}{k-1}\sum_{i=1}^{K}(\beta_i - \hat{b})(\beta_i - \hat{b})'\right]x(t)$$

is an estimate of the variance of $x(t)'\hat{b}$, $\rho(K, \alpha)$ satisfies

$$\int_0^1 P\{T_K(u) \leq \rho(K, \alpha)\}du = 1 - \alpha, \qquad (6.43)$$

where $T_K(u)$ denotes a random variable with noncentral t-distribution having $K - 1$ degrees of freedom and noncentrality parameter $\sqrt{K}\Phi^{-1}(1 - u)$, and Φ is the standard normal distribution function (i.e., $\rho(K, \alpha)$ is the $1 - \alpha$ quantile of the random variable $T_K(U)$, where U is uniform on [0,1]). Values of $\rho(K, \alpha)$ for $\alpha = 0.01, 0.05$, and 0.1 and $K = 3, 4, \ldots,$ and 20 are given in Table 6.3.

TABLE 6.3:   Values of $p(k,\alpha)$

	α		
k	0.01	0.05	0.10
3	13.929	5.840	3.771
4	10.153	5.262	3.662
5	9.178	5.222	3.756
6	8.903	5.331	3.905
7	8.889	5.496	4.072
8	8.994	5.684	4.245
9	9.160	5.881	4.417
10	9.358	6.080	4.587
11	9.574	6.279	4.753
12	9.800	6.475	4.916
13	10.031	6.669	5.074
14	10.265	6.859	5.229
15	10.498	7.045	5.380
16	10.730	7.228	5.527
17	10.961	7.407	5.671
18	11.189	7.583	5.812
19	11.417	7.755	5.590
20	11.637	7.924	6.084

Source: Shao, J. and Chen, L. (1997). *Statistics in Medicine*, 16, 1167–1173.

The $\hat{t}$ given in Equation 6.41 is an approximate $1 - \alpha$ lower prediction bound, provided that either

$$\min_{i=1,\ldots,K} X_i'X_i \longrightarrow \infty \qquad (6.44)$$

or

$$\sigma_e \to 0. \qquad (6.45)$$

Equation 6.44 is usually fulfilled if $n_i \to \infty$ for all i. Equation 6.45 is satisfied when the assay measurement error is small under controlled conditions in stability analysis. To show that

$$P\{\hat{t} \leq t_{true}\} \approx 1 - \alpha$$

holds under either conditions 6.44 or 6.45, we let F_{true} denote the distribution function of t_{true}. Thus, $\xi_u = F_{true}^{-1}(u)$ and $\sigma^2(t) = x(t)'\Sigma x(t)$. Note that

$$F_{true}(t) = P\{x(t)'\beta \leq \eta\} = \Phi\left(\frac{\eta - x(t)'b}{\sigma(t)}\right),$$

and

$$u = F_{true}(\xi_u) = \Phi\left(\frac{\eta - x(\xi_u)'b}{\sigma(\xi_u)}\right).$$

Hence,

$$\Phi^{-1}(1 - u) = \frac{x(\xi_u)'b - \eta}{\sigma(\xi_u)} \approx \frac{x(\xi_u)'b - \eta}{\sqrt{E[v(\xi_u)]}} \qquad (6.46)$$

since

$$E[v(t)] = \sigma^2(t) + \frac{\sigma_e^2}{K}\sum_{i=1}^{K} x(t)'(X_i'X_i)^{-1}x(t) \approx \sigma^2(t)$$

under either Equation 6.44 or 6.45. Therefore,

$$
\begin{aligned}
P\{t_{true} < \hat{t}\} &= \int_0^\infty P(t < \hat{t})dF_{true}(t) \\
&= \int_0^\infty P[(L(t) > \eta]dF_{true}(t) \\
&= \int_0^\infty P[(L(\xi_u) > \eta]du \\
&= \int_0^1 P\left\{\frac{x(\xi_u)'\hat{b} - \eta}{\sqrt{v(\xi_u)/K}} > \rho(K, \alpha)\right\} du \\
&\approx \int_0^1 P\{T_K(u) \leq \rho(K, \alpha)\}du \qquad (6.47) \\
&= \alpha,
\end{aligned}
$$

where the approximation in Equation 6.47 holds because Equation 6.46 and the last equality follow from Equation 6.43, the definition of $\rho(K, \alpha)$.

When K is large, we can remove the normality assumption on e_{ij}. In this case the result still holds. When K is large, we can also replace the normality assumption on β_i by the assumption that the distribution on β_i is from a location-scale family, that is,

$$P\{x(t)'\beta \leq \eta\} = \Psi\left(\frac{\eta - x(t)'b}{\sigma(t)}\right)$$

with a known distribution function Ψ. In such a case the result still holds with Φ replaced by Ψ in the definition of $\rho(k, \alpha)$ in Equation 6.43.

6.5.1 An Example

Consider the example given in Shao and Chow (1994), in which 300-mg tablets from $K = 5$ batches of a drug product were stored at room temperature in two types of containers (bottles and blister packages). The tablets were tested for potency at $t_j = 0, 3, 6, 9, 12$, and 18 months. The assay results are summarized in Table 5.6. Consider Model 6.31 with

$$x(t_{ij})' = (1, t_j, t_j w_j),$$

where $w_j = 0$ for the bottles and 1 for the blister packages. A test conducted by Shao and Chow (1994) shows a batch-to-batch variation at the 5% level of significance. Table 6.4 lists values of $L(t)$ given by Equation 6.42. For $\eta = 90\%$ and $\alpha = 0.05$, the labelled shelf-life is 27 months for bottles and 26 months for blister packages. The minimum approach suggested by the FDA gives a labelled shelf-life of 26 months for bottles. Thus, if the assurance level is $\alpha = 5\%$, the minimum approach is slightly too conservative. If the assurance level is $\alpha = 1\%$, the minimum approach gives a much too long labelled shelf-life.

TABLE 6.4:   Values of $L(t)$

	Bottle		Blister Package	
t **(in months)**	$\alpha = 0.01$	$\alpha = 0.05$	$\alpha = 0.01$	$\alpha = 0.05$
18	93.625	95.642	93.049	95.208
19	92.782	95.037	92.238	94.626
20	91.934	94.430	91.418	94.040
21	91.083	93.822	90.593	93.450
22	90.227	93.210	89.765	92.859
23	89.367	92.596	88.929	92.264
24	88.512	91.985	88.095	91.670
25	87.652	91.371	87.256	91.072
26	86.788	90.754	86.415	90.474
27	85.928	90.145	85.575	89.884
28	85.073	89.537	84.730	89.304

Source: Shao, J. and Chen, L. (1997). *Statistics in Medicine*, 16, 1167–1173.

The method by Shao and Chow (1994) and the method by Shao and Chen (1997) are in general not comparable since these were proposed under different perspectives. With $\varepsilon \leq 0.05$ in this example, the method by Shao and Chow is more conservative than the method by Shao and Chen. An advantage of Shao and Chen's prediction-bound approach is that it does not require the specification of an ε when it is not preassigned. In addition, Shao and Chen's prediction bound approach is applicable to any problem in which the data follow a linear random effects model and one is interested in estimating (or predicting) the intersection of the mean regression curve and a given curve. Examples of such applications include reliability analysis based on random effects degradation models (Lu and Meeker, 1993) and the determination of the regions for clean-up of contaminated soil in the U.S. Superfund program (Shao and Chen, 1997).

6.6 Concluding Remarks

For estimating drug shelf-life with random batches, in addition to Chow and Shao's method and the HLC method, Murphy and Weisman (1990) proposed the use of random slopes. Their idea is briefly outlined below. For the simple linear regression model, the slope β_i, $i = 1, \ldots, K$ is assumed to be i.i.d. as a normal random variable with mean zero and variance σ_β^2. Hence, the expected value of $MS(\beta)$ in the analysis of covariance table presented in Table 5.1 is then given by

$$E[MS(\beta)] = \sigma_e^2 + R\sigma_\beta^2,$$

where

$$R = \frac{1}{K-1} \left[S_{xx}(W) - \frac{\sum\limits_{i=1}^{K} S_{xx}^2(i)}{S_{xx}(W)} \right],$$

and $S_{xx}(i)$ and $S_{xx}(W)$ are as defined in Section 5.1. The same F statistic

$$F = \frac{MS(\beta)}{MSE}$$

can be applied to test the following hypotheses:

$$H_0 : \sigma_\beta^2 = 0 \quad \text{vs} \quad H_a : \sigma_\beta^2 > 0.$$

Depending on whether the null hypothesis above is rejected at the 0.25 level of significance or $MS(\beta)$ is larger than MSE, Murphy and Weisman (1990) proposed three methods to estimate shelf-life. Basically, their methods assume that the initial value of the strength at time zero is a fixed nonrandom quantity (i.e., 100% of label claim) for all three batches. Then, slope is use a to determine drug shelf-life. In other words,

Murphy and Weisman's method ignores the information observed at the initial time (i.e., time 0). As a result, their method suffers from the same drawbacks as the method suggested by Rahman (1992), which was discussed in Chapter 3. Even when the batch effect is assumed to be random, the same disadvantages remain in Murphy and Weisman's methods. It should be noted that Murphy and Weisman's methods are not based on the mean degradation line. Therefore, no probability statements can be made regarding the estimated shelf-life, observed strength, and lower specification limit. Moreover, in their simulation study the FDA's minimum approach was not considered as the referenced method for comparison. In addition, no coverage probabilities of their methods were reported.

For Chow and Shao's method, in practice, it may be difficult to choose an appropriate ϵ. Too large an ϵ can certainly increase the chance of the product being recalled prior to the expiration date, whereas too small an ϵ may increase the cost. By calculating $\tilde{t}$ for various values of ϵ, we can determine a labeled shelf-life by balancing the relative merits and disadvantages of having a longer labeled shelf-life against the risk of being recalled. In the situation where the final decision is made by the FDA, the research scientist or statistician may report $\tilde{t}$ for several values of ϵ in a reasonable range. When $\Sigma_\beta = 0$, both $\tilde{t}$ and $\hat{t}$ are valid labeled shelf-lives, but $\hat{t}$ is usually longer than $\tilde{t}$. When $\Sigma_\beta \neq 0$, however, $\hat{t}$ could be much too large. In practice, it is usually difficult to justify $\Sigma_\beta = 0$ since controlling the type II error for any procedure testing the null hypothesis H_0 is difficult owing to the complexity of the alternative hypothesis K_0. Thus, even if we cannot reject the hypothesis H_0, the use of $\hat{t}$ is still questionable. In this situation, however, the estimated shelf-lives determined by $\hat{t}$ and $\tilde{t}$ are close to each other because of the small contribution of Σ_β due to there being no evidence for the presence of batch-to-batch variability.

Lin (1990) investigated statistical analysis of stability data in terms of the intercept and slope of the degradation curve. Under the assumption of a fixed effects model, Lin described the possible situations for the slopes and intercepts encountered in the analysis of stability data: (a) common slope and common intercept, (b) different slopes and common intercept, (c) common slope and different intercepts, and (d) different slopes and different intercepts. As an alternative approach, Lin also described the possible use of a confidence limit (interval) approach for determining an expiration dating period when the drug characteristic increases or decreases with time. In addition, Lin pointed out some common deficiencies and concerns of statistical analyses that FDA reviewers often cite in their reviews of stability data submission.

Chapter 7

Stability Analysis with a Mixed Effects Model

As indicated earlier, a stability analysis is usually performed to characterize the degradation pattern or curve of a drug product by testing a limited number of batches under appropriate storage coniditions. Since different batches of a drug product may have different degradation patterns for various reasons (e.g., different stranghts, different package types, different storage conditions), both the FDA and ICH require that at least three batches, and preferably more, be tested to allow for some estimates of batch-to-batch variability and to test the hypothesis that a single expiration dating period period (shelf-life) is justifiable for all (future) batches. The expiration dating period is determined based on a statistical confidence (or prediction) interval analysis.

When there is no batch-to-batch variability, stability data from different batches can be pooled to determine a single shelf-life using the methods described in Chapter 3. When batch-to-batch variability is present, statistical methods for stability analysis as described in Chapters 5 (fixed batches) and 6 (random batches) are useful for determining a single shelf-life. In this chapter, as an alternative to the least sequares methods proposed by Chow and Shao (1991) and Shao and Chow (1994), we describe Chen, Hwang, and Tsong (1995) EM (Expectation and Maximization) algorithm procedure to obtain the maximum likelihood estimates of the regression coefficients of the fixed effects, random effects, and variance components under linear mixed effects. The likelihood ratio test of equal regression coefficients for the random effects components is used as a preliminary test of batch-to-batch variation for model selection.

In the next section a linear mixed effects model with replicates is introduced. Log-likelihood ratio tests under various hypotheses are given in Section 7.2. Section 7.3 provides details of the EM algorithm for obtaining maximum likelihood estimates of the parameters of the linear mixed effects model. An example is given in Section 7.4 to illustrate the use of the method proposed by Chen, Hwang, and Tsong (1995). A discussion is given in the last section to conclude this chapter.

7.1   Linear Mixed Effects Model

Chen, Hwang, and Tsong (1995) use the following linear mixed effects model to describe stability data collected from multiple batches:

$$y_{ijk} = \alpha + \beta x_{ij} + a_i + b_i x_{ij} + e_{ijk}, \qquad (7.1)$$

where y_{ijk} is the assay result (percent of label claim) from the kth replicate in the jth time point of the ith batch. $i = 1, \ldots, K$; $j = 1, \ldots, n_i$; $k = 1, \ldots, m_{ij}$. x_{ij} is the time of the stability sample corresponding to y_{ijk}, a_i is the ith batch effect (intercept), b_i is the effect of the degradation rate of the ith batch, and e_{ijk} is the random error in observing y_{ijk}. The random errors e_{ijk} are assumed to be independent and normally distributed with mean zero and variance σ^2. The random effects coefficients a_i and b_i are also assumed to be independent and normally distributed with mean 0 and variance σ_a^2 and σ_b^2, respectively. In addition, random variables e_{ijk}, a_i and b_i are assumed to be mutually independent. If $m_{ij} = 1$ for all i and j (i.e., there are no replicates), then Model 7.1 reduces to

$$y_{iy} = \alpha_i + \beta_i x_{ij} + e_{ij},$$

where $\alpha_i = \alpha + a_i$ and $\beta_i = \beta + b_i$. Under Model 7.1, the marginal mean of y_{ijk} is

$$E(y_{ijk}) = \alpha + \beta x_{ij} = \mu_{ij},$$

and the variance and covariance are given by

$$Var(y_{ijk}) = \sigma_a^2 + \sigma_b^2 x_{ij} + \sigma^2,$$

and

$$Cov(y_{ijk}, y_{ij'k'}) = \sigma_a^2 + \sigma_b^2 x_{ij} x_{ij'},$$

respectively. Model 7.1 represents a model of separate intercepts and separate slopes. If $a_1 = a_2 = \cdots = a_K = 0$ and $b_1 = b_2 = \cdots = b_K = 0$ (i.e., $\sigma_a^2 = 0$ and $\sigma_b^2 = 0$), then Model 7.1 has a common intercept and common slope. Similarly, if $b_1 = b_2 = \cdots = b_K = 0$ (i.e., $\sigma_b^2 = 0$), then Model 7.1 reduces to a common slope model. Furthermore, if $a_i = b_i$ for $i = 1, \ldots, K$, then Model 7.1 has common variance components (i.e., $\sigma_a^2 = \sigma_b^2$). This model was proposed by Chow and Wang (1994).

7.2   Model and Hypotheses

Chen, Hwang, and Tsong (1995) considered the following hypothesis tests for determining the poolability of different batches:

- Hypothesis 1: Test for equality of slopes.

- Hypothesis 2: Test for equality of intercepts given a common slope.

As a result, one of the following models can be selected for determining the degradation pattern of the drug product by testing the above hypotheses.

- Model 1: A common slope and common intercept if both hypotheses are not rejected

- Model 2: A common slope but different intercepts if the null hypothesis of equal intercepts is rejected but the hypothesis of equal slopes is not rejected

- Model 3: Separate slopes and separate intercepts if both hypotheses are rejected

As discussed in the previous chapters, if model 1 is selected, the shelf-life can be estimated as the time at which the 95% one-sided lower (or upper) confidence limit for the mean degradation curve intersects the acceptable lower (or upper) specification limit as given in the *USP-NF* (*USP-NF* 2000). If either model 2 or model 3 is selected, the FDA suggests obtaining the shelf-life of an individual batch and using the minimum of the obtained shelf-lives as the estimated shelf-life. In addition to the above models, Chen et al. also considered the following model

- Model 4: Common variance model

Model 4 assumes that $a_i = b_i$ for $i = 1, \ldots, K$, that is, $\sigma_a^2 = \sigma_b^2$.

7.3 Restricted Maximum Likelihood Estimation

To introduce the log-likelihood ratio tests proposed by Chen, Hwang, and Tsong (1995) under the above models, for simplicity, throughout this chapter we assume $m_{ij} = m$ for all replicates and $n_i = n$ for all i. Let $N = m \times n$,

$$y_i = (y_{i11}, \ldots, y_{i1m}, y_{i21}, \ldots, y_{i2m}, \ldots, y_{in1}, \ldots, y_{inm})'$$

be an N-vector of responses for the ith batch, X_i be an $N \times 2$ matrix whose first column is a vector of 1s and second column is

$$x_i = (x_{i1}, \ldots, x_{i1}, x_{i2}, \ldots, x_{i2}, \ldots, x_{in}, \ldots, x_{in})',$$

and D be a 2×2 diagonal matrix with $d_{11} = \sigma_a^2$ and $d_{22} = \sigma_b^2$. Then, the covariance matrix of y_i can be expressed as

$$V_i = \sigma^2 I + X_i D X_i'$$
$$= \sigma^2 I + \sigma_a^2 J + \sigma_b^2 x_i x_i',$$

where J is an $N \times N$ matrix containing 1s. The likelihood function for the data in the ith batch is then given by

$$L(y_i) = \frac{1}{(2\pi |V_i|)^{N/2}} \exp \left\{ -\frac{1}{2}(y_i - \mu_i)' V_i^{-1} (y_i - \mu_i) \right\},$$

where

$$\mu_i = (\mu_{i1}, \ldots, \mu_{in})'.$$

Thus, the log-likelihood is

$$LL(y_1, \ldots, y_K) = C - \frac{N}{2} \sum_{i=1}^{K} \log(|V_i|) \tag{7.2}$$

$$- \frac{1}{2} \sum_{i=1}^{K} (y_i - \mu_i)' V_i^{-1} (y_i - \mu_i).$$

Although maximum likelihood estimates of the parameters of the regression coefficients for fixed effects, random effects, and the variance components, can be obtained using numerical methods such as the familiar Newton-Raphson procedures, Chen, Hwang, and Tsong (1995) suggested using the EM algorithm of Dempster et al. (1977) for obtaining maximum likelihood estiamtes of the parameters of interest. This is because the EM estimates are well conditioned to lie within the parameter space, and the solution is robust to poor starting values. Details of the EM procedure are described in the next section.

Under Model 7.1, the significance of the random effects coefficients can be tested using the likelihood ratio test. Let $LL_{\max}$ be the maximum value of the log-likelihood in Equation 7.2 under a separate slopes and separate intercepts model and LL_{ab} be the maximum value of the log-likelihood with the constraints

$$a_1 = a_2 = \cdots = a_K = 0 \quad \text{and} \quad b_1 = b_2 = \cdots = b_K = 0$$

under Model 7.1 with a common slope and common intercept. Under the null hypothesis of Model 7.1, the likelihood ratio test

$$LR_{ab} = 2(LL_{\max} - LL_{ab})$$

has a chi-square distribution with $2K$ degrees of freedom. The likelihood ratio test LR_{ab} for testing a common slope model or the likelihood ratio statistic LR_a for testing a common intercept given a common slope can be computed similarly.

7.4 The EM Algorithm Procedure

The EM algorithm procedure consists of two steps: the E-step and the M-step. The E-step evaluates the expectation of the log-likelihood function, and the M-step maximizes the expectation. The two steps are repeated for obtaining maximum likelihood estimates of the parameters of interest. To apply the EM algorithm, for convenience's sake, we rewrite Model 7.1 as follows:

$$Y = X_1 \theta_1 + X_2 \theta_2 + X_3 \theta_3, \tag{7.3}$$

where θ_i is normal with mean vector zero and covariance matrix Σ_i, and $i = 1, 2, 3$. The relations between the above general form of Equation 7.3 and Model 7.1 are

$$Y' = (y_{111}, \ldots, y_{11m}, y_{121}, \ldots, y_{12m}, \ldots, y_{Kn1}, \ldots, y_{Knm}),$$
$$\theta_1' = (a_1, a_2, \ldots, a_K, b_1, b_2, \ldots, b_K),$$
$$\theta_2' = (e_{111}, \ldots, e_{11m}, e_{121}, \ldots, e_{12m}, \ldots, e_{Kn1}, \ldots, e_{Knm}),$$

where the covariance matrices are given by

$$\Sigma_1 \to \infty,$$
$$\Sigma_2 = diag(\sigma_a^2 I_K, \sigma_b^2 I_K),$$
$$\Sigma_3 = diag(\sigma^2 I_N, \sigma^2 I_N, \ldots, \sigma^2 I_N),$$

and the three design matrices are formed as

$$[X_1 | X_2 | X_3]$$
$$= \begin{bmatrix} 1 & x_1 & 1 & 0 & \cdots & 0 & x_1 & 0 & \cdots & 0 & I_N & 0 & \cdots & 0 \\ 1 & x_2 & 0 & 1 & \cdots & 0 & 0 & x_2 & \cdots & 0 & 0 & I_N & \cdots & 0 \\ \vdots & \vdots & \vdots & \vdots & \ddots & \vdots & \vdots & \vdots & \ddots & \vdots & \vdots & \vdots & \ddots & \vdots \\ 1 & x_K & 0 & 0 & \cdots & 0 & 0 & 0 & \cdots & x_K & 0 & 0 & \cdots & I_N \end{bmatrix}.$$

The joint normal distribution of Y and θ is then given by

$$\begin{bmatrix} Y \\ \theta \end{bmatrix} \sim \left(\begin{bmatrix} 0 \\ 0 \end{bmatrix}, \begin{bmatrix} Q & X\Sigma \\ \Sigma X' & \Sigma \end{bmatrix} \right), \tag{7.4}$$

where

$$X = (X_1, X_2, X_3),$$
$$\theta' = (\theta_1', \theta_2', \theta_3'),$$
$$\Sigma = diag(\Sigma_1, \Sigma_2, \Sigma_3),$$

and

$$Q = X\Sigma X'.$$

The conditional distribution of $\hat{\theta}$, given Y, is then given by

$$\hat{\theta}|Y \sim N(\theta, C), \tag{7.5}$$

where $\hat{\theta} = U'Y$, $U = Q^{-1}X\Sigma$, and $C = \Sigma - \Sigma X' Q^{-1} X\Sigma$.

Let

$$\Omega = Q_-^{-1} - Q_-^{-1} X_1 (X_1' Q_-^{-1} X_1)^{-1} X_1' Q_-^{-1},$$

where

$$Q_- = Q - X_1 \Sigma_1 X_1'.$$

Then, the $\hat{\theta}_i$'s and submatrices C_{ii}'s of C corresponding to Σ_i's in Equation 7.5 can be expressed as

$$\hat{\theta}_1 = Y'Q_-^{-1}X_1(X_1'Q_-^{-1}X_1)^{-1},$$
$$\hat{\theta}_2 = Y'\Omega X_2 \Sigma_2,$$
$$\hat{\theta}_3 = Y'\Omega X_3 \Sigma_3,$$
$$C_{11} = (X_1'Q_-^{-1}X_1)^{-1},$$
$$C_{22} = \Sigma_2 - \Sigma_2 X_2 \Omega X_2 \Sigma_2,$$
$$C_{12} = C_{11}X_1'Q_- X_2 \Sigma_2,$$
$$C_{33} = \Sigma_3 - \Sigma_3 X_3 \Omega X_3 \Sigma_3.$$

Estimation methods for the linear covariance components model were developed and illustrated by Dempster et al. (1981). These techniques include Bayesian estimation of fixed and random effects when the variances and covariances are known and point estimation of unknown variances and covariances using the EM algorithm. The EM algorithm proceeds by alternatively filling in values (E-step) for the random effects (more precisely, for the corresponding sufficient statistics), using conditional expectations given the observed Y and current estimated variances, and obtaining new estimates (M-step) from the filled-in data. This iterative procedure, in general, increases the likelihood at every step (Dempster et al. 1977). Conditions for convergence to a global maximum of the likelihood are complicated, but practical experience with variance component models has shown generally good results, although convergence may be very slow.

The E-step computes the conditional expectation of the sufficient statistic ($\theta_2'\theta_2$, $\theta_3'\theta_3$) given estimates of σ_a^2, σ_b^2, and σ^2 at the pth iteration. That is,

$$\tau_a^{(p)} = \hat{\theta}_{2a}'\hat{\theta}_{2a} + Tr(C_{221}),$$
$$\tau_b^{(p)} = \hat{\theta}_{2b}'\hat{\theta}_{2b} + Tr(C_{222}),$$
$$\tau^{(p)} = \hat{\theta}_3'\hat{\theta}_3 + Tr(C_{33}),$$

where $\hat{\theta}_{2a}$ and $\hat{\theta}_{2b}$ are the two subvectors of $\hat{\theta}_2$, and C_{221} and C_{222} are the two diagonal blocks of C_{22} corresponding to the two random components a_i and b_i. The M-step produces the maximum likelihood estimates of σ_a^2, σ_b^2, and σ^2 at the $(p+1)$th iteration. That is,

$$\sigma_a^{2, (p+1)} = \tau_a^{(p)}/K,$$
$$\sigma_b^{2, (p+1)} = \tau_b^{(p)}/K,$$
$$\sigma^{2, (p+1)} = \tau^{(p)}/(K \times N).$$

After the procedure converges, one may also obtain an estimate of fixed effect $\hat{\theta}_1$ and the conditional covariance matrix C_{11} in addition to the three variance component estimates.

The estimate (mean) of the drug characteristic at time t_i for the ith batch is

$$p(t_i) = \hat{\alpha} + \hat{\beta}t_i + \hat{a}_i + \hat{b}_i t_i \tag{7.6}$$

with the variance

$$
\begin{aligned}
V(p(t_i)) = {} & [V(\hat{\alpha}) + V(\hat{a}_i) + 2Cov(\hat{\alpha}, \hat{a}_i)] \\
& + 2[Cov(\hat{\alpha}, \hat{\beta}) + Cov(\hat{a}_i, \hat{b}_i)]t_i \\
& + [V(\hat{\beta}) + V(\hat{b}_i) + 2Cov(\hat{\beta}, \hat{b}_i)]t_i^2.
\end{aligned} \tag{7.7}
$$

It can be shown that in the balanced model, the variance and covariance terms are equal for all i in Equation 7.7. Thus, the variance of the predicted drug characteristic at time t for future batches is

$$
\begin{aligned}
V(p(t)) = {} & [V(\hat{\alpha}) + V(\hat{a}) + 2Cov(\hat{\alpha}, \hat{a})] \\
& + 2[Cov(\hat{\alpha}, \hat{\beta}) + Cov(\hat{a}, \hat{b})]t \\
& + [V(\hat{\beta}) + V(\hat{b}) + 2Cov(\hat{\beta}, \hat{b})]t^2.
\end{aligned} \tag{7.8}
$$

In the case of the unbalanced model, Chen, Hwang, and Tsong (1995) proposed the following for estimating the variance:

$$
\begin{aligned}
V(p(t)) = {} & [V(\hat{\alpha}) + \bar{V}(\hat{a}) + 2\overline{Cov}(\hat{\alpha}, \hat{a})] \\
& + 2[Cov(\hat{\alpha}, \hat{\beta}) + \overline{Cov}(\hat{a}, \hat{b})]t \\
& + [V(\hat{\beta}) + \bar{V}(\hat{b}) + 2\overline{Cov}(\hat{\beta}, \hat{b})]t^2,
\end{aligned} \tag{7.9}
$$

where

$$
\bar{V}(\hat{a}) = \frac{\sum_{ij} V(\hat{a}_i)}{\sum_j m_{ij}},
$$

$$
\bar{V}(\hat{b}) = \frac{\sum_{ij} V(\hat{b}_i)}{\sum_j m_{ij}},
$$

$$
\overline{Cov}(\hat{\alpha}, \hat{a}) = \frac{\sum_{ij} Cov(\hat{\alpha}, \hat{a}_i)}{\sum_j m_{ij}},
$$

$$
\overline{Cov}(\hat{\beta}, \hat{b}) = \frac{\sum_{ij} Cov(\hat{\beta}, \hat{b}_i)}{\sum_j m_{ij}},
$$

and

$$
\overline{Cov}(\hat{a}, \hat{b}) = \frac{\sum_{ij} Cov(\hat{a}_i, \hat{b}_i)}{\sum_j m_{ij}}.
$$

Without loss of generality, assume the drug characteristic decreases as time increases. Let η be a given lower specification limit. A 95% lower confidence bound for the predicted drug characteristic at t for future batches is

$$
\eta = \hat{\alpha} + \hat{\beta}t - t_{0.95}\sqrt{V(p(t))}. \tag{7.10}
$$

Thus, for a given η, an estimate of shelf-life is the solution of Equation 7.10.

7.5   An Example

Consider the example given in Chow and Shao (1991). The data set consists of 24 batches. The potency results were obtained at 0, 12, 24, and 36 months (Table 7.1). Based on the FDA-recommended approach, the data from individual batches support neither the hypothesis of common slope nor the hypothesis of common intercept. The minimum of the 24 shelf-life estimates is 26 months (the range of the 24 estimates is between 26 and 69). Table 7.2 contains the maximum likelihood estimates of α, β, σ_a^2, σ_b^2, and σ^2, the maximum value of the log-likelihood, and the 95% lower confidence limit estimate of shelf-life under model 1 (common slope and common intercept), model 2 (common slope but different intercepts), model 3 (separate slopes and separate intercepts), and model 4 (common variance model). The results indicate that the maximum values of log-likelihood under models 1, 2, and 3 are -196.14, -163.49, and -148.28, respectively. The maximum value under model 4 of a common variance component is -167.58, which is smaller than that under a common slope model.

TABLE 7.1:   Potency Assay Results (Percent of Label Claim)

	Age (Month)			
Batch	0	12	24	36
1	105	104	101	98
2	106	102	99	96
3	103	101	98	95
4	105	101	99	95
5	104	102	100	96
6	102	100	100	97
7	104	103	101	97
8	105	104	101	100
9	103	101	99	99
10	103	102	97	96
11	101	98	93	91
12	105	102	100	95
13	105	104	99	95
14	104	103	97	94
15	105	103	98	96
16	103	101	99	96
17	104	102	101	98
18	106	104	102	97
19	105	103	100	99
20	103	101	99	95
21	101	101	94	90
22	102	100	99	96
23	103	101	99	94
24	105	104	100	97

Source: Chow, S.C. and Shao, J. (1991). *Biometrics*, 47, 1071–1079.

TABLE 7.2: Maximum Likelihood Estimates and 95% Confidence Limits (Shelf-Life Estimates) for a Marketing Stability Data Set

Model	Coefficients		Variance Components			LL	Shelf-Life
	α	β	σ_a^2	σ_b^2	σ^2		
1	104.15(0.322)	−0.2198(0.0144)	0	0	3.5582	−196.14	59.65
2	104.15(0.365)	−0.2198(0.0089)	2.225	0	1.3800	−163.49	59.35
3	104.15(0.300)	−0.2198(0.0119)	1.536	0.0021	0.8985	−148.28	55.62
4	104.15(0.215)	−0.2198(0.0161)	0.004	0.004	1.5745	−167.58	53.75

Source: Chen, J.J, Hwang, J.S. and Tsong, Y. (1995). *Journal of Biopharmaceutical Statistics*, 5, 131–140.

The likelihood ratio test statistic LR_{ab} for the null hypothesis of a common intercept and common slope versus separate intercepts and separate slopes is

$$\chi_{46}^2 = 2(-148.278 + 196.142) = 95.728$$

with a p-value less than 0.0001. Thus, the null hypothesis is rejected. The likelihood ratio test LR_b for separate intercepts and a common slope versus separate intercepts and separate slopes is

$$\chi_{23}^2 = 2(-148.278 + 163.486) = 30.416$$

with a p-value of 0.138. If the level of significance is set to be 0.25, then the common slope model is rejected. Hence, model 3 is selected. The lower confidence limit shelf-life estimate is 55.62 months. For comparison, the likelihood ratio test statistic LR_a for a common intercept and common slope versus separate intercepts and a common slope is

$$\chi_{23}^2 = 2(-163.486 + 196.142) = 65.312$$

with a p-value less than 0.0001. Thus, a common intercept given a common slope model is rejected. The estimates of shelf-life are 59.65, 59.35, and 53.75 for models 1, 2, and 4, respectively.

7.6 Discussion

Chen, Hwang, and Tsong (1995) proposed a procedure for a marketing stability analysis using the EM algorithm. This procedure can be applied to either a balanced or an unbalanced design. The shelf-life estimate obtained from the EM approach is generally well behaved and lies within the range of shelf-life estimates of individual batches. The EM algorithm generally performs better than the commonly used Newton-Raphson procedure for unbalanced designs, although the convergence may be slow.

The simple linear mixed effects model presented in this chapter can be extended to a general linear models setting. For example, the fixed effects component $X_1\theta_1$ in Model 7.3 can include variables such as different dosage forms, strengths, and package types, in addition to the intercept and slope of the simple linear regression coefficients in Model 7.1. The random effects component $X_2\theta_2$ would include the coefficients of these added variables. The homogeneity of the degradation patterns across the dosages, strengths, and package types can be tested by the likelihood ratio procedure. However, the amount of data (number of batches and time points) required to support this model remains to be studied. Additional research is needed to determine relationships between the numbers of the random and fixed effects components, the number of batches, and the time points within each batch.

Chapter 8

Stability Analysis with Discrete Responses

For solid oral dosage forms such as tablets and capsules, the 1987 FDA stability guideline indicates that the characteristics of appearance, friability, hardness, color, odor, moisture, strength, and dissolution for tablets and the characteristics of strength, moisture, color, appearance, shape brittleness, and dissolution for capsules should be studied in stability studies. Some of these characteristics are measured based on a discrete rating scale. For example, an intensity scale of 0 (none) to 4 (severe) may be used for odor. A continuous response discredited by rounding is another example. The responses obtained from a discrete rating scale may be classified into acceptable (pass) and not acceptable (failure) categories, which results in binary stability data. Although in most stability studies, continuous responses such as potency are the primary concern, discrete responses such as appearance, color and odor should be considered for quality assurance or safety. For establishing drug shelf-life based on discrete responses, however, there is little discussion in either the FDA or the ICH stability guidelines.

The purpose of this chapter is to review statistical methods of estimating drug shelf-life with discrete stability data, according to the principle of shelf-life estimation for continuous data as described in the 1987 and 1998 FDA stability guidelines (Chow and Shao, 2003). This is useful for quality assurance of the drug product prior to the expiration date established based on a primary drug characteristic such as the strength (potency) of the drug product. In the next section we will consider the binary case without batch-to-batch variation. Section 8.2 considers shelf-life estimation in the binary case when batch-to-batch variation is present. A statistical test for batch-to-batch variation is proposed in Section 8.3. In Section 8.4 we consider an example to illustrate the proposed procedure. Section 8.5 discusses methods for ordinal responses.

8.1 Binary Data Without Batch-To-Batch Variation

Suppose there are K batches of a drug product in a stability study and that from the ith batch, y_{ij}, $j = 1, \ldots, n_i$, are binary responses observed at some time points. When there is no batch-to-batch variation, we assume that y_{ij}'s are independent and follow the following logistic regression model:

$$E(y_{ij}) = \psi(\beta' x_{ij}), \qquad (8.1)$$
$$Var(y_{ij}) = \tau(\beta' x_{ij}),$$

where $i = 1, \ldots, K$; $j = 1, \ldots, n_i$,

$$\psi(z) = \frac{e^z}{1 + e^z}$$
$$\tau(z) = \psi(z)[1 - \psi(z)],$$

and x_{ij} is a p-vector of covariates (time and other covariates such as the bottle size or container type), β is a p-vector of unknown parameters, and β' is its transpose. Typically, $x'_{ij} = (1, t_{ij})$, $(1, t_{ij}, t_{ij}^2)$, $(1, t_{ij}, w_{ij}t_{ij})$, or $(1, t_{ij}, w_{ij}, w_{ij}t_{ij})$, where t_{ij} is the jth time point for batch i and w_{ij} is a vector of covariates such as the bottle size or container type.

We now define the true (unknown) shelf-life of the drug product. Let $x(t)$ be x_{ij} with t_{ij} replaced by t and $w_{ij} = w$, a fixed particular value. The mean drug characteristic at time t (with other covariates fixed at a particular value) is $\psi(\beta'x(t))$ under Model 8.1. Assume the mean drug characteristic decreases as t increases; that is, $\beta'x(t)$ is a decreasing function of t. Since $\psi(z)$ is a strictly increasing function of z, the true shelf-life t^* satisfies $\beta'x(t^*) = \psi^{-1}(\eta)$, where η is the approved specification limit.

Under Model 8.1, $\hat{\beta}$, the maximum likelihood estimator of β, can be obtained by solving the following equation:

$$\sum_{i=1}^{K}\sum_{j=1}^{n_i} x_{ij}[y_{ij} - \psi(\beta'x_{ij})] = 0.$$

When the total number of responses $N = \sum_{i=1}^{K} n_i$ is large, $\hat{\beta} - \beta$ is approximately distributed as $N(0, V)$, where

$$V = \left[\sum_{i=1}^{K}\sum_{j=1}^{n_i} x_{ij}x'_{ij}\tau(\beta'x_{ij})\right]^{-1},$$

see, for example, Shao (1999, Section 4.4–4.5). Consequently, an approximate 95% lower confidence bound for $\beta'x(t)$ is given by

$$L(t) = \hat{\beta}'x(t) - z_{0.95}\sqrt{x(t)'\hat{V}x(t)}, \tag{8.2}$$

where z_a is the $100a$th percentile of the standard normal distribution and $\hat{V}$ is V with β replaced by $\hat{\beta}$. Following the same principle of shelf-life for continuous data as described in the 1987 FDA stability guideline and the ICH Q1A (R2) guideline for stability, we propose the following estimated shelf-life:

$$\hat{t}^* = \inf\{t : L(t) \le \psi^{-1}(\eta)\}, \tag{8.3}$$

which satisfies

$$\begin{aligned}
P_Y\left(\hat{t}^* > t^*\right) &\le P_Y(L(t^*) > \psi^{-1}(\eta)) \\
&= P_Y(L(t^*) > \beta'x(t^*)) \\
&\approx 5\%.
\end{aligned}$$

That is $\hat{t}^*$ is an approximate 95% lower confidence bound for the true shelf-life t^*, where P_Y is the probability related to the responses $y'_{ij}s$, and the approximation is based on the normal approximation to the distribution of $\hat{\beta}$ according to the central limit theorem. If $L(t)$ in Equation 8.2 is strictly decreasing in t, then $\hat{t}^*$ satisfies $L(t^*) = \psi^{-1}(\eta)$.

Unlike the continuous responses, a large number of observations are required for binary responses. In an application (such as a new drug application) with a small K, a large number of time points or some replicates at each time point are recommended. In sample size determination one may carry out some simulation studies using some initial guessing values of parameters.

8.2 The Case of Random Batches

When there is batch-to-batch variation, the parameter vector β in Model 8.1 takes different values for different batches and, thus, should be denoted by $\beta_i, i = 1, \ldots, K$. Some researchers (e.g., Ruberg and Hsu, 1992) considered β_i's as unknown fixed effects in estimating shelf-life. The fixed-effect approach, however, may not provide a shelf-life estimator that is applicable to all future batches (of the same drug product) based on stability data from the K batches. A more reasonable approach is to consider the K batches as a random sample from a population of all future batches (Shao and Chow, 1994). Consequently, β_i's are random effects. Hence, the following mixed-effect model is considered:

$$E(y_{ij}|\beta_i) = \psi(\beta'x_{ij}), \qquad (8.4)$$
$$Var(y_{ij}|\beta_i) = \tau(\beta'x_{ij}),$$

where $i = 1, \ldots, K; j = 1, \ldots, n_i$, $\beta'_i s$ are independently distributed as $N(\beta, \Sigma)$, where ψ and τ are the same as those given in Model 8.1; $E(y_{ij}|\beta_i)$ and $Var(y_{ij}|\beta_i)$ are, respectively, the conditional expectation and variance of y_{ij}, given that the random effect β_i, $\beta = E(\beta_i)$ is an unknown covariance matrix. If $\Sigma = 0$, then there is no batch-to-batch variation, and Model 8.4 reduces to Model 8.1.

Let $\psi[\beta'_{future}x(t)]$ be the mean degradation at time t for a future batch of the drug product. The true shelf-life for this batch is then given by

$$t^*_{future} = \inf\{t : \beta'_{future}x(t) \le \psi^{-1}(\eta)\},$$

which is a random variable since β_{future} is random. Consequently, a shelf-life estimator should be a 95% lower prediction bound (instead of a lower confidence bound) for t^*_{future}.

Since β_i's are unobserved random effects, the prediction bound has to be obtained based on the marginal model specified by

$$E(y_{ij}) = E[E(y_{ij}|\beta_i)] = E[\psi(\beta'_i x_{ij})],$$

and

$$
\begin{aligned}
Var(y_{ij}) &= Var[E(y_{ij}|\beta_i)] + E[Var(y_{ij}|\beta_i)] \\
&= Var[\psi(\beta_i' x_{ij})] + E[\tau(\beta_i' x_{ij})].
\end{aligned}
$$

However, neither $E(y_{ij})$ nor $Var(y_{ij})$ is an explicit function of $\beta_i' x_{ij}$, so an efficient estimator of β, such as the maximum likelihood estimator, is difficult to compute. Furthermore, when k is small (which is typically the case in stability analysis for a new drug application), the computation of the maximum likelihood estimator requires an iteration process that may not converge.

Alternatively, the following method is proposed for a small k such as 3. Assume that n_i's are large so that Model 8.1 can be fitted within each fixed batch. For each fixed i, let $\hat{\beta}_i$ be a solution of the equation

$$
\sum_{j=1}^{n_i} x_{ij}[y_{ij} - \psi(\beta_i' x_{ij})] = 0.
$$

That is, $\hat{\beta}_i$ is the maximum likelihood estimator of β_i based on the data observed from the ith batch, given β_i. For large n_i, $\hat{\beta}_i$ is approximately distributed as $N[\beta_i, V_i(\beta_i)]$, depending on β_i, where

$$
V_i(\beta_i) = \left[\sum_{j=1}^{n_i} x_{ij} x_{ij}' \tau(\beta_i' x_{ij}) \right]^{-1}.
$$

Unconditionally, $\hat{\beta}_i$ is approximately distributed as $N(\beta, D_i)$, where

$$
\begin{aligned}
D_i &= E[Var(\hat{\beta}_i|\beta_i)] + Var[E(\hat{\beta}_i|\beta_i)] \\
&\approx E[V_i(\beta_i)] + Var(\beta_i) \\
&= E[V_i(\beta_i)] + \Sigma.
\end{aligned}
$$

Let

$$
\hat{\beta} = \frac{1}{K} \sum_{i=1}^{K} \hat{\beta}_i.
$$

Then, $\hat{\beta}$ is approximately distributed as $N(\beta, K^{-1}D)$, where

$$
D = \frac{1}{K} \sum_{i=1}^{K} D_i.
$$

Define

$$
v(t) = x(t)' \left[\frac{1}{K-1} \sum_{i=1}^{K} (\hat{\beta}_i - \hat{\beta})(\hat{\beta}_i - \hat{\beta})' \right] x(t).
$$

Then,

$$\frac{\hat{\beta}'x(t) - \psi^{-1}(\eta)}{\sqrt{v(t)/K}}$$

is approximately distributed as the noncentral t distribution with $K - 1$ degrees of freedom and the noncentrality parameter of

$$\frac{\hat{\beta}'x(t) - \psi^{-1}(\eta)}{\sqrt{x(t)'\Sigma x(t)}}.$$

Following the idea of Shao and Chen (1997), Chow and Shao (2003) proposed the following approximate 95% lower prediction bound for t^*_{future} as an estimated shelf-life:

$$\hat{t}^*_{future} = \inf\left\{t : L(t) \le \psi^{-1}(\eta)\right\}, \tag{8.5}$$

where

$$L(t) = \hat{\beta}'x(t) - \rho_{0.95}(K)\sqrt{v(t)/K}, \tag{8.6}$$

and $\rho_a(K)$ satisfies

$$\int_0^1 P\left\{T_K(u) \le \rho_a(K)\right\} du = a.$$

$T_K(u)$ denotes a random variable, with the noncentral t distribution having $K - 1$ degrees of freedom and the noncentrality parameter of $\sqrt{K}\Phi^{-1}(1 - u)$; Φ is the standard normal distribution function.

Let F_{shelf} be the distribution function of the random true shelf-life $\hat{t}^*_{future}$. Then

$$\begin{aligned}
F_{shelf}(t) &= P(t^*_{future} \le t) \\
&= P(\psi(\beta'_{future}x(t) \le \eta)) \\
&= P(\beta'_{future}x(t) \le \psi^{-1}(\eta)) \\
&= \Phi\left(\frac{\psi^{-1}(\eta) - \beta'x(t)}{\sqrt{x(t)\Sigma x(t)}}\right).
\end{aligned}$$

Let $\xi_u = F^{-1}_{shelf}(u)$. Then

$$\begin{aligned}
u &= F_{shelf}(\xi_u) \\
&= \Phi\left(\frac{\psi^{-1}(\eta) - \beta'x(\xi_u)}{\sqrt{x(\xi_u)\Sigma x(\xi_u)}}\right)
\end{aligned}$$

and, thus, the noncentrality parameter of

$$\frac{\hat{\beta}'x(\xi_u) - \psi^{-1}(\eta)}{\sqrt{v(\xi_u)/K}}$$

is

$$\sqrt{K}\,\Phi^{-1}(1-u) = \frac{\hat{\beta}'x(\xi_u) - \psi^{-1}(\eta)}{\sqrt{x(\xi_u)\Sigma x(\xi_u)/K}}.$$

Then, for $\hat{t}^*_{future}$ as defined in Equations 8.5 and 8.6, we have

$$P\left(t^*_{future} < \hat{t}^*_{future}\right) = \int_0^\infty P_Y\left(t < \hat{t}^*_{future}\right) dF_{shelf}(t)$$

$$= \int_0^\infty P_Y\left\{L(t) > \psi^{-1}(\eta)\right\} dF_{shelf}(t)$$

$$= \int_0^1 P_Y\left\{L(\xi_u) > \psi^{-1}(\eta)\right\} du$$

$$= \int_0^1 P_Y\left\{\frac{\hat{\beta}'x(\xi_u) - \psi^{-1}(\eta)}{\sqrt{v(\xi_u)/K}} > \rho_{0.95}(K)\right\} du$$

$$\approx \int_0^1 P\left\{T_K(u) > \rho_{0.95}(K)\right\} du$$

$$= 0.05.$$

This shows that $\hat{t}^*_{future}$ in Equation 8.5 is indeed an approximate 95% lower prediction bound for the true shelf-life t^*_{future}. Note that $\hat{t}^*_{future}$ in Equation 8.5 has the same form as the $\hat{t}^*$ in Model 8.3 except that $L(t)$ in Model 8.2 is replaced by a more conservative bound in Equation 8.6 that incorporates the batch-to-batch variability.

8.3   Testing for Batch-To-Batch Variation

If the batch-to-batch variability is not statistically significant, it would be advantageous to combine the data from different batches and apply the shelf-life estimation procedure given in the previous section. However, combining the data from different batches should be supported by a preliminary test for batch similarity. For continuous responses, the 1987 FDA stability guideline recommends that a preliminary test for batch-to-batch variation be performed at the 25% level of significance as suggested by Bancroft (1964).

Chow and Shao (2003) proposed some tests for batch-to-batch variation to determine which of the methods described in the previous sections should be applied. Testing for batch-to-batch variation is equivalent to testing the following hypotheses:

$$H_0 : \Sigma = 0 \quad \text{versus} \quad H_a : \Sigma \neq 0.$$

First, consider the case where $x_{ij} = x_j$ and $n_i = n$ for all i (i.e., the stability designs for all batches are the same). From the results described in the previous section, under H_0, approximately

$$[V_0(\hat{\beta})]^{-1/2}(\hat{\beta}_i - \beta) \sim N(0, I_p),$$

where I_p is the identity matrix of order p and

$$V_0(\beta) = \left[\sum_{j=1}^{n} x_j x_j' \tau(\beta' x_j) \right]^{-1}.$$

Since $\hat{\beta}_i$'s are independent, under H_0, approximately

$$T_0 = \sum_{i=1}^{K} (\hat{\beta}_i - \hat{\beta})'[V_0(\hat{\beta})]^{-1}(\hat{\beta}_i - \hat{\beta}) \sim \chi^2_{p(K-1)}, \qquad (8.7)$$

where χ_r^2 denotes the chi-square distribution with r degrees of freedom. Under H_a, since

$$Var(\hat{\beta}_i) = \Sigma + E[V_i(\beta_i)].$$

$E(T_0)$ is much larger than $p(K-1)$, which is $E(T_0)$ under H_0. Therefore, a large value of T_0 indicates that H_a is true. Chow and Shao (2003) proposed that the p-value for testing batch-to-batch variation is then given by

$$1 - \chi^2_{p(K-1)}(T_0).$$

According to the 1987 FDA stability guideline (for continuous responses), to obtain an estimated shelf-life we can apply the method in Section 8.1 when the p-value is larger than or equal to 0.25 and apply the method in Section 8.2 when the p-value is smaller than 0.25.

Next, consider the general case where x_{ij}'s depend on i. Let $c = (c_1, \ldots, c_K)$ be a constant vector satisfying $\sum_{i=1}^{K} c_i = 0$. From the result in Section 8.2, under H_0, approximately

$$\left[\sum_{i=1}^{K} c_i V(\hat{\beta}) \right]^{-1/2} \sum_{i=1}^{K} c_i \hat{\beta}_i \sim N(0, I_p),$$

where

$$V_i(\beta) = \left[\sum_{j=1}^{n_i} x_{ij} x_{ij}' \tau(\beta_i' x_{ij}) \right]^{-1}.$$

Then, approximately

$$T = \sum_{i=1}^{K} c_i \hat{\beta}_i' \left[\sum_{i=1}^{K} c_i V_i(\hat{\beta}) \right]^{-1} \sum_{i=1}^{K} c_i \hat{\beta}_i \sim \chi^2_p$$

under H_0. Our proposed p-value for testing batch-to-batch variation is then

$$1 - \chi^2_p(T).$$

The constant vector c can be chosen as follows. If $K = 3$, we can choose $c = (1, 1, -2)$. If K is even, we can choose $c = (1, -1, 1, -1, \ldots, 1, -1)$. If K is odd and $K \geq 5$, we can choose $c = (1, 1, -2, 1, -1, \ldots, 1, -1)$.

8.4 An Example

A stability study was conducted on the dosage form of tablets of a drug product. Tablets from three batches were stored at room temperature (25°C) in two types of containers. In addition to the potency test, the tablets were tested for odor at 0, 3, 6, 9, 12, 18, 24, 30, and 36 months. At each time point, five independent assessments were performed. The results of the odor intensity tests were expressed as either "acceptable" (denoted by 0) or "not acceptable" (denoted by 1). Table 8.1 displays the data from the odor intensity tests.

TABLE 8.1:   Test Results for Odor Intensity

			Sampling Time (Months)								
Package	Batch	Replicate	0	3	6	9	12	18	24	30	36
Bottle	1	1	0	0	0	0	0	1	0	0	0
		2	0	0	0	0	0	0	0	0	1
		3	0	0	0	0	0	0	0	0	1
		4	0	0	0	0	0	0	0	0	0
		5	0	0	0	0	0	0	0	0	0
	2	1	0	0	0	0	0	0	0	0	1
		2	0	0	0	0	0	0	0	0	0
		3	0	0	0	0	1	0	0	0	0
		4	0	0	0	0	0	0	0	0	1
		5	0	0	0	0	0	0	0	0	0
	3	1	0	0	0	0	0	0	0	0	1
		2	0	0	0	0	0	0	0	0	0
		3	0	0	0	0	0	0	1	0	1
		4	0	0	0	0	0	0	0	0	0
		5	0	0	0	0	0	0	0	0	0
Blister	1	1	0	0	0	0	0	0	0	1	0
		2	0	0	0	0	0	0	0	0	0
		3	0	0	0	0	0	0	0	0	0
		4	0	0	0	0	0	0	0	0	0
		5	0	0	0	0	0	0	0	0	0
	2	1	0	0	0	0	0	0	0	1	0
		2	0	0	0	0	0	0	0	0	0
		3	0	0	0	0	0	0	0	0	1
		4	0	0	0	0	0	0	0	0	0
		5	0	0	0	0	0	0	0	0	0
	3	1	0	0	0	0	0	0	1	0	0
		2	0	0	0	0	0	0	0	0	0
		3	0	0	0	0	0	0	0	0	1
		4	0	0	0	0	0	0	0	0	1
		5	0	0	0	0	0	0	0	0	0

With $x'_j = (1, t_j, w_j t_j)$, where w_j is the indicator for container type, the statistic T_0 given in Equation 8.7 is equal to 0.5035 based on the data in Table 8.1, which results in a p-value of 0.9978, that is, the batch-to-batch variation is not significant. Thus, the shelf-life estimation procedure described in Section 8.1 should be applied by combining data in different batches. $L(t)$ in Equation 8.2 was computed. For a given η, the intersect of the horizontal line and $L(t)$ gives the shelf-life of the drug product. For $\eta = 90\%$, the estimated shelf-life is about 21 months for container type 1 and 22 months for container type 2.

8.5 Ordinal Responses

In this section we consider the situation where y_{ij} is an ordinal response with more than two categories. We introduce the following three approaches.

8.5.1 Generalized Linear Models

Suppose a parametric model (conditional on the covariate) can be obtained for the response y_{ij}. For example, y_{ij} follows a binomial distribution or a (truncated) Poisson distribution, given x_{ij}. Then, Model 8.1 (which is a generalized linear model) still holds with a proper modification of the variance function τ. For example, if y_{ij} is binomial with values $0, 1, \ldots, m$, then $\tau(z) = m\psi(z)[1 - \psi(z)]$. Consequently, the results described in the previous sections can still be applied with the modification of the function τ. Under this approach, it is assumed that the mean of the discrete response is an appropriate summary measure for the stability analysis.

8.5.2 Threshold Approach (Multivariate Generalized Models)

Suppose the ordinal response y is a categorized version of a latent continuous variable U. For example, if y is a grouped continuous response, then U may be the unobserved underlying continuous variable. Suppose the relationship between y and U is determined by

$$y = r \quad \text{if and only if} \quad \theta_r < U \leq \theta_{r+1}, \quad r = 0, 1, \ldots, m,$$

where $\theta_0 = -\infty$, $\theta_m = \infty$, and θ_r, $r = 1, \ldots, m - 1$ are unknown parameters. Assume further that the latent variable U and the covariate x follow a linear regression model,

$$U = -\beta'x + \epsilon,$$

where ϵ is a random error with distribution function F. Then,

$$P(y \leq r | x) = F(\theta_{r+1} + \beta'x), \quad r = 0, 1, \ldots, m.$$

This is referred to as the threshold approach (Fahrmeir and Tutz, 1994). If $F(z) = 1/(1 + e^{-z/\sigma})$ (the logistic distribution with mean 0 and variance $\sigma^2 \pi^2/3$), then

$$P(y \leq r|x) = \frac{e^{(\theta_{r+1} + \beta'x)/\sigma}}{1 + e^{(\theta_{r+1} + \beta'x)/\sigma}} = \psi(\tilde{\theta}_{r+1} + \tilde{\beta}'x),$$

where $\tilde{\theta}_r = \theta_r/\sigma$ and $\tilde{\beta} = \beta/\sigma$.

Consider the case of no batch-to-batch variation. Suppose the shelf-life of the drug product is defined to be the time interval that the mean of U remains above the specification $\psi^{-1}(\eta)$. Let $y_{ij}^{(r)} = I(y_{ij} \leq r)$, where $I(\cdot)$ is the indicator function. Then, an extension of Model 8.1 is

$$E(y_{ij}^{(r)}) = \psi(\tilde{\beta}'x_{ij}), \quad i = 1, \ldots, K, \tag{8.8}$$

$$Var(y_{ij}^{(r)}) = \tau(\tilde{\beta}'x_{ij}), \quad j = 1, \ldots, n_i,$$

$r = 0, 1, \ldots, m - 1$. Since $y_{ij}^{(r)}$, $r = 0, 1, \ldots, m$ are dependent, Equation 8.8 is a multivariate generalized linear model. Maximum likelihood estimation can be carried out as described in Section 3.4 of Fathrmeir and Tutz (1994). Let $L(t)$ be an approximate 95% lower confidence bound for $\tilde{\beta}'x(t)$ based on the maximum likelihood estimation. Then, the estimated shelf-life can still be defined by Equation 8.3. A combination of this approach and the method in Section 8.2 can be used to handle the case where random batch-to-batch variation is present.

8.5.3 Binary Approach

The threshold approach involves some complicated computation, since a multivariate generalized linear model has to be fitted. A simple but not very efficient approach is to binarize the ordinal responses. Let r_0 be a fixed threshold and define $\tilde{y}_{ij} = 0$ if $y_{ij} \leq r_0$ and 1 otherwise. If the ordinal response y follows the model described in the threshold approach, then

$$P(y \leq r_0|x) = F(\theta_{r_0+1} + \beta'x) = \psi(\tilde{\theta}_{r_0+1} + \tilde{\beta}'x)$$

and Model 8.1 holds for the summary data set $\{\tilde{y}_{ij}\}$. Thus, we can apply the methods described in the previous sections.

8.6 Concluding Remarks

In practice, the appearance, color, and odor of a drug product may not have a direct impact on the potency (strength) of a drug product. However, changes in appearance, color, and odor may be an indication that a significant change in potency has occurred. If this occurs, it is suggested that a retest of stability of the same batch be conducted to confirm that the potency of the drug product remains within approved specifications prior to its expiration dating period. As indicated earlier, stability tests for appearance, color, and odor (discrete responses) are often conducted for quality assurance and quality control. The possible causes (related or unrelated to the degradation of the drug product) of significant changes in appearance, color, or odor should be identified, controlled, and removed before the drug product is released for marketplace.

Unlike stability tests for potency, we consider an approximate 95% lower prediction bound for t^*_{future} rather than an approximate 95% lower confidence bound for t^*_{future} as an estimated shelf-life for some secondary characteristics of drug products such as appearance, color, and odor. These characteristics may be related to one another. Individual estimates of drug shelf-lives based on these discrete responses may provide a much shorter drug shelf-life. In this case it is suggested that either (a) the quality of the drug product in terms of the discrete response be improved or (b) the product specification of the discrete response be adjusted to achieve the established drug shelf-life estimated based on the strength of the drug product.

Chapter 9

Stability Analysis with Multiple Components

In the previous chapters we only considered a drug product with a single active ingredient. In practice, while most drug products consist of a single active ingredient, some drug products contain multiple active ingredients (see e.g., Pong and Raghavarao (2001) Chow, Pong, and Chang, 2006). For example, Premarin (conjugated estrogens, USP) is known to contain at least five active ingredients: estrone, equilin, 17α-dihydroequilin, 17α-estradiol, and 17β-dihydroequiliin. Other examples include combinational drug products, such as the traditional Chinese medicines (see, e.g., Stefan and Chantal, 2005; Chow, Pong, and Chang, 2006). For a drug product with multiple active ingredients (or components), an ingredient-by-ingredient (or component-by-component) stability analysis may not be appropriate, since these active ingredients may have some unknown interactions.

In the next section the basics of obtaining an estimate of drug shelf-life for drug products with multiple components is described. The model and assumptions are given in Section 9.2. In Section 9.3 we introduce the statistical method proposed by Chow and Shao (2007) for estimating drug shelf-life for drug products with multiple components. An example concerning a traditional Chinese medicine is given in Section 9.4 to illustrate the described statistical method. A brief discussion is given in the last section of this chapter.

9.1 Basic Idea

Let $y(t, k)$ be the potency of the kth ingredient at time t after the manufacture of a given drug product, $k = 1, \ldots, p$. For ingredient k, its shelf-life is the time interval at which $E[y(t, k)]$ (the expectation of $y[t, k]$) remains within a specified limit, whereas the shelf-life for the drug product may be the time interval at which $E[f(y(t, 1), \ldots, y(t, p))]$ remains within the specified limits, where f is a function (such as a linear combination of $y[t, 1], \ldots, y[t, p]$) that characterizes the impact of all active ingredients. In general, f is a vector-valued function with a dimension $q \leq p$.

If data are observed from $y(t, 1), \ldots, y(t, p)$ and the function f in the previous discussion is a known function, then stability analysis can be made by using the transformed data $z(t) = f[y(t, 1), \ldots, y(t, p)]$. If the dimension of f is 1, then $z(t)$ can be treated as a single ingredient. If the dimension of f is $q > 1$, then one may define the shelf-life to be the minimum of the shelf-lives $\tau_1, \ldots, \tau_q$, where τ_h is the shelf-life when the hth component of $z(t)$ is treated as a single ingredient. One special case is where f is the identity function so that the shelf-life is the minimum of all shelf-lives corresponding to different ingredients $y(t, k)$, $k = 1, \ldots, p$.

In practice, however, f is typically unknown. Although the best way to estimate f is to fit a model between the y and z variables, it requires data observed from both y and z, which is not a common practice in the pharmaceutical industry, because the variable z is not clearly defined in many problems, such as the traditional Chinese medicines (see, e.g., Chow, Pong, and Chang 2006). Chow and Shao (2007) assumed that the components of z are linear combinations of the components of y and proposed a method to establish the shelf-life. Note that Chow and Shao's approach is basically an application of the factor model in multivariate analysis (see, e.g., Johnson and Wichern, 1998).

9.2   Models and Assumptions

Let $y(t)$ denote the p dimensional vector whose kth component is the potency of the kth ingredient at time t after the manufacture of a given drug product, $k = 1, \ldots, p$. We assume that the drug potency is expected to decrease with time. If $p = 1$, that is, $y(t)$ is univariate, the current established procedure to determine a shelf-life is to use the time at which a 95% lower confidence bound for the mean degradation curve $E[y(t)]$ intersects the acceptable lower product specification limit as specified in the 1987 FDA stability guideline (see also ICH Q1A (R2), 2003). Let η be the vector whose kth component is the lower product specification limit as specified in the USP/NF for the kth component of $y(t)$.

Assume that, for any t,

$$y(t) - E[y(t)] = LF_t + \varepsilon_t, \tag{9.1}$$

where L is a $p \times q$ nonrandom unknown matrix of full rank, F_t and ε_t are unobserved independent random vectors of dimensions q and p, respectively, $E(F_t) = 0$, $\mathrm{Var}(F_t) = I_q$ (the identity matrix of order q), $E(\varepsilon_t) = 0$, $\mathrm{Var}(\varepsilon_t) = \Psi$, and Ψ is an unknown diagonal matrix of order p. Note that Model 9.1 with the assumptions on F_t and ε_t is the so-called orthogonal factor model (see Johnson and Wichern, 1998). If ε_t is treated as a random error, then Model 9.1 assumes that the p dimensional ingredient vector $y(t)$ is governed by a q dimensional unobserved vector F_t. Normally q is much smaller than p.

Let $z(t) = (L'L)^{-1}L'[y(t) - \eta]$. It follows from Model 9.1 that

$$z(t) - E[z(t)] = F_t + (L'L)^{-1}L'\varepsilon_t. \tag{9.2}$$

If L is known, then Model 9.2 suggests performing a stability analysis based on the transformed data observed from $z(t)$. In practice, since L is unknown, if we can estimate L based on Model 9.1 and the observed data from $y(t)$, then we can carry out a stability analysis using the transformed $z(t)$ with L replaced by its estimate.

Let $x(t)$ be an s dimensional covariate vector associated with $y(t)$ at time t. For example, $x(t) = (1, t)'$ $(s = 2)$ or $x(t) = (1, t, t^2)'$ $(s = 3)$. We assume the following model at any time t:

$$E[y(t) - \eta] = Bx(t), \quad \text{Var}[y(t)] = \Sigma, \quad i = 1, \ldots, m, \quad j = 1, \ldots, n, \tag{9.3}$$

where B is a $p \times s$ matrix of unknown parameters and $\Sigma > 0$ is an unknown $p \times p$ positive definite covariance matrix. Since $z(t) = (L'L)^{-1}L'[y(t) - \eta]$, it follows from Model 9.3 that

$$E[z(t)] = \gamma' x(t), \quad i = 1, \ldots, m, \quad j = 1, \ldots, n, \tag{9.4}$$

where $\gamma = B'L(L'L)^{-1}$.

9.3 Shelf-Life Determination

Suppose we independently observe data y_{ij}, $i = 1, \ldots, m$, $j = 1, \ldots, n$, where y_{ij} is the jth replicate of $y(t_i)$ and $t_1, \ldots, t_m$ are designed time points for the stability analysis. Define

$$x_i = x(t_i), \quad z_{ij} = (L'L)^{-1}L'(y_{ij} - \eta), \quad i = 1, \ldots, m, \quad j = 1, \ldots, n \tag{9.5}$$

First consider the case of $q = 1$; that is, z_{ij} in Model 9.5 is univariate. If z_{ij}'s are observed, then an approximate 95% lower confidence bound for $E[z(t)] = \gamma' x(t)$ is

$$l(t) = \hat{\gamma}' x(t) - t_{0.95,mn-s} \hat{\sigma} \sqrt{D(t)} \tag{9.6}$$

where $\hat{\gamma}$ is the least squares estimator of γ in Model 9.4 based on data from z_{ij}'s and x_i's, $\hat{\sigma}^2$ is the usual sum of squared residuals divided by its degrees of freedom $mn - s$, $t_{0.95,mn-s}$ is the 95th percentile of the t-distribution with degrees of freedom $mn - s$, and

$$D(t) = \left[n \sum_{i=1}^{m} x(t)' x_i x_i' x(t) \right]^{-1}.$$

Hence, if z_{ij}'s are observed, a shelf-life according to the 1987 FDA stability guideline is

$$\tau = \inf\{t : l(t) \leq 0\}. \tag{9.7}$$

In our problem y_{ij}'s, not z_{ij}'s, are observed. Hence, the lower confidence bound $l(t)$ in Equation 9.6 needs to be modified. Since $\gamma' = (L'L)^{-1}L'B$, we can obtain an estimator of γ in two steps. At the first step we use Model 9.3, observed data y_{ij}'s and x_i's, and the multivariate linear regression (see, e.g., Johnson and

Wichern, 1998) to obtain a least squares estimator $\hat{B}$ of B. At the second step we consider the orthogonal factor model of Model 9.1 and apply the method of principal components (see, e.g., Johnson and Wichern, 1998) to obtain an estimator $\hat{L}$ of L, using $y_{ij} - \eta - \hat{B}x_i, i = 1, \ldots, m, j = 1, \ldots, n$. More precisely, $\hat{L}$ is the normalized eigenvector corresponding to the largest eigenvalue of the sample covariance matrix based on $y_{ij} - \hat{B}x_i, i = 1, \ldots, m, j = 1, \ldots, n$. Let $\hat{\gamma} = \hat{B}'\hat{L}(\hat{L}'\hat{L})^{-1}$. The lower confidence bound in Equation 9.6 is modified to

$$l(t) = \hat{\gamma}'x(t) - t_{0.95, mn-s}\sqrt{x(t)'Vx(t)}, \tag{9.8}$$

where V is the jackknife variance estimator of $\hat{\gamma}$ (see, e.g., Shao and Tu, 1995); that is,

$$V = \frac{mn-1}{mn}\sum_{i=1}^{m}\sum_{j=1}^{n}(\hat{\gamma}_{i,j} - \hat{\gamma})(\hat{\gamma}_{i,j} - \hat{\gamma})',$$

where $\hat{\gamma}_{i,j}$ is the estimator of γ calculated using the same method as in the calculation of $\hat{\gamma}$ but with the (i, j)th data point deleted.

The result for $q = 1$ is sufficient for applications with a small or moderate p. When p is large, we propose the following procedure with $1 < q < p$. Let $\hat{B}$ be defined as before, λ_k be the kth largest eigenvalue of the sample covariance matrix based on $y_{ij} - \eta - \hat{B}x_i, i = 1, \ldots, m, j = 1, \ldots, n$, and e_k be the normalized eigenvector corresponding to λ_k. Then, our estimator $\hat{L}$ of L is the $p \times q$ matrix whose kth column is $\lambda_k e_k, k = 1, \ldots, q$. Our estimator of γ is still $\hat{\gamma} = \hat{B}'\hat{L}(\hat{L}'\hat{L})^{-1}$, which is an $s \times q$ matrix. Let $\hat{\gamma}_k$ be the kth column of $\hat{\gamma}, k = 1, \ldots, q$,

$$l_k(t) = \hat{\gamma}_k'x(t) - t_{1-0.05/q, mn-s}\sqrt{x(t)'V_kx(t)}, \tag{9.9}$$

and

$$V_k = \frac{mn-1}{mn}\sum_{i=1}^{m}\sum_{j=1}^{n}\left(\hat{\gamma}_{k,i,j} - \hat{\gamma}_k\right)\left(\hat{\gamma}_{k,i,j} - \hat{\gamma}_k\right)',$$

where $\hat{\gamma}_{k,i,j}$ is the same as $\hat{\gamma}_k$ but calculated with the (i, j)th data point deleted. Then, $l_k(t), k = 1, \ldots, q$ are approximate 95% simultaneous lower confidence bounds for $\zeta_k(t), k = 1, \ldots, q$, where $\zeta_k(t)$ is the kth component of $E[z(t)] = \gamma'x(t)$. An approximate level 95% shelf-life for the drug product (when the sample size mn is large) is

$$\tau = \min_{k=1,\ldots,q} \tau_k,$$

where each τ_k is defined by the right-hand side of Equation 9.7 with $l(t)$ replaced by $l_k(t)$ and is a shelf-life for the kth component of z with confidence level $(1 - 0.05/q)\%$.

9.4   An Example

To illustrate the proposed method for determining the shelf-life of a drug product with multiple active ingredients, consider a stability study conducted for a traditional

TABLE 9.1: List of Components of a Traditional Chinese Medicine

Component	Formulation
HE	60 mg
B	25 mg
C	25 mg
Excipient	90 mg
Total	200 mg

Chinese herbal medicine, which is newly developed for treatment of patients with rheumatoid arthritis. This medicine contains three active botanical components, namely *Herba epimedii* (HE), B extract, and C extract. Each of the three components has been used as an herbal remedy since ancient times and is well documented in the Chinese Pharmacopeia. The proportions of each components are summarized in Table 9.1.

To establish a shelf-life for this product, a stability study was conducted for a time period of 18 months under a testing condition of 25°C/60% relative humidity. The lower product specification limit for each component is 90%.

Stability data (percent of label claim) at each sampling time point for the three components are given in Table 9.2.

Since $p = 3$, we consider that $q = 1$. Using the proposed procedure described in the previous sections, we obtain $l(t)$ in Equation 9.8 for various t (month), which are given in Table 9.3.

Hence, the estimated shelf-life for this product is 27 months.

TABLE 9.2: Stability Data of the Traditional Chinese Medicine

Component	Sampling Time Point (Month)					
	0	3	6	9	12	18
HE	99.6	97.5	96.8	96.2	94.8	95.3
	99.7	98.3	97.0	96.0	95.1	94.8
	100.2	99.0	98.2	97.1	95.3	94.6
B	99.5	98.4	96.3	95.4	93.2	91.0
	100.5	98.5	97.4	94.9	94.5	92.1
	99.3	99.0	97.3	95.0	93.1	91.5
C	100.0	99.5	98.9	98.2	97.9	97.5
	99.8	99.4	99.0	98.5	98.0	97.9
	101.2	99.9	100.3	99.5	98.9	98.0

TABLE 9.3: $\ell(+)$ Values for Various t

t	19	20	21	22	23	24	25	26	27	28
$l(t)$	4.97	4.36	3.75	3.14	2.52	1.90	1.28	0.66	0.03	−0.60

9.5 Discussion

This chapter introduces the method for determining the shelf-life of a drug product with p active ingredients proposed by Chow and Shao (2007). Basically, Chow and Shao (2007) assume that these active ingredients are linear combinations of q factors. Since these factors are chosen using principal components, the first factor can be viewed as the primary active factor, and the second factor can be viewed as the secondary active factor. Chow and Shao (2007) assume that active ingredients decrease with time. If one or more ingredients increase with time, then a transformation such as $g(y) = -y$ or $g(y) = 1/y$ may be applied. If p is small or moderate, then $q = 1$ is recommended. If p is large, then adding a few more factors may be considered. Since the principal components are orthogonal, adding more factors will not affect the previous selected factors (except that $t_{0.95,mn-s}$ is changed to $t_{1-0.05/q,mn-s}$) so that one can compare the results in a sensitivity analysis. Finally, adding more factors always results in a more conservative procedure.

Note that in their proposed approach, Chow and Shao (2007) assume that there is no significant toxic degradant in the test drug product with multiple components. This is a reasonable assumption for most traditional Chinese medicines since multiple ingredients are used to reduce toxicities when used in conjunction with primary therapy. However, when toxic degradation products are detected, special attention should be paid to: (a) identity (chemical structure), (b) cross reference to information about biological effects and the significance of the concentration likely to be encountered, and (c) indications of pharmacological action or inaction as indicated in the FDA guidelines for stability analysis. Chow and Shao's approach is useful when different ingredients degrade not independently of each other, which is the case for most traditional Chinese medicines. If multiple ingredients degrade independently, then an ingredient-by-ingredient analysis may be appropriate. If Chow and Shao's approach is applied, it is suggested that q be selected as $q = 1$ or $q = 2$ factors that are ingredients having the most variability.

Chapter 10

Stability Analysis with Frozen Drug Products

Unlike most drug products, some drug products must be stored at specific temperatures, such as $-20°C$ (frozen temperature), $5°C$ (refrigerator temperature), and $25°C$ (room temperature), to maintain stability until use (Mellon, 1991). Drug products of this kind are usually referred to as *frozen* products. Unlike the other drug products, a typical shelf-life statement for frozen drug products usually consist of multiple phases with different storage temperatures. For example, a commonly adopted shelf-life statement for frozen products could be 24 months at $-20°C$ followed by 2 weeks at $5°C$. As a result, the drug shelf-life is determined based on a two-phase stability study. However, no discussion of the statistical methods for estimating two-phase shelf-life is available in either the FDA stability guidelines or the ICH stability guidelines. Mellon (1991) suggested that data obtained from the two-phase stability study be analyzed separately to obtain a combined shelf-life for the frozen products. Mellon's method does not account for the fact that stability at the second phase may depend on the stability at the first phase. That is, an estimated shelf-life at the second phase following 3 months of the first phase may be longer than that following 6 months of the first phase. To overcome this problem, Shao and Chow (2001a) proposed a method for a two-phase stability study using a two-phase linear regression based on the statistical principle described in both the FDA and ICH stability guidelines.

In the next section, the concept of a two-phase stability study is introduced, followed by statistical methods for two-phase shelf-life estimation (Section 10.2). An example is given in Section 10.3 to illustrate the use of the described method. Section 10.4 provides a brief discussion of two-phase shelf-life estimation in stability analysis.

10.1 Two-Phase Stability Study

In the pharmaceutical industry two-phase stability studies are usually conducted to characterize degradation of frozen drug products. The first phase is to determine drug shelf-life under a frozen storage condition such as $-20°C$, and the second phase is to estimate drug shelf-life under a refrigerated condition. A first-phase stability study is usually referred to as a frozen study, and a second-phase stability study is known as a thawed study. The frozen study is usually conducted like a regular long-term stability study, except that the drug is stored in frozen conditions. In other words, stability testing will be normally conducted at 3-month intervals during the first year, 6-month intervals during the second year, and annually thereafter. Stability testing for the thawed study follows the stability testing for the frozen study. It may be performed at 1-week (or 1-day) intervals up to several weeks.

Concerns in the design and analysis of two-phase stability studies for frozen products are as follows (Mellon, 1991):

- When are the best times to assay?

- How many assays should be made?

- Does frozen time affect thawed degradation?

- Does concentration level at time points during the frozen state (i.e., strength) affect degradation during the thawed state?

- Do different lots affect degradation?

Mellon (1991) also provided a number of approaches that may be used for the analysis of frozen drug products:

- Cell means model

- Estimation of trends separately for each lot and temperature

- One grand regression, including lot and temperature

- Regression of frozen and thawed data separately

With separate analyses for frozen and thawed studies, Mellon (1991) suggested that the following approximate confidence intervals be obtained:

- Bonferroni intervals

- Confidence intervals based on asymptotic theory

- Confidence intervals using bootstrap or jackknife procedures

- Confidence intervals using the Satterthwaite approximation

The most important issue for determining shelf-life for frozen drug products is that the same acceptable lower specification limit cannot be used for both frozen and thawed studies. The strength of a frozen drug product at the time it is to be administered at room temperature is very likely to be below the acceptable lower specification limit if it is used to establish the shelf-life for the frozen drug product. Second, determination of the shelf-life for refrigerated and ambient conditions should be analyzed separately from that for frozen conditions. The shelf-life for frozen conditions, measured in months, is usually much longer than that under refrigerated or ambient conditions, measured in either weeks or days. Therefore, the design for the thawed study should have different measures for time points than the frozen study. If one uses only one regression model to fit the data from both states, the resulting estimated shelf-lives under refrigerated and ambient conditions may not be reliable, owing to rapid degradation and a much shorter time interval for the thawed study.

As indicated earlier, unlike the usual stability studies, the stability at the second phase (thawed study) may depend on the stability at the first phase (frozen study). Thus, it is very likely that an estimated shelf-life from the thawed study following 3 months of frozen storage may be different from that following 6 months of frozen storage. As a result, design strategies for selecting the sampling time points as well as the sample size in the second phase need to be further studied. Shao and Chow (2001a) suggested having shorter sampling time intervals for the second phase in later months to have a more accurate assessment of the degradation of the drug products. In other words, collect assay more frequently in later weeks.

10.2 Stability Data and Model

For estimating the first-phase shelf-life, we have the following stability data

$$y_{ik} = \alpha + \beta t_i + e_{ik},$$

where y_{ik} is the drug characteristic of interest (e.g., potency, dissolution, etc.), $i = 1, \ldots, I \geqslant 2$; $k = 1, \ldots, K_i \geqslant 1$; α and β are unknown parameters, and e_{ik}'s are independent and identically distributed random errors with mean 0 and variance $\sigma_i^2 > 0$. Typically, $t_i = 0, 3, 6, 9, 12,$ and 18 months. Thus, the total number of data for the first phase is

$$n_1 = \sum_i K_i,$$

which equals IK if $K_i = K$ for all i. Since stability studies are usually conducted under well-controlled conditions, the error variance σ_i^2 is expected to be small. Thus, the sample size n_1 is often not very large.

At time t_i, $K_{ij} \geq 1$ second-phase stability data are collected at time intervals t_{ij}, $j = 1, \ldots, J \geq 2$. The total number of data for the second phase is

$$n_2 = \sum_i \sum_j K_{ij},$$

which equals IJK if $K_{ij} = K$ for all i and j. Data from the two phases are independent. Typically,

$$t_{ij} = t_i + s_j$$

where s_j is time in the second phase and $s_j = 1, 2, 3$ days (or weeks) and so on. Since the degradation lines for the two phases intersect, the intercept of the second-phase degradation line at time t is $\alpha + \beta t$. Let $\gamma(t)$ be the slope of the second-phase degradation line at time t. Then, at time t_i, $i = 1, \ldots, I$, we have second-phase stability data

$$y_{ijk} = \alpha + \beta t_i + \gamma(t_i)s_j + e_{ijk},$$

where e_{ijk}'s are independent and identically distributed random errors with mean 0 and variance $\sigma_{2i}^2 > 0$. For simplicity, we assume that $\gamma(t)$ is a polynomial in t with unknown coefficients. Typically, $\gamma(t)$ could be

- $\gamma(t) = \gamma_0$ (common slope model)

- $\gamma(t) = \gamma_0 + \gamma_1 t$ (linear trend model)

- $\gamma(t) = \gamma_0 + \gamma_1 t + \gamma_2 t^2$ (quadratic trend model)

In general, $\gamma(t)$ could be expressed as

$$\gamma(t) = \sum_{h=0}^{H} \gamma_h t^h, \tag{10.1}$$

where γ_h's are unknown parameters and

$$H + 1 < \sum_j K_{ij}, \quad \text{for all } i \text{ and } H < I.$$

In the pharmaceutical industry drug products are usually manufactured in different batches and the FDA requires testing of at least three batches, and preferably more, in a stability analysis to account for batch-to-batch variation. Thus, typically,

$$K_i = K_{ij} = K$$

is the number of batches tested in a stability analysis. According to the FDA, if the p-value in testing batch-to-batch variation is larger than 0.25, then batch-to-batch variation can be ignored and the analysis can be done by treating data from different batches as replicates; otherwise, the shelf-life for the drug product is the minimum of the shelf-lives of different batches, where each shelf-life for a batch is obtained using data from the given batch only.

10.2.1 Estimating the First-Phase Shelf-Life

Following the 1987 FDA stability guideline and the ICH Q1A (R2) guideline for stability, the first-phase shelf-life can be estimated based on the first-phase data, that is, $\{y_{ik}\}$ as the time point at which the lower product specification limit intersects the 95% lower confidence bound of the mean degradation curve (FDA, 1987; ICH Q1A [R2], 2003). We first consider the case where there is no batch-to-batch variation (or the batch-to-batch variation can be ignored). Let $\hat{\alpha}$ and $\hat{\beta}$ be the least squares estimators of α and β based on the first phase data. Now, let

$$L(t) = \hat{\alpha} + \hat{\beta}t - t_{0.95, n_1 - 2}\sqrt{v(t)}$$

be the 95% lower confidence bound for $x + \beta t$, where $t_{0.95,m}$ is the 95th quantile of the t-distribution with m degrees of freedom and

$$v(t) = \hat{\sigma}_1^2 \left[\frac{1}{n_1} + \frac{(t - \bar{t})^2}{\sum_i K_i (t_i - \bar{t})^2} \right],$$

$$\bar{t} = \frac{1}{n_1} \sum_i K_i t_i,$$

and

$$\hat{\sigma}_1^2 = \frac{1}{n_1 - 2} \sum_{i,k} (y_{ik} - \hat{\alpha} - \hat{\beta} t_i)^2,$$

which is the usual error variance estimator based on residuals. Suppose the lower limit for the drug characteristic is η (we assume that $x + \beta t$ decreases as t increases). Then, the first phase shelf-life is the first solution of $L(t) = \eta$, that is

$$\hat{t} = \inf\{t : L(t) \leq \eta\}.$$

The first-phase shelf-life is constructed so that

$$P\{\hat{t} \leq \text{ the true first-phase shelf-life}\} = 95\% \qquad (10.2)$$

assuming that e_{ik}'s are normally distributed. Without the normality assumption, the above result approximately holds for small σ_1^2/n_1 (large sample size n_1 or small error variance σ_1^2). In practice, $\hat{t}$ may be much larger than the study range, the maximum of t_i's in the study. The validity of the shelf-life $\hat{t}$ depends on the validity of the statistical model assumption for t-values beyond the study range. Usually, the FDA allows an estimated shelf-life beyond the study range for up to 6 months. Unless the drug product is relatively stable, the FDA may question an estimated shelf-life of more than 6 months beyond the study range.

If there is batch-to-batch variation, which is a fixed (nonrandom) effect, then the shelf-life is

$$\hat{t} = \min_k \hat{t}_k,$$

where $\hat{t}_k$ is the shelf-life obtained using data from the kth batch and the method described in Chapter 3. When there is batch-to-batch variation, the methods described in Chapter 5 (fixed batch effects) and Chapter 6 (random batch effects) are useful for establishing the first-phase shelf-life.

10.2.2   Estimating the Second-Phase Shelf-Life

We now consider the second-phase shelf-life, under Equation 10.1. Again, we first consider the case where there is no batch-to-batch variation (or the batch-to-batch variation can be ignored). Since the slope of the second-phase degradation line may vary with t, we estimate the slope at time t_i using the second-phase data at t_i:

$$b_i = \frac{\sum_{j,k}(s_j - \bar{s}) y_{ijk}}{\sum_{j,k}(s_j - \bar{s})^2},$$

where s_j's are the second-phase time intervals, and $\bar{s}$ is the average of s_j's. Since $H < I$, the unknown parameters γ_h in Equation 10.1 can be estimated by the least squares estimators under the following model:

$$b_i = \sum_{h=0}^{H} \gamma_h t_i^h + error, \tag{10.3}$$

that is, the parameter vector $\gamma = (\gamma_0, \gamma_1, \ldots, \gamma_H)'$ can be estimated by

$$\hat{\gamma} = (W'W)^{-1} W'b, \tag{10.4}$$

where A' is the transpose of a vector or matrix A, that is,

$$b = (b_1, b_2, \ldots, b_I)',$$
$$W' = [l(t_1), \ldots, l(t_I)],$$

and

$$l(t) = (1, t, \ldots, t^H)'.$$

Consequently, the slope $\gamma(t)$ can be estimated by

$$\hat{\gamma}(t) = [l(t)]'\hat{\gamma} = [l(t)]'(W'W)^{-1}W'b.$$

The covariance matrix of $\hat{\gamma}$ is given by

$$V(\hat{\gamma}) = (W'W)^{-1} W' \Sigma W (W'W)^{-1},$$

where Σ is an $I \times I$ diagonal matrix whose ith diagonal element is

$$\frac{\sigma_{2i}^2}{\sum_{j,k}(s_j - \bar{s})^2}.$$

Each σ_{2i}^2 can be estimated by

$$\hat{\sigma}_{2i}^2 = \frac{1}{\sum_j K_{ij} - (H+2)} \sum_{j,k} [y_{ijk} - \bar{y}_i - b_i(s_j - \bar{s})]^2,$$

where $\bar{y}_i$ is the average of y_{ijk} over all j and k. Under the assumption of equal variances (i.e., $\sigma_{2i}^2 = \sigma_2^2$) for all i, σ_2^2 can be estimated by the combined estimator

$$\hat{\sigma}_2^2 = \frac{1}{n_2 - I(H+2)} \sum_{i,j,k} [y_{ijk} - \bar{y}_i - b_i(s_j - \bar{s})]^2. \tag{10.5}$$

Let $\hat{\Sigma}$ be the same as Σ but with σ_{2i}^2 replaced by $\hat{\sigma}_{2i}^2$ or $\hat{\sigma}_2^2$. Then $\hat{\gamma}$ and its estimated covariance matrix $(W'W)^{-1} W' \hat{\Sigma} W (W'W)^{-1}$ can be used to form approximate t-tests to select a polynomial model in Equation 10.1. The variance of $\hat{\gamma}$ can be estimated by

$$\hat{V}[\hat{\gamma}(t)] = [l(t)]'(W'W)^{-1} W' \hat{\Sigma} W (W'W)^{-1}[l(t)].$$

For fixed t and s, let

$$v(t, s) = v(t) + \hat{V}[\hat{\gamma}(t)]s^2,$$

and

$$L(t, s) = \hat{\alpha} + \hat{\beta}t + \hat{\gamma}(t)s - t_{0.95, n_1 + n_2 - 2 - I(H+2)} \sqrt{v(t, s)}.$$

For any fixed t less than the first-phase true shelf-life, that is, t satisfying $\alpha + \beta t > \eta$, the second-phase shelf-life can be estimated as

$$\hat{s}(t) = \inf\{s \geq 0 : L(t, s) \leq \eta\}$$

If $L(t, s) < \eta$ for all s, then $\hat{s}(t) = 0$. That is, if the drug product is taken out of the first-phase storage condition at time t, then the estimated second-phase shelf-life is $\hat{s}(t)$. The justification for $\hat{s}(t)$ is that for any t satisfying $\alpha + \beta t > \eta$,

$$P\{\hat{s}(t) \leq \text{ the true second phase shelf-life}\} \approx 95\%$$

for large sample sizes or small error variances.

In practice, the time at which the drug product is taken out of the first-phase storage condition may be unknown. In such a case we may apply the following method to assess the second-phase shelf-life. Select a set of time intervals

$$t_l < \hat{t}, \quad l = 1, \ldots, L$$

and construct a table (or a figure) for $[t_l, \hat{s}(t_{l+1})]$, $l = 1, \ldots, L$ (see, for example, Table 10.1). If a drug product is taken out of the first-phase storage condition at time t_0, which is between t_l and t_{l+1}, then its second-phase shelf-life is $\hat{s}(t_{l+1})$. If there is batch-to-batch variation, which is a fixed effect, we can apply the method to the data from each batch and use the FDA's approach by taking the minimum. Extensions to the case of random batch-to-batch variation require further study.

10.3 An Example

To illustrate the application of the method for determining the shelf-life for the frozen product described in the previous section, we consider a stability test conducted for characterizing the degradation of a frozen product at a pharmaceutical company. The pharmaceutical company wants to establish a shelf-life for the product at a frozen condition of $-20°C$ followed by a shelf-life at a refrigerated condition of $5°C$. As a result, the stability study consisted of a frozen phase at $-20°C$ and a refrigerated phase (or thawed phase) at $5°C$. During the frozen phase the product stored at $-20°C$ was tested at 0, 3, 6, 9, 12, and 18 months. At each sampling time point of the frozen phase, the product was tested at 1, 2, 3, and 5 weeks at the refrigerated condition of $5°C$. Three batches of the drug product were used at each sampling time point (i.e., $K_i = K_{ij} = 3$). Assay results in percent of label claim (i.e., labeled concentration at $g/50$ ml) are given in Table 10.2. The p-values from F-tests of batch-to-batch

TABLE 10.1:   Lower Confidence Bounds and Estimated Shelf-Lives

| | First Phase | | | Second Phase | | | |
| | | | | Common Slope Model | | Linear Trend Model | |
Month t	$\hat{y}(t)$	$[\hat{V}(\hat{y}(t))]^{1/2}$	$L(t)$	Day s	$L(t, s)$	Day s	$L(t, s)$
1	−0.2056	0.0037	99.70	40	90.14	42	90.21
2	−0.2087	0.0034	99.41	39	90.05	41	90.08
3	−0.2118	0.0031	99.11	38	90.21	39	90.19
4	−0.2149	0.0029	98.82	37	90.15	38	90.06
5	−0.2180	0.0027	98.52	36	90.09	36	90.15
6	−0.2211	0.0025	98.22	35	90.02	35	90.01
7	−0.2242	0.0024	97.93	33	90.19	33	90.10
8	−0.2273	0.0024	97.34	32	90.13	31	90.19
9	−0.2304	0.0024	97.34	31	90.06	30	90.03
10	−0.2335	0.0025	97.02	29	90.23	28	90.12
11	−0.2366	0.0027	96.72	28	90.17	26	90.21
12	−0.2397	0.0029	96.42	27	90.10	25	90.05
13	−0.2428	0.0031	96.11	26	90.04	23	90.16
14	−0.2459	0.0034	95.81	24	90.20	22	90.01
15	−0.2490	0.0037	95.50	23	90.14	20	90.14
16	−0.2521	0.0040	95.20	22	90.07	19	90.01
17	−0.2552	0.0043	94.89	20	90.23	17	90.17
18	−0.2583	0.0046	94.59	19	90.16	16	90.06
19	−0.2614	0.0050	94.28	18	90.09	14	90.25
20	−0.2645	0.0054	93.97	17	90.02	13	90.17
21	−0.2676	0.0057	93.67	15	90.18	12	90.09
22	−0.2707	0.0061	93.36	14	90.11	11	90.03
23	−0.2738	0.0064	93.05	13	90.04	9	90.28
24	−0.2769	0.0068	92.75	11	90.20	8	90.24
25	−0.2800	0.0072	92.44	10	90.13	7	90.20
26	−0.2831	0.0076	92.13	9	90.05	6	90.18
27	−0.2862	0.0080	91.82	7	90.21	5	90.16
28	−0.2893	0.0083	91.52	6	90.14	4	90.15
29	−0.2924	0.0087	91.21	5	90.06	3	90.15
30	0.2955	0.0091	90.90	3	90.22	2	90.15
31	−0.2986	0.0095	90.60	2	90.14	1	90.15
32	−0.3017	0.0099	90.30	1	90.07	0	90.30
33	−0.3048	0.0103	89.98				

Source: Shao, J. and Chow, S.C. (2001a). *Statistics in Medicine*. 20, 1239–1248.

variation for two phases of data are larger than 0.25. Thus, we combined three batches in the analysis. As discussed in the previous section, if batch-to-batch variation is significant, we should consider the minimum of within-batch shelf-lives.

Assay results indicate that the degradation of the drug product at the refrigerated phase is faster at later sampling time points of the frozen phase. Assuming the linear trend model,

$$\gamma(t) = \gamma_0 + \gamma_1 t$$

TABLE 10.2: Stability Data

Time		Stability Data	
t (months)	s (days)	**Frozen Condition**	**Refrigerated Condition**
0		100.0, 100.1, 100.1	
3		99.2, 99.0, 99.1	
	7		97.7, 97.4, 97.4
	14		96.5, 96.2, 95.9
	21		94.8, 94.5, 94.6
	35		91.7, 91.8, 91.4
6		98.2, 98.2, 98.1	
	7		96.7, 96.5, 96.7
	14		95.4, 95.2, 95.1
	21		93.6, 93.5, 93.6
	35		90.3, 90.6, 90.5
9		97.5, 97.4, 97.5	
	7		95.9, 95.6, 95.7
	14		94.5, 94.3, 94.1
	21		92.7, 92.5, 92.5
	35		89.3, 89.5, 89.3
12		96.4, 96.5, 96.5	
	7		94.5, 94.7, 94.8
	14		92.8, 93.3, 93.4
	21		91.3, 91.4, 91.5
	35		88.2, 88.2, 87.9
18		94.4, 94.6, 94.5	
	7		92.6, 92.6, 92.4
	14		91.0, 91.1, 90.6
	21		88.9, 89.0, 89.0
	35		85.1, 85.6, 85.6

Source: Shao, J. and Chow, S. C. (2001a). *Statistics in Medicine*, 20, 1239–1248.

which is the case where $H = 1$ in Equation 10.1. Using Equation 10.4, we obtain the following estimates:

$$\hat{\gamma}_0 = -0.2025 \quad \text{and} \quad \hat{\gamma}_1 = -0.0031$$

with estimated standard errors 0.0037 and 0.0004, respectively, where the combined estimator given in Equation 10.5 was used to estimate the second-phase error variances. This results in, the null hypothesis that $\gamma_1 = 0$ (which leads to the common slope model $\gamma(t) = \gamma_0$) is rejected based on an approximate t-test with significance level ≤ 0.0001.

For convenience's sake, Table 10.1 lists the lower confidence bounds $L(t)$ and $L(t, s)$, where $L(t)$ is obtained by using the FDA's method and ignoring the data from the second-phase, and $L(t, s)$ is computed by using the method described in the previous section. For comparison, the results are given under both common slope and linear trend models. The acceptable lower product specification limit in this example is 90%. Thus, the first-phase (frozen) shelf-life for the drug product is given as 32 months. Since the study range for the first-phase is 18 months, a first-phase shelf-life of 24 months can be granted from the FDA. A longer first-phase shelf-life might be granted since the drug product is very stable.

Note that the results for $L(t, s)$ in Table 10.1 are given at s's such that $L(t, s) > 90\%$, whereas $L(t, s + 1) < 90\%$. Thus, the s's listed in Table 10.1 are the second-phase (refrigerated) shelf-lives for various t's. For example, if the drug product comes out of the frozen storage condition at month 10, its refrigerated shelf-life is 29 days under the common slope model or 28 days under the linear trend model. In practice, if we are unable to determine which of the two methods (common slope and linear trend) is significantly better, we could adopt a conservative approach by selecting the shorter of the shelf-lives computed under the two methods. A higher-degree polynomial model does not always produce shorter shelf-lives at the second phase. In our example the linear trend model gave shorter shelf-lives in later months but longer second-phase shelf-lives in early months. This is because under the common slope model, the estimated common slope is the average of the five estimated slopes (in different months) that are getting more negative with time t since $\hat{\gamma}_1 < 0$.

10.4   Discussion

In practice, the assumption of simple linear regression in two phases of stability studies may not be appropriate. For example, there may be acceleration in decay with s in the second phase. Shao and Chow's method can be easily extended to the case where polynomial regression models are considered in both phases. For example, suppose that in the second phase

$$y_{ijk} = \alpha + \beta t_i + \gamma(t_i)s_j + \rho(t_i)s_j^2 + e_{ijk},$$

where both $\gamma(t)$ and $\rho(t)$ are polynomials. At month t_i, let b_i and c_i be the least squares estimates of the coefficients of the linear and quadratic terms in the second-phase quadratic model. Then, estimators of $\gamma(t)$ and $\rho(t)$ can be obtained using

$$b_i = \gamma(t_i) + error,$$

and

$$c_i = \rho(t) + error,$$

which is similar to Equation 10.3. Usual model diagnostic methods may be applied to select an adequate model. Further research on models more complicated than polynomials is needed.

The selection of sampling time points in the second-phase stability test depends on how fast the degradation would be. It may be a good idea to have shorter time intervals for the second phase in later months. For example, we may collect assays more frequently in later weeks. Design strategies on how to select the sampling time points as well as the sample sizes in the second phase need to be further studied as well.

Chapter 11

Stability Testing for Dissolution

In the previous chapters we focused on stability testing for strength (potency) and some discrete drug characteristics such as odor, color, and hardness of oral solid dosage form. In practice, it is also important to make sure other characteristics such as dissolution will remain within approved specification limits prior to the drug expiration date. In the pharmaceutical industry *in vitro* dissolution testing is one of the primary *United States Pharmacopeia-National Formulary (USP-NF)* tests that is often performed to ensure that a drug product meets the *USP-NF* standards for identity, strength, purity, stability, and reproducibility. As indicated in Chapter 1, dissolution failure was one of the top 10 reasons for drug recalls in the fiscal year of 2004 (it becomes the top two reasons for drug recalls in the fiscal year of 2005). As a result, dissolution testing at various critical stages of the manufacturing process for in-process controls, at the end of manufacturing process (final product) for quality assurance, and at commercial marketing for stability plays an important role to ensure that the drug product meets the *USP-NF* standards prior to its expiration date. In addition, after a drug is approved for commercial use, there may be changes with respect to chemistry, manufacturing, and controls. Before the postapproval change formulation can be approved, the Food and Drug Administration (FDA) requires that similarity in dissolution profiles between postchange and prechange formulations be established. In this chapter dissolution testing including the *USP-NF* dissolution test and stability testing for dissolution profiles are discussed.

The rest of this chapter is organized as follows. In the next section sampling plans, testing procedures, and acceptance criteria for *USP-NF* dissolution testing are briefly outlined. Also included in this section are the assessment of the probability lower bound of the *USP-NF* dissolution test (Chow and Shao, 2002b) and a set of recently proposed three-stage sequential dissolution test rules (Tsong et al., 2004). In Section 11.2 the concept of assessing similarity between dissolution profiles is introduced. The f_2 similarity factor recommended by the FDA is also discussed in this section. Statistical methods based on a time series model and a set of hypotheses for the similarity factors for assessment of similarity between dissolution profiles are discussed in Section 11.3. Section 11.4 illustrates the described statistical methods through numerical examples. Concluding remarks are given in the last section of this chapter.

11.1   *USP-NF* Dissolution Testing

Dissolution testing is typically performed by placing a dosage unit in a transparent vessel containing a dissolution medium. A variable-speed motor rotates a cylindrical basket containing the dosage unit. The dissolution medium is then analyzed to determine the percentage of the drug dissolved. Dissolution testing is usually performed on six units simultaneously. The dissolution medium is routinely sampled at various predetermined time intervals to form a dissolution profile.

11.1.1   Sampling Plan and Acceptance Criteria

Dissolution testing usually involves a sampling plan and a set of acceptance criteria for passage of the test. In what follows, commonly considered sampling plan and acceptance criteria are briefly described.

11.1.1.1   *USP-NF* Three-Stage Acceptance Criteria

As indicated in Chapter 1, the *USP-NF* recommends a three-stage sampling plan and acceptance criteria for dissolution testing. For the first stage (S_1), six dosage units are tested. The requirement for the first stage is met if each unit is not less than $Q + 5\%$. If the product fails to pass S_1, an additional six units are tested at the second stage (S_2). The product is considered to have passed S_2 if the average of the 12 units from S_1 and S_2 is equal to or greater than Q and if no unit is less than $Q - 15\%$. If the product fails to pass both S_1 and S_2, an additional 12 units are tested at a third stage (S_3). If the average of all 24 units from S_1, S_2, and S_3 is equal to or greater than Q, no more than two units are less than $Q - 15\%$, and no unit is less than $Q - 25\%$, the product has passed the *USP-NF* dissolution test. To provide a better understanding, the three-stage acceptance rule is summarized in Figure 11.1.

11.1.1.2   Three-Stage Sequential Dissolution Testing Rules

Tsong et al. (1995) criticized the *USP-NF* three-stage acceptance criteria for lacking discriminating capability. In other words, the *USP-NF* three-stage acceptance criteria often fail to reject batches or lots with large percentages below specifications. As an alternative, Tsong et al. (2004) proposed a three-stage sequential dissolution testing procedure based on the concept of O'Brien and Fleming's (1979) alpha spending function. Their proposed procedure is described below:

- At the first stage, based on a sample of six dosage units, accept the batch if $\bar{y}_1 - Q > A_1$, where $A_1 = C(\alpha_1)(s_1\sqrt{6}) - s_1 z_{0.1}$, $\bar{y}_1$, and s_1 are the sample mean and sample standard deviation of the six dosage units, and $C(\alpha_k) = t_{1-\alpha_k, n_k-1}(-\sqrt{n_k}z_{1-p})$ is the $(1 - \alpha_k)$th quantile of a noncentral t-distribution with $n_k - 1$ degrees of freedom and a noncentrality parameter of $-\sqrt{n_k}z_{1-p}$. Note that p is the regulatory required percent of dosage units that dissolve at least Q percent of the label claim at a specific time. Otherwise, move to the second stage and sample six more dosage units.

- At the second stage, accept the batch if $\bar{y}_2 - Q > A_2$, where $A_2 = C(\alpha_2)(s_2\sqrt{12}) - s_2 z_{0.1}$ and $\bar{y}_2$ and s_2 are the sample mean and sample standard deviation of the 12 dosage units; otherwise, move to the third stage and sample 12 more dosage units.

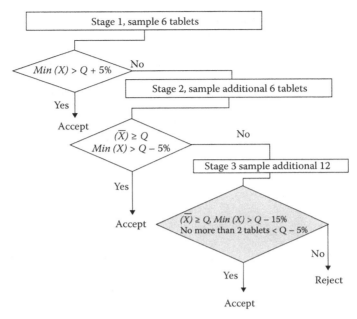

Figure 11.1: *USP-NF* three stage acceptance rule. Source: Tsong, Chen, and Shah (2004).

- At the third stage, accept the batch if $\bar{y}_3 - Q > A_3$, where $A_3 = C(\alpha_3)(s_3\sqrt{24}) - s_3 z_{0.1}$ and $\bar{y}_3$ and s_3 are the sample mean and sample standard deviation of the 24 dosage units; otherwise reject the batch.

Note that α_k can be selected like O'Brien and Fleming's alpha spending function (Lan and DeMets, 1983; O'Brien and Fleming, 1979). To provide a better understanding, the three-stage sequential dissolution testing rules are summarized in Figure 11.2.

11.1.1.3 Remarks

The European Pharmacopeia uses the same criteria as *USP-NF*. The Japanese (Katori et al., 1998), however, suggests a sampling plan based on a sampling mean that accepts the unit when $\bar{y} - 0.8226s > Q$ and $\min(y) \geq Q - 10\%$ in the first stage with six dosage units, where $\bar{y}$ and s are the sample mean and sample standard deviation, respectively. If the unit is not accepted at the first stage, another six dosage units are sampled, and the batch is accepted at the second stage if $\bar{y} - 0.5184s > Q$ and $\min(y) \geq Q - 10\%$. The plan rejects the batch otherwise.

11.1.2 Probability Lower Bound

Under the three-stage sampling plan and acceptance criteria, it is of interest to evaluate the probability of passing the *USP-NF* dissolution test, denoted by P_{DT}. Knowledge of P_{DT} is useful in establishing a set of in-house specification limits for dissolution

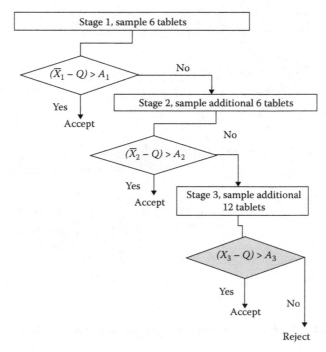

Figure 11.2:   Three-stage group sequential testing method. Source: Tsong, Chen, and Shah (2004).

testing. Bergum (1990) provided the following lower bound for P_{DT}:

$$P_B = P_{Q-15}^{24} + 24 P_{Q-15}^{23} (P_{Q-25} - P_{Q-15})$$
$$+ 276 P_{Q-15}^{22} (P_{Q-25} - P_{Q-15})^2 - P(\bar{y}_{24} \le Q),$$

where

$$P_x = P(y \ge x),$$

and $\bar{y}_{24}$ is the average of the dissolution testing results from all three stages. Although Bergum's probability lower bound is easy to compute, it is somewhat conservative. Alternatively, Chow, Shao, and Wang et al. (2002b) proposed a better probability lower bound, which is described below.

Let y_i, $i = 1, \ldots, 6$, be the dissolution testing results from the first stage, y_i, $i = 7, \ldots, 12$, be the dissolution testing results from the second stage, y_i, $i = 13, \ldots, 24$, be the dissolution testing results from the third stage, and $\bar{y}_k$ be the average of $y_1, \ldots, y_k$. Let S_i denote the event that the ith stage of a k-stage *USP-NF* test is passed. Also, let C_{ij} be the event that the jth criterion at the ith stage is met, where $j = 1, \ldots, m_i$ and $i = 1, \ldots, k$. Then,

$$S_i = C_{i1} \cap C_{i2} \cap \cdots \cap C_i m_i, \quad i = 1, \ldots, k$$

and the event of passing the *USP-NF* test is

$$S_1 \cup S_2 \cup \cdots \cup S_k.$$

Define the following events:

$$
\begin{aligned}
S_1 &= \{y_i \geq Q + 5, \ i = 1, \ldots, 6\}, \\
C_{21} &= \{y_i \geq Q - 15, \ i = 1, \ldots, 12\}, \\
C_{22} &= \{\bar{y}_{12} \geq Q\}, \\
C_{31} &= \{y_i \geq Q - 25, \ i = 1, \ldots, 24\}, \\
C_{32} &= \{\text{no more than two } y_i\text{'s} < Q - 15\}, \\
C_{33} &= \{\bar{y}_{24} \geq Q\}, \\
S_2 &= C_{21} \cap C_{22}, \\
S_3 &= C_{31} \cap C_{32} \cap C_{33}.
\end{aligned}
$$

Then,

$$P_{DT} = P(S_1 \cup S_2 \cup S_3).$$

Bergum's lower bound P_B is obtained by using the inequalities

$$P(S_1 \cup S_2 \cup S_3) \geq P(S_3),$$

and

$$P(C_{31} \cap C_{32} \cap C_{33}) \geq P(C_{31} \cap C_{32}) - P(C_{33}^c),$$

where A^c denotes the complement of the event A and the fact that $S_3 = C_{31} \cap C_{32} \cap C_{33}$ and y_i's are independent and identically distributed. When the probability $P(C_{33}^c)$ is not small, these inequalities are not sharp enough. From the equation

$$P_{DT} = P(S_3) + P(S_2 \cap S_3^c) + P(S_1 \cap S_2^c) + P(S_1 \cap S_2^c \cap S_3^c),$$

a lower bound for P_{DT} can be obtained by deriving lower bounds for $P(S_3)$, $P(S_2 \cap S_3^c)$, and $P(S_1 \cap S_2^c \cap S_3^c)$. We take Bergum's bound P_B as the lower bound for $P(S_3)$. For $P(S_1 \cap S_2^c)$, consider that

$$
\begin{aligned}
P(S_2 \cap S_3^c) &= P(C_{21} \cap C_{22} \cap S_3^c) \geq P(C_{21} \cap S_3^c) - P(C_{22}^c) \\
&= P(C_{21} \cap C_{31}^c) + P(C_{21} \cap C_{31} \cap C_{32}^c) + P(C_{21} \cap C_{31} \cap C_{32} \cap C_{33}^c) \\
&\quad - P(\bar{y}_{12} < Q) \geq P(C_{21} \cap C_{31}^c) + P(C_{21} \cap C_{31} \cap C_{32}^c) - P(\bar{y}_{12} < Q).
\end{aligned}
$$

Since y_i's are independent and identically distributed,

$$
\begin{aligned}
P(C_{21} \cap C_{31}^c) &= P(C_{21}) - P(C_{21} \cap C_{31}) \\
&= P_{Q-15}^{12} - P_{Q-15}^{12} P_{Q-25}^{12}.
\end{aligned}
$$

Note that

$$P(C_{21} \cap C_{31} \cap C_{32}) = P(y_i \geq Q - 15, \ i = 1, \dots, 24)$$
$$+ 12P \left(\begin{matrix} y_i \geq Q - 15, \ i = 1, \dots, 23 \\ Q - 25 \leq y_{24} < Q - 15 \end{matrix} \right)$$
$$+ \binom{12}{2} P \left(\begin{matrix} y_i \geq Q - 15, \ i = 1, \dots, 22 \\ Q - 25 \leq y_i < Q - 15, \ i = 23, 24 \end{matrix} \right)$$
$$= P_{Q-15}^{24} + 12 P_{Q-15}^{23}(P_{Q-25} - P_{Q-15}) + 66 P_{Q-15}^{22}(P_{Q-25} - P_{Q-15})^2.$$

Hence,

$$P\left(C_{21} \cap C_{31} \cap C_{32}^c\right) = P\left(C_{21} \cap C_{31}\right) - P\left(C_{21} \cap C_{31} \cap C_{32}\right)$$
$$= P_{Q-15}^{12} P_{Q-25}^{12} - P_{Q-15}^{24} - 12 P_{Q-15}^{23}(P_{Q-25} - P_{Q-15})$$
$$- 66 P_{Q-15}^{22}(P_{Q-25} - P_{Q-15})^2.$$

Thus, a lower bound for $P(S_2 \cap S_3^c)$ is

$$P_C = P_{Q-15}^{12} - P_{Q-15}^{24} - 12 P_{Q-15}^{23}(P_{Q-25} - P_{Q-15})$$
$$- 66 P_{Q-15}^{22}(P_{Q-25} - P_{Q-15})^2 - P(\bar{y}_{12} < Q).$$

This lower bound is good when $P(\bar{y}_{12} < Q)$ is small. However,

$$P\left(S_2 \cap S_3^c\right) = P[C_{21} \cap C_{22} \cap (C_{31} \cap C_{32} \cap C_{33})^c]$$
$$\geq P\left(C_{21} \cap C_{22} \cap C_{33}^c\right)$$
$$\geq P\left(C_{22} \cap C_{33}^c\right) - P\left(C_{21}^c\right)$$
$$= P(\bar{y}_{12} \geq Q, \bar{y}_{24} < Q) - \left(1 - P_{Q-15}^{12}\right).$$

The previous two inequalities provide an accurate lower bound if $P(C_{21}^c)$ and $P(C_{21} \cap C_{31}^c \cup C_{32}^c)$ are small. Thus, a better lower bound for $P(S_2 \cap S_3^c)$ is the larger of P_C and

$$P_D = P(\bar{y}_{12} \geq Q, \bar{y}_{24} < Q) - \left(1 - P_{Q-15}^{12}\right).$$

For $P(S_1 \cap S_2^c \cap S_3^c)$, consider that

$$P\left(S_1 \cap S_2^c \cap S_3^c\right) = P\left(S_1 \cap C_{21}^c \cap S_3^c\right) + P\left(S_1 \cap C_{21} \cap C_{22}^c \cap S_3^c\right)$$
$$\geq P\left(S_1 \cap C_{21}^c \cap S_3^c\right)$$
$$= P\left(S_1 \cap C_{21}^c \cap C_{31}^c\right) + P\left(S_1 \cap C_{21}^c \cap C_{31} \cap C_{32}^c\right)$$
$$+ P\left(S_1 \cap C_{21}^c \cap C_{31} \cap C_{32}^c \cap C_{33}^c\right)$$
$$\geq P\left(S_1 \cap C_{21}^c \cap C_{31}^c\right) + P\left(S_1 \cap C_{21}^c \cap C_{31} \cap C_{32}^c\right)$$
$$= P\left(S_1 \cap C_{21}^c \cap C_{31}^c\right) + P\left(S_1 \cap C_{21}^c \cap C_{31}\right)$$
$$- P\left(S_1 \cap C_{21}^c \cap C_{31} \cap C_{32}\right)$$
$$= P\left(S_1 \cap C_{21}^c\right) - P\left(S_1 \cap C_{31} \cap C_{32}\right)$$
$$+ P(S_1 \cap C_{21} \cap C_{31} \cap C_{32}).$$

Since

$$P\left(S_1 \cap C_{21}^c\right) = P(S_1) - P(S_1 \cap C_{21})$$
$$= P_{Q+5}^6 - P_{Q+5}^6 P_{Q-15}^6,$$

then

$$P(S_1 \cap C_{31} \cap C_{32}) = P\left(\begin{array}{l} y_i \geq Q+5, \ i = 1, \ldots, 6 \\ y_i \geq Q-15, \ i = 7, \ldots, 24 \end{array}\right)$$
$$+ 18P\left(\begin{array}{l} y_i \geq Q+5, \ i = 1, \ldots, 6 \\ y_i \geq Q-15, \ i = 7, \ldots, 23 \\ Q-25 \leq y_{24} < Q-15 \end{array}\right)$$
$$+ \binom{18}{2} P\left(\begin{array}{l} y_i \geq Q+5, \ i = 1, \ldots, 6 \\ y_i \geq Q-15, \ i = 7, \ldots, 22 \\ Q-25 \leq y_i < Q-15, \ i = 23, 24 \end{array}\right)$$
$$= P_{Q+5}^6 P_{Q-25}^{18} + 18 P_{Q+5}^6 P_{Q-15}^{17}(P_{Q-25} - P_{Q-15})$$
$$+ 153 P_{Q+5}^6 P_{Q-15}^{16}(P_{Q-25} - P_{Q-15})^2,$$

and

$$P(S_1 \cap C_{21} \cap C_{31} \cap C_{32}) = P\left(\begin{array}{l} y_i \geq Q+5, \ i = 1, \ldots, 6 \\ y_i \geq Q-15, \ i = 7, \ldots, 24 \end{array}\right)$$
$$+ 12P\left(\begin{array}{l} y_i \geq Q+5, \ i = 1, \ldots, 6 \\ y_i \geq Q-15, \ i = 7, \ldots, 23 \\ Q-25 \leq y_{24} < Q-15 \end{array}\right)$$
$$+ \binom{12}{2} P\left(\begin{array}{l} y_i \geq Q+5, \ i = 1, \ldots, 6 \\ y_i \geq Q-15, \ i = 7, \ldots, 22 \\ Q-25 \leq y_i < Q-15, \ i = 23, 24 \end{array}\right)$$
$$= P_{Q+5}^6 P_{Q-15}^{18} + 12 P_{Q+5}^6 P_{Q-15}^{17}(P_{Q-25} - P_{Q-15})$$
$$+ 66 P_{Q+5}^6 P_{Q-15}^{16}(P_{Q-25} - P_{Q-15})^2.$$

A lower bound for $P(S_1 \cap S_2^c \cap S_3^c)$ is given by

$$P_E = P_{Q+5}^6 - P_{Q+5}^6 P_{Q-15}^6 - 6 P_{Q+5}^6 P_{Q-15}^{17}(P_{Q-25} - P_{Q-15})$$
$$- 87 P_{Q+5}^6 P_{Q-15}^{16}(P_{Q-25} - P_{Q-15})^2.$$

Combining these results, the following lower bound for P_{DT} can be obtained

$$P_{DT} = \max(0, P_B) + \max(0, P_C, P_D) + P_E.$$

This lower bound is given in terms of six probabilities: P_{Q+5}, P_{Q-15}, P_{Q-25}, $P(\bar{y}_{12} < Q)$, $P(\bar{y}_{24} < Q)$, and $P(\bar{y}_{12} \geq Q, \bar{y}_{24} < Q)$. If y_i is normally distributed with mean μ and variance σ^2, then

$$P_x = 1 - \Phi\left(\frac{x - \mu}{\sigma}\right),$$
$$P(\bar{y}_k < Q) = \Phi\left(\frac{\sqrt{k}(Q - \mu)}{\sigma}\right).$$

and

$$P(\bar{y}_{12} \geq Q, \quad \bar{y}_{24} < Q) = P(\bar{y}_{24} < Q) - P(\bar{y}_{12} < Q, \ \bar{y}_{24} < Q)$$
$$= \Phi \left(\frac{\sqrt{24}(Q - \mu)}{\sigma} \right) - \Psi(Q - \mu, \ Q - \mu),$$

where $x = Q + 5$, $Q - 15$, or $Q - 25$; $k = 12$ or 24; Φ is the standard normal distribution function; and Ψ is the bivariate normal distribution with mean 0 and covariance matrix

$$\frac{\sigma^2}{24} \begin{pmatrix} 2 & 1 \\ 1 & 1 \end{pmatrix}.$$

If μ and σ^2 are unknown, they can be estimated using data from previously sampled test results. Chow, Shao, and Wang et al. (2002b) studied the probability lower bound through a simulation study. The results indicate that the probability lower bound for P_{DT} described above is better than Bergum's lower bound and is a very accurate approximation to P_{DT} when $\sigma \geq 5$. More details regarding the simulation results can be found in Chow, Shao, and Wang et al. (2002b). Note that an improved method for estimation of the probability lower bound of passing the dissolution test was recently proposed by Wang (2007).

11.2   Dissolution Profile Testing

To characterize the dissolution profile, dissolution testing is typically performed by placing a dosage unit in a transparent vessel containing a dissolution medium. The dissolution medium is routinely sampled at various predetermined time points, for example, at 15, 30, 45, 60, and 120 minutes or until asymptote is reached. For a drug product, the curve of the mean dissolution rate over time is referred to as its *dissolution profile*.

11.2.1   Scale-Up and Postapproval Changes

After a drug is approved for commercial marketing, there may be some changes with respect to chemistry, manufacturing, and controls. Before the postchange formulation can be approved for commercial use, its quality and performance need to be demonstrated to show similarity to the prechange formulation. Because drug absorption depends on the dissolved state of drug products, *in vitro* dissolution testing is believed to provide a rapid assessment of the rate and extent of drug release. As a result, Leeson (1995) suggested that *in vitro* dissolution testing be used as a substitute for *in vivo* bioequivalence studies to assess equivalence between the postchange and prechange formulations. For this purpose, the FDA has issued several guidances regarding scale-up and postapproval changes (see, e.g., SUPAC-IR, 1995; SUPAC-MR,

1997, and SUPAC-SS, 1997). These SUPAC guidances make recommendations to the sponsors of new drug applications, abbreviated new drug applications, and abbreviated antibiotic drug applications regarding the following four types of postapproval changes:

- The components or compositions

- The site of manufacture

- The scale-up/scale-down of manufacture

- The manufacturing process and equipment of a formulation

If dissolution profile similarity is demonstrated between the prechange drug product and the postchange formulation, *in vivo* bioequivalence testing can be waived for most changes.

11.2.2 Dissolution Profile Comparison

In practice, depending on the study objectives, we may want to compare dissolution profiles of a given batch of a drug product at two different sampling time intervals, different batches of a drug product at a given time sampling time interval, or two drug products at a given sampling time interval. For comparison of dissolution profiles, several methods have been proposed in the literature. These methods include model-dependent and model-independent methods, which are described below.

11.2.2.1 Model-Dependent Methods

For comparison of dissolution profiles between a test product and a reference product, Tsong et al. (1995) suggested modeling the dissolution profile curves based on previously approved batches for the reference drug product. For example, Langenbucher (1972) indicated that the dissolution profile can be approximated by Weibull distribution after linearization. Dawoodbhai et al. (1991) considered the Gompertz model to characterize the dissolution profile, while Pena Romero et al. (1991) introduced the use of the logistic curve, which was found useful in the study of water uptake and force development in an optimized prolonged release formulation. Kervinen and Yliruusi (1993) proposed three control factor models for S-shaped dissolution profile curves to connect the curves with the physical phenomenon of dissolution. Tsong et al. (1995) indicated that we can consider a sigmoid curve or a probit model to describe the dissolution profile. After a parametric model is chosen, unknown parameters in the model are estimated based on dissolution data. If the estimated parameters of the test product fall within some predetermined specifications, we can conclude that the two drug products have similar dissolution profiles.

This approach may sound reasonable, but, how do we select an appropriate model for the dissolution profile? Does the dissolution profile of the test product follow the same model as that of the reference product? These questions remain unsolved, which limits the application of this approach.

11.2.2.2   Mondel-Independent Methods

Some model-independent approaches such as the analysis of variance, the analysis of covariance, and the split-plot analysis are considered in assessing similarity between the two dissolution profiles. However, these methods are not appropriate because dissolution testing results over time are not independent owing to the nature of dissolution testing. As an alternative, Tsong et al. (1996) proposed a multivariate analysis by considering the distance between the mean dissolution rates of the two drug products at two time points. The idea is to use a Hotelling T^2 statistic to construct a 90% confidence region for the difference in dissolution means of two batches of the reference product at the two time points. This confidence region is then used as equivalence criteria for assessing similarity between dissolution profiles. For a given test product, if the constructed confidence region for the difference in dissolution means between the test product and the reference product at the two time points is within the equivalence region, we can conclude that the two dissolution profiles are similar at the two time points. One disadvantage of this method is that it is impossible to visualize the confidence region when there are more than two time points.

Another model-independent approach is to consider the method of analysis of variance with repeated measures proposed by Gill (1988). The idea is to consider a nested model with a covariate error structure to account for correlation between observations (repeated measurements). This model is useful in: (a) comparing dissolution rates between drug products at given time points, (b) detecting time effect within treatment, and (c) comparing mean dissolution rate changes from one time point to another. One disadvantage of Gill's method is that the dissolution profiles can only be compared at each time point when there is a significant treatment-by-time interaction.

11.2.3   Concept of Similarity

Unlike the concept of equivalence in *in vivo* bioequivalence testing, there exists no universal definition for similarity in *in vitro* dissolution testing. Many criteria for assessment of similarity have been proposed in the literature (see, e.g., Moore and Flanner, 1996; Chow and Ki, 1997; Tsong et al., 1996). These criteria are briefly described below.

11.2.3.1   Local/Global Similarity

Since dissolution profiles are curves over time, we introduce the concepts of local similarity and global similarity. Two dissolution profiles are said to be locally similar at a given time point if their difference or ratio at the given time point is within some equivalence limits, denoted by (δ_L, δ_U). Two dissolution profiles are considered globally similar if their differences or ratios are within the equivalence limits (δ_L, δ_U) across all time points. Note that global similarity is also referred to as uniformly similar by many researchers. Chow and Ki (1997) suggested the following equivalence limits

$$\delta_L = \frac{Q - \delta}{Q + \delta},$$

and

$$\delta_U = \frac{Q + \delta}{Q - \delta},$$

where Q is the desired mean dissolution rate of a drug product as specified in the *USP-NF* individual monograph, and δ is a meaningful difference of scientific importance in mean dissolution profiles of two drug products under consideration. In practice, δ is usually determined by a pharmaceutical scientist. For more details regarding the derivation of the equivalence limits, see Chow and Ki (1997). It should be noted that the concept of global similarity or uniformly similarity was not adopted by the FDA due to its over-stringency.

11.2.3.2 The f_2 Similarity

Instead of comparing two dissolution profiles, we may consider a statistic that measures the closeness of two dissolution profiles. Such a statistic is called a similarity factor. Two dissolution profiles are considered to be similar if the similarity factor is within some specified equivalence limits. Moore and Flanner (1996) considered a similarity factor that is referred to as the f_2 similarity factor. The f_2 similarity factor can be described as follows. Since the FDA SUPAC guidances require that all profiles be conducted on at least 12 individual dosage units, we assume that at each time point and for each drug product, there are n individual dosage units. Let y_{hti} be the cumulative percent dissolved for dosage unit i at the tth sampling time point for drug product h, where $i = 1, \ldots, n$, $t = 1, \ldots, T$; and $h = 1, 2$. Denote the average cumulative percent dissolved at the tth time point for product h as

$$\bar{y}_{ht} = \frac{1}{n} \sum_{i=1}^{n} y_{hti},$$

and the sum of squares of difference in average cumulative percent dissolved between the two drug products over all sampling time points as

$$D = \sum_{t=1}^{T} (\bar{y}_{1t} - \bar{y}_{2t})^2.$$

The f_2 similarity factor proposed by Moore and Flanner (1996) is then defined to be the logarithmic reciprocal square root transformation of one plus the mean squared (the average sum of squares) difference in observed average cumulative percent dissolved between the two products over all sampling time points, that is,

$$f_2 = 50 \times \log_{10} \left(\frac{100}{\sqrt{1 + \frac{D}{T}}} \right)$$

$$= 100 - 25 \log_{10} \left(1 + \frac{D}{T} \right), \tag{11.1}$$

where $\log_{10}$ denotes the logarithmic base 10 transformation. f_2 is a strictly decreasing function of D. If there is no difference in average cumulative percent at all sampling time points ($D = 0$), f_2 reaches its maximum value 100. A large value of f_2 indicates the similarity of the two dissolution profiles. For immediate-release solid dosage forms, the FDA suggests that two dissolution profiles are considered to be similar if the f_2 similarity factor is between 50 and 100 (SUPAC-IR, 1995).

The use of the f_2 similarity factor has been discussed and criticized by many researchers (see, e.g., Liu et al., 1997; Shah et al., 1998; Tsong et al., 1996; Ma et al., 1999). Chow and Shao (2002b) pointed out two main problems in using the f_2 similarity factor for assessing similarity between the dissolution profiles of two drug products. The first problem is its lack of statistical justification. Since f_2 is a statistic and, thus, a random variable, $P(f_2 > 50)$ may be quite large when the two dissolution profiles are not similar. However, $P(f_2 > 50)$ can be very small when the two dissolution profiles are similar. Suppose the expected value $E(f_2)$ exists and that we can find a 95% lower confidence bound for $E(f_2)$, which is denoted by $\tilde{f}_2$. A reasonable modification to the approach of using f_2 is to replace f_2 by $\tilde{f}_2$; that is, two dissolution profiles are considered to be similar if $\tilde{f}_2 > 50$. However, Liu et al. (1997) indicated that the distribution of f_2 is very complicated and almost intractable because of the unnecessary logarithmic reciprocal square root transformation, which they believe exists just to make the artificial acceptable range from 50 to 100. Since f_2 is a strictly decreasing function of D/T, $f_2 > 50$ if and only if $D/T < 99$. Hence, an alternative method is to consider a 95% upper confidence bound $\tilde{D}/T$ for $E(D)/T$ as the similarity factor; that is, two dissolution profiles are similar if $\tilde{D}/T < 99$. Note that

$$\frac{E(D)}{T} = \frac{1}{T} \sum_{t=1}^{T} \mu_{Dt}^2 + \frac{1}{nT} \sum_{t=1}^{T} \sigma_{Dt}^2,$$

where $\mu_{DT} = E(y_{1ti} - y_{2ti})$ is the difference between two dissolution profiles at the ith time point, and $\sigma_{Dt}^2 = Var(y_{1ti} - y_{2ti})$. However, the construction of an upper confidence bound for $E(D)/T$ is still a difficult problem, and more research is required.

The second problem with using the f_2 similarity factor is that the f_2 similarity factor assesses neither local similarity nor global similarity, owing to the use of the average of $(\bar{y}_{1t} - \bar{y}_{2t})^2$, $t = 1, \dots, T$. To illustrate the problem, let us consider the following example. Suppose n is large enough so that

$$D \approx E(D) = \sum_{t=1}^{T} \mu_{Dt}^2,$$

and that T is an even integer, $\mu_{Dt} = 0$ when $t = 1, \dots, T/2$, and $|\mu_{Dt}| = 10$ when $t = T/2 + 1, \dots, T$. Since

$$\frac{E(D)}{T} = \frac{10^2}{2} = 50 < 99,$$

the value of f_2 is less than 50. However, the two dissolution profiles are not globally similar, if a difference of ± 10 is considered to be large enough for nonsimilarity. At time points $t = T/2 + 1, \dots, T$, the two dissolution profiles are not locally similar either.

11.2.3.3 Other Similarity Factors

Moore and Flanner (1996) suggested a different similarity factor, f_1, which is defined as the ratio of the sum of absolute mean differences and the sum of cumulative dissolution of the reference product. As a result, it is a reference-dependent metric and is more complicated in its interpretation. Besides, statistical properties of f_1 are not well understood.

Since the f_2 similarity factor is derived based on the average squared mean difference, it can be expressed as a function g as follows:

$$g\left[\sum_{t=1}^{T} w_t(\bar{y}_{1t} - \bar{y}_{2t})^2\right],$$

where w_t is a prespecified weight at time point t. For the f_2 similarity factor, equal weights are applied, that is, $w_t = w = 1/T$. Alternatively, unlike the f_2 factor, Ma et al. (2000) considered the so-called g_1 similarity factor, which is a function of the absolute mean difference between two dissolution profiles. In other words, they considered

$$g\left(\frac{1}{T}\right)\sum_{t=1}^{T} w_t |\bar{y}_{1t} - \bar{y}_{2t}|,$$

where w_t is a prespecified weight at time point t. For example, we may consider the area between two time points. That is, we may use

$$w_1 = \frac{t_1}{2},$$
$$w_2 = \frac{t_3 - t_1}{2},$$
$$\vdots$$
$$w_k = \frac{t_{k+1} - t_{k-1}}{2},$$
$$\vdots$$
$$w_T = \frac{t_T - t_{T-1}}{2}.$$

As a result, if we consider average distance, this leads to the criterion that is defined as follows:

$$g_1 = \frac{1}{T}\sum_{t=1}^{T} |\bar{y}_{1t} - \bar{y}_{2t}|.$$

The possible values of g_1 range from 0 to 100. Unlike the f_2 factor, this range is based on the original untransformed scale. Because g_1 is a simple linear monotone function of the difference cumulated at each time point, the interpretation of g_1 is straightforward and easily understood by chemists, pharmacologists, and nonstatisticians. However, with the difference squared in f_2, the similarity factor is more sensitive than g_1 for large profile differences at a single time point.

11.2.3.4 Remarks

Shah et al. (1998) and Chow and Shao (2002), pointed out that the similarity factors introduced by Moore and Flanner (1996) are statistics. Thus, they are random variables, which suffer from some deficiencies. To overcome this problem, we may modify the criteria as follows. For the same dosage unit, we consider

$$y_{ht} = (y_{ht}, y_{ht2}, \dots, y_{htn})'$$

with mean

$$\mu_h = (\mu_{h1}, \mu_{h2}, \dots, \mu_{hn})'$$

and covariance matrix Σ_h, $h = 1, 2$. Since y_{hti} is the cumulative percent dissolved, we have

$$y_{ht1} \leq y_{ht2} \leq \dots \leq y_{htn},$$

and

$$\mu_{h1} \leq \mu_{h2} \leq \dots \leq \mu_{hn}.$$

Now, we redefine the f_2 similarity factor as follows:

$$f_2 = 50 \times \log_{10} \left(\frac{100}{\sqrt{1 + \frac{W}{T}}} \right)$$

$$= 100 - 25 \log_{10} \left(1 + \frac{W}{T} \right), \tag{11.2}$$

where

$$W = \sum_{t=1}^{T} (\mu_{1t} - \mu_{2t})^2.$$

Similarly, g_1 is redefined as

$$g_1 = \frac{1}{T} \sum_{t=1}^{T} |\mu_{1t} - \mu_{2t}|. \tag{11.3}$$

The newly defined f_2 and g_1 similarity factors are no longer statistics, but are unknown parameters.

11.3 Statistical Methods for Assessing Similarity

In this section we introduce statistical methods based on the concept of local and global similarity and the f_2 similarity factor for assessing similarity between two dissolution profiles. Other similarity factors such as the g_1 similarity factor are also considered. For the concept of local and global similarity, a time series model proposed by Chow and Ki (1997) and Chow and Shao (2002b) are considered. For similarity factors f_2 and g_1, the method of hypotheses testing proposed by Ma et al. (2000) is discussed.

11.3.1 Time Series Model

For assessing dissolution profiles between two drug products Chow and Ki (1997) considered the ratio of the dissolution results of the ith dosage unit at the tth time point, denoted by

$$R_{ti} = \frac{y_{1ti}}{y_{2ti}}, \tag{11.4}$$

which can be viewed as a measure of the relative dissolution rate at the ith dissolution medium (or the ith location) and the tth time point. In the following discussion, R_{ti} can be replaced by the difference $y_{1ti} - y_{2ti}$ or the log-transformation on y_{1ti}/y_{2ti}. Since the dissolution results at the tth time point depend on the results at the previous time point $t - 1$, the dissolution results are correlated over time. To account for this correlation, Chow and Ki (1997) considered the following autoregressive time series model for R_{ti}:

$$R_{ti} - \gamma_t = \phi(R_{(t-1)i} - \gamma_{t-1}) + \epsilon_{ti}, \quad i = 1, \dots, n, \quad t = 2, \dots, T, \tag{11.5}$$

where $\gamma_t = E(R_{ti})$ is the mean relative dissolution rate at the tth time point, $|\phi| < 1$ is an unknown parameter, and ϵ_{ti}'s are independent and identically normally distributed with mean 0. When γ_t does not vary with t, Chow and Ki (1997) derived a method of assessing similarity between dissolution profiles of two drug products based on Equation 11.5 and methods from time series analysis. Consider the case where $\gamma_t = \gamma$ for all t. Chow and Ki proposed to construct a 95% confidence interval (L, U) for γ and then compare (L, U) with the similarity limits (δ_L, δ_U). If (L, U) is within (δ_L, δ_U), we claim that the two dissolution profiles are similar. Under Model 11.5, we have

$$E(R_{ti}) = \gamma \quad \text{and} \quad Var(R_{ti}) = \sigma_R^2,$$

where

$$\sigma_R^2 = \frac{\sigma_\epsilon^2}{1 - \phi^2}.$$

Let

$$\hat{\gamma} = \frac{1}{nT} \sum_{t=1}^{T} \sum_{i=1}^{n} R_{ti}.$$

Then

$$E(\hat{\gamma}) = \gamma,$$

and

$$Var(\hat{\gamma}) = \frac{1}{n} Var\left(\frac{1}{T} \sum_{t=1}^{T} R_{ti}\right)$$

$$= \frac{\sigma_R^2}{nT}\left(1 + 2\sum_{t=1}^{T} \frac{T-t}{T}\phi^t\right).$$

If σ_R^2 and ϕ are known, a 95% confidence interval for γ is (L, U) with

$$L = \hat{\gamma} - z_{0.975} \sqrt{Var(\hat{\gamma})},$$

and

$$U = \hat{\gamma} + z_{0.975} \sqrt{Var(\hat{\gamma})},$$

where z_a is the $100 \times a$th percentile of the standard normal distribution. When σ_R^2 and ϕ are unknown, we replace them in L and U by their estimators

$$\sigma_R^2 = \frac{1}{nT - 1} \sum_{t=1}^{T} \sum_{i=1}^{n} (R_{ti} - \hat{\gamma})^2,$$

and

$$\hat{\phi} = \frac{\sum_{t=1}^{T-1} \sum_{i=1}^{n} (R_{ti} - \hat{\gamma})(R_{(t+1)i} - \hat{\gamma})}{\sum_{t=1}^{T} \sum_{i=1}^{n} (R_{ti} - \hat{\gamma})^2},$$

respectively, which results in an approximate 95% confidence interval (L, U) for γ since $\hat{\sigma}_R^2$ and $\hat{\phi}$ are consistent estimators as $nT \to \infty$.

In practice, however, γ_t usually varies with t. Chow and Ki (1997) considered a random effects model for γ_t, in which γ_t's are assumed to be independent and normally distributed random variables with mean γ and variance σ_γ^2. For a fixed t, we have

$$R_{ti}|\gamma_t \sim N(\gamma_t, \sigma_R^2),$$

and

$$\bar{R}_t|\gamma_t \sim N\left(\gamma_t, \frac{1}{n}\sigma_R^2\right).$$

Hence, unconditionally,

$$\bar{R}_t \sim N\left(\gamma, \frac{1}{n}\sigma_R^2 + \sigma_\gamma^2\right),$$

where $\bar{R}_t$ is the average of $R_{ti}, i = 1, \dots, n$. Then,

$$\gamma_t|\bar{R}_t \sim N\left(\frac{\sigma_R^2/n}{\sigma_R^2/n + \sigma_\gamma^2} + \frac{\sigma_\gamma^2}{\sigma_R^2/n + \sigma_\gamma^2}\bar{R}_t, \frac{\sigma_\gamma^2\sigma_R^2}{\sigma_R^2 + n\sigma_\gamma^2}\right).$$

Therefore, when γ, σ_R^2, and σ_γ^2 are unknown, we replace γ with $\hat{\gamma}$ and σ_R^2 with its unbiased estimator,

$$\sigma_R^2 = \frac{1}{T(n-1)} \sum_{t=1}^{T} \sum_{i=1}^{n} (R_{ti} - \bar{R}_t)^2.$$

Note that since

$$s^2 = \frac{1}{T-1} \sum_{t=1}^{T} (\bar{R}_t - \hat{\gamma})^2$$

is an unbiased estimator of $\sigma_R^2/n + \sigma_\gamma^2$, σ_γ^2 can be estimated by

$$\hat{\sigma}_\gamma^2 = \max\{0,\ s^2 - \hat{\sigma}_R^2/n\}.$$

As a result, the prediction limits are given by

$$L_t = \bar{R}_t - \min\left\{1,\ \frac{\hat{\sigma}_R^2}{ns^2}\right\}(\bar{R}_t - \hat{\gamma}) - z_{0.975}\frac{\hat{\sigma}_\gamma\hat{\sigma}_R}{s\sqrt{n}},$$

and

$$U_t = \bar{R}_t - \min\left\{1,\ \frac{\hat{\sigma}_R^2}{ns^2}\right\}(\bar{R}_t - \hat{\gamma}) + z_{0.975}\frac{\hat{\sigma}_\gamma\hat{\sigma}_R}{s\sqrt{n}}.$$

Thus, the two dissolution profiles are considered to be locally similar at the time point if (L_t, U_t) is within (δ_L, δ_U). It should be noted that, under the random effects model approach, it is difficult to assess the global similarity between two dissolution profiles. We consider instead the similarity of the two dissolution profiles at all time points considered in dissolution testing. That is, we construct the simultaneous prediction interval $(\tilde{L}_t, \tilde{U}_t)$, $t = 1, \ldots, T$, satisfying

$$P\left(\tilde{L}_t < \gamma_t < \tilde{U}_t,\ t = 1, \ldots, T | \bar{R}_t, t = 1, \ldots, T\right) \geq 95\%,$$

and consider the two dissolution profiles to be globally similar when $(\tilde{L}_t, \tilde{U}_t)$ is within (δ_L, δ_U) for $t = 1, \ldots, T$. Using Bonferroni's method, simultaneous prediction intervals can be constructed using the limits

$$\tilde{L}_t = \bar{R}_t - \min\left\{1,\ \frac{\hat{\sigma}_R^2}{ns^2}\right\}(\bar{R}_t - \hat{\gamma}) - z_{1-0.025/T}\frac{\hat{\sigma}_\gamma\hat{\sigma}_R}{s\sqrt{n}},$$

and

$$\tilde{U}_t = \bar{R}_t - \min\left\{1,\ \frac{\hat{\sigma}_R^2}{ns^2}\right\}(\bar{R}_t - \hat{\gamma}) + z_{1-0.025/T}\frac{\hat{\sigma}_\gamma\hat{\sigma}_R}{s\sqrt{n}}.$$

11.3.1.1 Remarks

Note that the random effects approach described above has two disadvantages. First, there are often deterministic and monotone trends in γ_t, $t = 1, \ldots, T$, which cannot be appropriately described by random effects. Second, it is difficult to use the random effects approach to assess the global similarity between two dissolution profiles, since the global similarity should also apply to time points that are not in the sample.

Alternatively, Chow and Shao (2002b) proposed a method based on the time series model (Model 11.5) and the following polynomial model for γ_t:

$$\gamma_t = \beta_0 + \beta_1 x_t + \beta_2 x_t^2 + \cdots + \beta_p x_t^p, \tag{11.6}$$

where x_t is the value of the tth time point and β_j's are unknown parameters. The variances $\sigma_t^2 = Var(\epsilon_{ti})$, $t = 1, \ldots, T$ may depend on t. Let

$$\bar{R} = \begin{pmatrix} \bar{R}_1 \\ \bar{R}_2 \\ \vdots \\ \bar{R}_T \end{pmatrix}, \quad X = \begin{pmatrix} 1 & x_1 & x_1^2 & \cdots & x_1^p \\ 1 & x_2 & x_2^2 & \cdots & x_2^p \\ \vdots & \vdots & \vdots & \ddots & \vdots \\ 1 & x_T & x_T^2 & \cdots & x_T^p \end{pmatrix}, \quad \beta = \begin{pmatrix} \beta_0 \\ \beta_1 \\ \vdots \\ \beta_p \end{pmatrix},$$

and $\bar{R}_t$ be the average of R_{ti}, $i = 1, \ldots, n$, where R_{ti} is given by Equations 11.4 and Model 11.5. Then, the ordinary least squares estimator of β is

$$\hat{\beta}_{OLS} = (X'X)^{-1}X'\bar{R},$$

which is unbiased with

$$Var(\hat{\beta}_{OLS}) = \frac{1}{n}(X'X)^{-1}X'V_\epsilon X(X'X)^{-1},$$

where V_ϵ is a $T \times T$ matrix whose element of the tth row and sth column is $\phi^{|t-s|}\sigma_t\sigma_s$. Let

$$\hat{\sigma}_t^2 = \frac{1}{n-1}\sum_{i=1}^{n}(R_{ti} - \bar{R}_t)^2,$$

and

$$\hat{\phi} = \frac{\sum_{i=1}^{T-1}\sum_{i=1}^{n} z_{ti} z_{(t+1)i}}{\sum_{i=1}^{T-1}\sum_{i=1}^{n} z_{ti}^2},$$

where

$$z_{ti} = R_{ti} - (1, x_t, \ldots, x_t^p)\hat{\beta}_{OLS}.$$

Then, approximately (when nT is large),

$$\frac{\hat{\beta}_{OLS} - \beta}{\left[\frac{1}{n}(X'X)^{-1}X'\hat{V}_\epsilon X(X'X)^{-1}\right]^{1/2}} \sim N(0, I_{p+1}),$$

where $\hat{V}_\epsilon$ is V_ϵ with σ_t replaced by $\hat{\sigma}_t$ and ϕ replaced by $\hat{\phi}$ and I_{p+1} is the identity matrix of order $p + 1$. Consequently, an approximate $(1 - \alpha) \times 100\%$ confidence interval for the mean relative dissolution rate at time x (which may be different from any of x_t, $t = 1, \ldots, T$) is $[L(x), U(x)]$ with

$$L(x) = \hat{\beta}'_{OLS}a(x) - z_{1-\alpha/2}\sqrt{n^{-1}a(x)'(X'X)^{-1}X'\hat{V}_\epsilon X(X'X)^{-1}a(x)},$$

and

$$U(x) = \hat{\beta}'_{OLS}a(x) + z_{1-\alpha/2}\sqrt{n^{-1}a(x)'(X'X)^{-1}X'\hat{V}_\epsilon X(X'X)^{-1}a(x)},$$

where $a(x) = (1, x_t, \ldots, x_t^p)'$ and $z_{1-\alpha/2}$ is the $(1 - \alpha/2)$th percentile of the standard normal distribution. Thus, two dissolution profiles are locally similar at time x if $[L(x), U(x)]$ is within (δ_L, δ_U).

To assess the global similarity of two dissolution profiles, we can consider the simultaneous confidence intervals $[\tilde{L}(x), \tilde{U}(x)]$, where $\tilde{L}(x)$ and $\tilde{U}(x)$ are the same as $L(x)$ and $U(x)$, respectively, with $z_{1-\alpha/2}$ replaced by

$$\sqrt{\chi^2_{1-\alpha/2, p+1}},$$

which is the square root of the $(1 - \alpha/2)$th percentile of the chi-sqaure distribution with $p - 1$ degrees of freedom. Thus, two dissolution profiles are globally similar if $[\tilde{L}(x), \tilde{U}(x)]$ is within (δ_L, δ_U) for all possible time values x. The ordinary least squares estimator $\hat{\beta}_{OLS}$ may be improved by the generalized least squares estimator

$$\hat{\beta}_{GLS} = (X'\hat{V}_\epsilon^{-1}X)^{-1}X'\hat{V}_\epsilon^{-1}\bar{R},$$

which has its asymptotic variance–covariance matrix (see, e.g., Fuller, 1996)

$$\frac{1}{n}(X'V_\epsilon^{-1}X)^{-1}.$$

Therefore, the previously described procedures for assessing similarity can be modified by replacing $\hat{\beta}_{OLS}$ and $(X'X)^{-1}X'\hat{V}_\epsilon X(X'X)^{-1}$ with $\hat{\beta}_{GLS}$ and $(X'V_\epsilon^{-1}X)^{-1}$, respectively. This method may result in shorter confidence intervals and, thus, a more efficient statistical method for assessing similarity.

11.3.2 Hypotheses Testing for Similarity Factors

Under Equations 11.2 and 11.3, it can be easily verified that the maximum of f_2 is 100. When $\mu_{it} = \mu_{2t}$ for all time points, two dissolution profiles are similar if f_2 is greater than some allowable lower limit, for example, θ_0. As a result, the hypothesis for evaluation of dissolution profile similarity based on f_2 is a one-sided hypothesis expressed as

$$H_0 : f_2 \leq \theta_0 \quad \text{or} \quad H_0 : f_2 < \theta_0. \tag{11.7}$$

Similarly, based on g_1, the dissolution profiles between the two drug products are concluded similarly if g_1 is smaller than an allowable upper limit, for example, δ_0.

The corresponding hypothesis can be formulated as

$$H_0 : g_1 \leq \delta_0 \quad \text{or} \quad H_0 : g_1 < \delta_0. \tag{11.8}$$

Under these hypotheses, statistical tests can be derived. As discussed earlier, f_2 and g_1 can be estimated, respectively, by

$$\hat{f}_2 = 100 - 25 \log_{10} \left(1 + \frac{D}{T} \right), \tag{11.9}$$

and

$$\hat{g}_1 = \frac{1}{T} \sum_{t=1}^{T} |\bar{y}_{1t} - \bar{y}_{2t}|, \tag{11.10}$$

where D, $\bar{y}_{1t}$, and $\bar{y}_{2t}$ are as defined before.

Let $x = y_1 - y_2$, where $y_h = \frac{1}{n} \sum_{i=1}^{n} y_{hi}$ and $h = 1, 2$. We assume that $y_{hi} = (y_{h1i}, y_{h2i}, \ldots, y_{hTi})'$ follows a multivariate normal distribution with mean vector $\mu_h = (\mu_{h1}, \mu_{h2}, \ldots, \mu_{hT})'$ and covariance matrix Σ_h for $h = 1, 2$. Then, x also follows an T-variate normal distribution with mean $\mu_d = \mu_1 - \mu_2$ and covariance matrix

$$\Sigma_d = \frac{1}{n} [\Sigma_1 + \Sigma_2].$$

The expected value of D is given by

$$E(D) = \mu_D^2 + \sigma_D^2,$$

where

$$\mu_D^2 = \mu_d' \mu_d = \sum_{t=1}^{T} \mu_{Dt}^2,$$

$$\sigma_D^2 = \frac{1}{T} \sum_{t=1}^{T} \sigma_{Dt}^2,$$

$$\mu_{Dt} = \mu_{1t} - \mu_{2t},$$

$$\sigma_{Dt}^2 = \sigma_{1t}^2 + \sigma_{2t}^2,$$

where σ_{ht}^2 is the tth diagonal element of Σ_h for $t = 1, \ldots, n; h = 1, 2$. The expected value of $\hat{f}_2$ can then be approximated by Taylor's expansion about $E(D)$ as follows

$$E(\hat{f}_2) \approx 100 - 25 \log_{10} \left[1 + \frac{E(D)}{T} \right]$$

$$= 100 - 25 \log_{10} \left[1 + \frac{W}{T} + \sigma_D^2 \right]$$

$$\leq f_2.$$

Because $\sigma_D^2 \geq 0$, $E(\hat{f}_2) \leq f_2$. As a result, Shah et al. (1998) indicated that $\hat{f}_2$ is conservative in evaluation of dissolution profile similarity. However, as $T \to \infty$, $\hat{f}_2$ is asymptotically unbiased for f_2. The expected value of $\hat{g}_1$ is given as follows (see, also Ma et al., 1999)

$$E(\hat{g}_1) = \frac{1}{T} \sum_{t=1}^{T} \{\mu_{Dt}[2\Phi(\sqrt{n}\mu_{Dt}/\sigma_{Dt}) - 1] + 2(\sigma_{Dt}/\sqrt{n})\phi(\sqrt{n}\mu_{Dt}/\sigma_{Dt})\},$$

and its bias is

$$E(\hat{g}_1) - g_1 = \frac{1}{T} \sum_{t=1}^{T} \{\mu_{Dt}[2\Phi(\sqrt{n}\mu_{Dt}/\sigma_{Dt}) - 1]$$
$$+ 2(\sigma_{Dt}/\sqrt{n})\phi(\sqrt{n}\mu_{Dt}/\sigma_{Dt}) - |\mu_{Dt}|\}, \quad (11.11)$$

where Φ and ϕ are the cumulative probability function and probability density function, respectively, of the standard normal random variable.

From Equation 11.11, it can be easily verified that $\hat{g}_1$ is also a consistent estimator of g_1. The bias of $\hat{g}_1$ goes to zero as the number of dosage units becomes large. Under the normality assumption, the sampling distribution function of $\hat{f}_2$ is complex, and its expected value does not have a closed form. In addition, the asymptotic distribution of $\hat{f}_2$ is also complicated by the fact that the Taylor series expansion of $\log_{10}(1 + D/T)$ converges only if $D/T < 1$. However, the absolute value is not continuous at zero. Both distribution and asymptotic distribution of $\hat{g}_1$ depend on not only the sign of μ_{Dt} but also on whether μ_{Dt} is zero. As a result, the distribution of $\hat{g}_1$ is complex.

Because of the complexity of the distributions and because $\hat{f}_2$ and $\hat{g}_1$ are model dependent, Ma et al. (2000) suggested relaxing the normality assumption and applying nonparametric bootstrap method (Efron, 1993) to evaluate the sampling distributions and for testing the hypotheses described in Equations 11.7 and 11.8. Ma et al. (2000) suggested the following bootstrap procedure.

- Step 1: Based on dosage unit, generate bootstrap samples

$$y_{ht}^* = (y_{ht1}^*, \ldots, y_{htT}^*)'$$

by sampling with replacement from T-variate vectors of the observed cumulative percent dissolved:

$$y_{ht} = (y_{ht1}, \ldots, y_{htT})'.$$

Sampling should be performed independently and separately for samples of dosages units from both drug products.

- Step 2: Calculate $\hat{f}_2^*$ and $\hat{g}_1^*$ based on the bootstrap sample according to Equations 11.9 and 11.10, respectively.

- Step 3: Repeat steps 1 and 2 many times (e.g., $B = 3,000$ times).

- Step 4: The $(1 - \alpha) \times 100\%$ lower confidence limit for f_2, L_f is the $\alpha \times 100\%$ quantile of the bootstrap values of $\hat{f}_2^*$. The $(1 - \alpha) \times 100\%$ lower confidence limit for g_1, U_g is the $(1 - \alpha) \times 100\%$ quantile of the bootstrap values of $\hat{g}_1^*$.

- Step 5: Based on criterion f_2, reject the null hypothesis in Equation 11.7 at the α level of significance and conclude that the dissolution profiles of the two drug products are similar if $L_f > \theta_0$. Based on criterion g_1, reject the null hypothesis in Equation 11.8 at the α level of significance and conclude that the dissolution profiles of the two drug products are similar if $U_g < \delta_0$.

This confidence limit approach for assessing similarity was also proposed by Shah et al. (1998) based on $\hat{f}_2$ using the bootstrap method. Bias and potential correction of bias and the correct hypotheses that $\hat{f}_2$ is testing for were also described and discussed by Shah et al. (1998).

11.4 Numerical Examples

We will now consider two examples, one from Tsong et al. (1996) and the other from Shah et al. (1998), to illustrate the use of the statistical methods under a time series model and under a hypothesis testing framework as described in the previous section, respectively.

11.4.1 Example 1: Time Series Model

Consider the dissolution data given by Tsong et al. (1996) and Chow and Ki (1997). For illustration, we consider dissolution data from the new lot as the dissolution data of the test product (y_{1ti}'s) and dissolution data from lot 1 as the dissolution data of the reference product (y_{2ti}'s). Based on the dissolution data, which are listed in Table 11.1, D and the f_2 similarity factor are 193.3 and 63.6, respectively. Since the f_2 similarity factor is between 50 and 100, the two dissolution profiles are considered to be similar according to the criterion of using the f_2 similarity factor.

Consider the ratio of the dissolution data R_{ti} under Model 11.5, Figure 11.2 shows that γ_t varies with t, and a quadratic relationship between γ_t and the time (hour) is revealed. Using the random effects approach of Chow and Ki (1997) and the data in Table 11.1, prediction bounds L_t and U_t and simultaneous prediction bounds $\tilde{L}_t$ and $\tilde{U}_t$ are summarized in Table 11.2. To apply the method introduced in the previous section, we choose a quadratic model for γ_t, that is, $p = 2$ in Equation 11.3. The corresponding confidence bounds $L(x_t)$ and $U(x_t)$ and the simultaneous confidence bounds $\tilde{L}(x_t)$ and $\tilde{U}(x_t)$ are also given in Table 11.2. These confidence and prediction bounds are shown in Figure 11.3 for comparison. Suppose the desired mean dissolution rate is $Q = 75\%$ and $\delta = 5\%$. Then, the equivalence limits for similarity are $(\delta_L, \delta_U) = (87.5, 114.3\%)$, which are plotted in Figure 11.3. Under these equivalence limits, it can be seen from Table 11.2 or Figure 11.3 that the two

TABLE 11.1: Dissolution Data (Percent Label Claim)

Product	Location	Times (hours)						
		1	2	3	4	6	8	10
Test	1	34	45	61	66	75	85	91
	2	36	51	62	67	83	85	93
	3	37	48	60	69	76	84	91
	4	35	51	63	61	79	82	88
	5	36	49	62	68	79	81	89
	6	37	52	65	73	82	93	95
	7	39	51	61	69	77	85	93
	8	38	49	63	66	79	84	90
	9	35	51	61	67	80	88	96
	10	37	49	61	68	79	91	91
	11	37	51	63	71	83	89	94
	12	37	54	64	70	80	90	93
Reference	1	50	56	68	73	80	86	87
	2	43	48	65	71	77	85	92
	3	44	54	63	67	74	81	82
	4	48	56	64	70	81	84	93
	5	45	56	63	69	76	81	83
	6	46	57	64	67	76	79	85
	7	42	56	62	67	73	81	88
	8	44	54	60	65	72	77	83
	9	38	46	54	58	66	70	76
	10	46	55	63	65	73	80	85
	11	47	55	62	67	76	81	85
	12	48	55	62	66	73	78	85

Source: Chow, S.C. and Shao, J. (2002b). *Journal of Biopharmaceutical Statistics*, 12, 311–321.

TABLE 11.2: Analysis of Similarity

Time (hours)	x_t	1	2	3	4	6	8	10
Mean (%)	$\bar{R}_t$	0.81	0.93	1.00	1.02	1.06	1.08	1.08
Lower bound (%)	L_t	78.2	89.3	95.6	97.3	101.9	103.5	103.6
	$\tilde{L}_t$	76.7	87.8	94.1	95.8	100.4	102.0	102.1
	$L_t(x_t)$	80.0	87.9	93.9	98.4	104.2	105.9	102.4
	$\tilde{L}_t(x_t)$	77.7	86.1	92.2	96.6	102.1	103.8	99.9
Upper bound (%)	U_t	86.5	97.7	103.9	105.7	110.2	111.8	111.9
	$\tilde{U}_t$	88.0	99.2	105.4	107.2	111.7	113.3	113.4
	$U(x_t)$	88.3	94.4	100.1	105.1	111.6	113.3	111.3
	$\tilde{U}(x_t)$	90.7	96.2	101.8	106.9	113.7	115.4	113.8

Source: Chow, S.C. and Shao, J. (2002b). *Journal of Biopharmaceutical Statistics*, 12, 311–321.

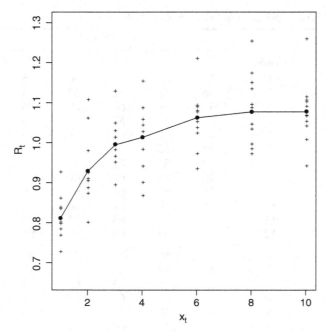

Figure 11.3: Dissolution data and their average over locations. Source: Chow and Shao (2002b).

dissolution profiles are not globally similar, although they are locally similar when $x_t \geq 2$. The simultaneous bounds obtained using the method proposed by Chow and Shao (2002b) and the random effect approach are similar. However, the bounds obtained using Chow and Shao (2002b)'s method apply to all time points between 1 and 10, whereas the bounds obtained using the random effects approach only apply to time points 1 to 4, 6, 8, and 10.

11.4.2   Example 2: Hypotheses Testing

Consider the dissolution data listed in Table 11.3 taken from Shah et al. (1998). For illustration, we consider the cumulative percent dissolved of a prechange batch and the postchange batch 4. The coefficients of variation at 30, 60, 90, and 180 minutes are 6.76%, 4.27%, 3.76%, and 2.87%, respectively, for the prechange batch and are 15.24%, 4.77%, 3.76%, and 2.87%, respectively, for the postchange batch. The mean cumulative percents dissolved are the same for both postchange batch 4 and the prechange batch at all sampling time points except for 30 minutes, where a mean difference of 19.84% is observed. To have a stable and precise confidence limit, the bootstrap procedure is carried out with 10,000 bootstrap samples. Table 11.4 presents the results of the samples. The results indicate that the bootstrap sampling distributions of $\hat{f}_2$ and $\hat{g}_1$ are skewed to the right (i.e., their medians are smaller than the means). However, the bootstrap distribution of g_1 is more skewed than that

TABLE 11.3: Cumulative Percent Dissolved of Prechange and Five Postchange Batches

Time in Minutes	Prechange Batch				Postchange Batch 1				Postchange Batch 2			
	30	60	90	180	30	60	90	180	30	60	90	180
Tablets												
1	36.1	56.6	80.0	93.3	38.75	61.79	85.14	100.20	48.0	60.0	84.0	103.0
2	33.0	59.5	80.8	95.7	36.16	61.21	84.25	97.30	52.0	75.0	89.0	99.0
3	35.7	62.3	83.0	97.1	38.49	63.89	84.94	96.39	48.0	60.0	83.0	101.0
4	32.1	62.3	81.3	92.8	37.27	62.52	85.65	95.47	53.0	70.0	93.0	103.0
5	36.1	53.6	72.6	88.8	48.12	77.18	95.32	99.30	45.0	60.0	84.0	105.0
6	34.1	63.2	83.0	97.4	48.45	80.62	95.05	98.94	48.0	66.0	90.0	103.0
7	32.4	61.3	80.0	96.8	41.08	67.62	84.94	99.03	51.0	71.0	91.0	100.0
8	39.6	61.8	80.4	98.6	39.64	63.68	80.73	95.63	49.0	63.0	89.0	104.0
9	34.5	58.0	76.9	93.3	36.06	61.59	82.22	96.12	44.0	60.0	84.0	103.0
10	38.0	59.2	79.3	94.0	36.69	63.60	84.50	98.42	53.0	68.0	81.0	104.0
11	32.2	56.2	77.2	96.3	39.95	67.98	87.40	98.10	49.0	63.0	86.0	105.0
12	35.2	58.0	76.7	96.8	43.41	74.07	93.95	97.80	52.0	68.0	87.0	104.0
Mean	34.9	59.5	79.3	95.1	40.30	67.10	87.00	97.70	49.3	65.3	86.8	102.8
S.D.	2.4	2.8	3.0	2.7	4.30	6.60	5.00	1.60	3.0	5.1	3.7	1.9

(Continued)

TABLE 11.3: Cumulative Percent Dissolved of Prechange and Five Postchange Batches (Continued)

Time in Minutes	Postchange Batch 3				Postchange Batch 4				Postchange Batch 5			
	30	60	90	180	30	60	90	180	30	60	90	180
Tablets												
1	28.7	48.2	63.8	85.6	17.1	58.6	80.0	93.3	41.5	78.0	86.4	98.3
2	26.4	53.1	68.3	90.6	16.0	59.5	80.8	95.7	43.7	78.3	85.9	96.4
3	25.4	52.4	70.0	89.5	12.7	62.3	83.0	97.1	46.3	78.3	86.9	96.4
4	23.2	49.5	65.5	92.2	15.1	62.3	81.3	92.8	44.0	79.9	88.6	96.0
5	25.1	50.7	68.0	87.6	14.1	53.6	72.6	88.8	42.6	73.2	81.4	95.5
6	28.7	54.1	70.8	93.6	12.1	63.2	83.0	97.4	44.4	78.4	86.2	98.4
7	23.5	50.3	66.1	85.1	14.4	61.3	80.0	96.8	43.0	79.0	87.5	99.5
8	26.2	50.6	67.7	88.0	19.6	61.8	80.4	98.6	44.4	79.6	87.3	99.9
9	25.0	49.1	63.6	85.8	14.5	58.0	76.9	93.3	44.8	78.7	86.9	97.8
10	24.9	49.5	66.7	86.6	14.0	59.2	79.3	94.0	41.7	76.9	84.5	100.0
11	30.4	53.9	70.4	89.9	18.2	56.2	77.2	96.3	42.3	77.0	81.9	97.9
12	22.0	46.3	63.0	88.7	13.2	58.0	76.7	96.8	42.0	78.2	92.4	100.3
Mean	25.8	50.6	67.0	88.6	15.1	59.5	79.3	95.1	43.4	78.0	86.3	98.6
S.D.	2.5	2.4	2.7	2.7	2.3	2.8	3.0	2.7	1.5	1.7	2.9	2.1

Source: Ma, M.C., Wang, B.B.C., Liu, J.P., and Tsong, Y. (2000). *Journal of Biopharmaceutical Statistics*, 10, 229–249.

TABLE 11.4: Results of Bootstrap for Comparison of Dissolution Profiles Between the Prechange Batch and Postchange Batch 4

Statistics	$\hat{f}_2$	$\hat{g}_1$
Observed value	50.07	4.96
Bootstrap mean	49.99	5.63
Bootstrap median	49.97	5.55
5% quantile	48.39	4.97
95% quantile	51.64	6.53
p-value for normality	0.011	0.001

Number of bootstrap samples = 10,000
Source: Ma, M.C., Wang, B.B.C., Liu, J.P., and Tsong, Y. (2000).
Journal of Biopharmaceutical statistics, 10, 229–249.

of $\hat{f}_2$. The corresponding p-values for normality are 0.011 and 0.0001 for $\hat{f}_2$ and $\hat{g}_1$, respectively. This implies that asymptotic normal approximation to the sampling distribution for both statistics computed from this data set is not adequate.

The observed estimates of f_2 and g_1 from the samples are 50.07 and 4.95, respectively, with the means of 49.99 and 5.63, respectively, from their bootstrap distributions. The 95% lower confidence limit for $\hat{f}_2$ can be estimated by the 5% quantile of the bootstrap sampling distribution, which is 48.39. Because 48.39 is smaller than the allowable lower limit of 50, based on $\hat{f}_2$ and at the 5% significance level the dissolution profile of postchange batch 4 is not similar to that of the prechange batch. The 95% quantile of the bootstrap sampling distribution of $\hat{g}_1$ is 6.53%. Based on $\hat{g}_1$ and at the 5% level of significance, the dissolution profiles of postchange batch 4 and the prechange batch would be claimed similar since the 95% upper confidence limit of 6.53% is smaller than the allowable upper limit of 10%. Table 11.5 summarizes

TABLE 11.5: Results of Comparisons of Dissolution Profiles Based on Data in Table 11.3[a]

Comparison[b]	Statistics	Observed Value	Bootstrap Mean	Quantile 5%	Quantile 95%	Similarity
T_1 vs. R	$\hat{f}_2$	60.035	60.425	53.219	—	Yes
	$\hat{g}_1$	5.825	5.854	—	8.001	Yes
T_2 vs. R	$\hat{f}_2$	51.082	51.022	48.275	—	No
	$\hat{g}_1$	8.850	8.877	—	10.296	No
T_3 vs. R	$\hat{f}_2$	51.184	51.197	48.342	—	No
	$\hat{g}_1$	4.960	5.625	—	6.533	Yes
T_5 vs. R	$\hat{f}_2$	48.052	48.022	46.022	—	No
	$\hat{g}_1$	9.375	9.381	—	10.513	No

[a] The number of bootstrap samples = 10,000
[b] T: postchange batch, and R: prechange batch
Source: Ma, M.C., Wang, B.B.C., Liu, J.P., and Tsong, Y. (2000). *Journal of Biopharmaceutical Statistics*, 10, 229–249.

the individual results for comparisons of the dissolution profiles of all postchange batches in Table 11.3 versus the prechange batch. From Table 11.5, for this data set, the bootstrap means are close to their observed values. In general, for f_2, the bootstrap mean is smaller than the observed values, while the observed value of g_1 is smaller than the bootstrap means. At the 5% level of significance and an allowable lower limit of 50, for f_2. Only the dissolution profiles of postchange batch 1 are concluded to be similar to that of the prechange batch. For g_1, the dissolution profiles of postchange batches 1 and 4 are both declared to be similar to that of the prechange batch at the 5% level of significance and an allowable upper limit of 10%. This example demonstrates the sensitivity of $\hat{f}_2$ with a large difference at a single time point and the liberalism of $\hat{g}_1$ in concluding similarity in such cases.

11.5   Concluding Remarks

In comparison to f_2, g_1 has a direct interpretation on the untransformed data. As shown in Ma et al. (1999), g_1 is independent of the number of sampling time points. If one directly tests the hypotheses Equations 11.7 and 11.8 by the observed values of f_2 or g_1, respectively, the type I error rate could be large when the within-time-point variation is large. For example, at the 5% level of significance and with $\delta = 10$, the empirical type I error rate from using the observed value of f_2 as the test statistic is at least 44%. Therefore, Shah et al. (1998) suggested the employment of the confidence limit approach via the bootstrap method. The confidence interval approach is more important when the observed value is less than 60. From the simulation, the bootstrap confidence interval procedure based on g_1 slightly inflates the size.

As discussed in the previous sections, $\hat{f}_2$ underestimates f_2. The magnitude of the underestimate bias is reduced when the difference between the two profiles increases (i.e., δ increases). For example, for 12 tablets, it is reduced to less than -1.5 when $\delta \geq 6$. This property is helpful for testing Equation 11.7 with $\theta_0 = 50$. Shah et al. (1998) have proposed a bias correction estimated by

$$D' = D - \frac{1}{n(n-1)} \sum_{t=1}^{T} \sum_{i=1}^{n} (y_{hti} - \bar{y}_{ht})^2$$

and f_2 is estimated by

$$\hat{f}'_2 = 100 - 25 \log_{10} \left(1 + \frac{D'}{T} \right).$$

The distribution of $\hat{f}'_2$ is even more complicated than that of $\hat{f}_2$. Ma et al. (2000) conducted a simulation under the same conditions and combinations as described in the previous sections. The results indicated that when $\delta = 0$, $\hat{f}'_2$ still underestimates f_2, but the magnitude of underestimation reduces to within 9. However, when $\delta > 0$, $\hat{f}'_2$ overestimates f_2 by a range of 0.33 to 9.5. As a result, the magnitude of overestimation

TABLE 11.6: Empirical Proportion of Negative Bias Correction Factor[a]

Number of Sampling Time Points	Dissolution Difference at Each Time Point		
	$\delta = 0$	$\delta = 3$	$\delta = 6$
3	63.6	20.8	2.5
4	63.3	19.4	2.1
5	63.0	18.4	2.0
6	62.9	17.9	1.9
7	63.1	17.4	1.8

[a]Based on 10,000 simulated samples

Source: Shah et al. (1998). *Pharmaceutical Research*, 15, 889–896.

by $\hat{f}'_2$ (of f_2) is much larger than $\hat{f}_2$. The reason is that D' can be negative. When D' is negative, D' is set to be 0 and $\hat{f}'_2$ is 100. Table 11.6 provides proportions of D' smaller than 0 from the simulation. For $\delta = 0$, the proportion of D' less than zero is about 63%. Even when $\delta = 6$, about 2% of D' can be negative. Hence, until the issue of handling a negative D' for estimation of f_2 is resolved, $\hat{f}'_2$ should be used cautiously.

If the coefficient of variation (CV) is 10%, even when $\delta = 0$, insufficient power is provided by both procedures if the number of dosage units is small. Power increases drastically if the number of dosage units increases from 6 to 12. Therefore, we recommend that the number of dosage units be at least 12 for comparison of dissolution profiles. As demonstrated in the simulation, when the number of dosage units is 12, the proposed confidence limit approach based on $\hat{f}_2$ not only can control the type I error rate under 5%, but also provides adequate power when the two dissolution profiles are similar. Although the confidence limit approach based on $\hat{f}_2$ is a bit conservative, it is an adequate method for assessing similarity of dissolution profiles between the changed and unchanged drug products. In comparison to g_1, f_2 is more sensitive to large differences at a few time points.

Chapter 12

Current Issues and Recent Developments

As indicated in the previous chapters, the shelf-life of a drug product is estimated based on stability data collected from long-term stability studies. These long-term stability studies are usually conducted under certain assumptions about controlled design factors such as storage conditions (e.g., temperature and humidity) and package types. These assumptions include, but are not limited to: (a) the drug product is expected to degrade linearly over time, (b) the drug is stored under ambient storage conditions such as 25°C and 60% relative humidity, and (c) the components and composition remain unchanged. Once these assumptions are met, an optimal design is selected to achieve the best precision for estimating the drug shelf-life. In practice, these assumptions may not be met, especially when: (a) there are postapproval changes in components and composition, site, batch size (scale-up or scale-down), or manufacturing equipment and processes, and (b) the drug product is shipped to foreign markets or different regions. In this case regulatory agencies may require additional stability data to be collected to support the changes.

In the next section regulatory requirements for stability testing for scale-up and postapproval changes are discussed. Also included in this section is the impact of scale-up and postapproval changes on dissolution testing. Section 12.2 introduces the approach for classification of countries according to climatic zones I, II, III, and IV (Dietz et al., 1993; Grimm, 1998). Recent development on proposed optimality criteria for choosing a stability design is discussed in Section 12.3. Section 12.4 discusses issues of stability analysis. These issues include the use of overage, assay sensitivity, assessment of multiple drug characteristics, batch similarity, deviation from linearity, and sample size. Section 12.5 includes software for stability analysis (i.e., the SAS/PC program system STAB) developed by the FDA.

12.1 Scale-Up and Postapproval Changes

Between 1995 and 1997, the FDA published a number of guidances regarding scale-up and postapproval changes. We will refer to these guidances as SUPAC guidances. SUPAC guidances include guidances for scale-up and postapproval changes for immediate-release solid oral dosage forms (SUPAC-IR, 1995), modified-release solid oral dosage forms (SUPAC-MR, 1997), and nonsterile semisolid dosage forms (SUPAC-SS, 1997). SUPAC initiatives are aimed at lowering the regulatory burden in the industry by providing informal and nonbinding communication that currently represents the best scientific judgement of the FDA. SUPAC guidances provide recommendations to pharmaceutical sponsors of new drug applications, abbreviated new drug applications, and abbreviated antibiotic drug applications who intend to make postapproval changes to: (a) the components and composition of the drug, (b) the manufacturing process or equipment, (c) the batch size, or (d) the site of manufacturing. For each of these four categories, the SUPAC guidances define what constitutes the difference between levels 1, 2, and 3 changes, which testing methods should be employed to support each level of changes. and which type of filing documentation will be required in accordance with the levels of change. The levels of change are generally categorized according to the probable impact on the drug due to the postapproval changes. In other words, a level 1 change will be a minor change, a level 2 change will be a moderate change, and a level 3 change will be a major or significant change.

In this section we will discuss the impact of SUPAC guidances on stability testing and dissolution testing with respect to postapproval changes. Owing to the similarity of these guidances, for illustration, we will focus on SUPAC-IR as an example. Impacts on stability testing with respect to postapproval changes in; (a) the components and composition of the drug, (b) the manufacturing process or equipment, (c) the batch size, and (d) the site of manufacturing are summarized below.

12.1.1 Components and Composition

For changes in components and composition, the SUPAC guidances define level 1 changes as changes that are unlikely to have detectable impact on formulation quality and performance. As a result, one batch on long-term stability testing is required, and the stability data should be reported in an annual report. The SUPAC guidances define changes that could have a significant impact on formulation quality and performance as level 2 changes. Thus, SUPAC guidances require that one batch with 3 months' accelerated stability testing and one batch on long-term stability testing be conducted. Level 3 changes are changes that are likely to have a significant impact on formulation quality and performance. Hence, SUPAC guidances recommend that one batch with 3 months' accelerated stability data be reported in a supplement and one batch on long-term stability data reported in an annual report if a significant body of information is available. A significant body of information on the stability of the drug product is likely to exist after five years of commercial experience for new molecular entities or three years of commercial experience for new dosage forms. If a significant body of information is not available, it is suggested that up to three batches with 3 months' accelerated stability data be reported in a supplement and one batch on long-term stability data be reported in an annual report.

12.1.2 Manufacturing Site

Site changes consist of changes in location of the site of manufacture for both company-owned and contract manufacturing facilities. The SUPAC guidances indicate that level 1 changes consist of site changes within a single facility, where the same equipment, standard operating procedures (SOPs), environmental conditions such as temperature and humidity and controls, and personnel common to both manufacturing sites are used and where no changes are made to manufacturing batch records, except for administrative information and the location of the facility. Level 2 changes in manufacturing site consist of site changes within a contiguous campus or between facilities in adjacent city blocks, where the sample equipment, SOPs, environmental conditions and controls, and personnel common to both manufacturing sites are used and where no changes are made to the manufacturing batch records, except for administrative information and the location of the facility. No stability testing is required for level 1 and level 2 changes in manufacturing site.

Level 3 changes consist of changes in manufacturing site to a different campus. A different campus is defined as one that is not on the same original contiguous site or where the facilities are not in adjacent city blocks. To qualify as a level 3 change, the same equipment, SOPs, environmental conditions, and controls should be used in the manufacturing process at the new site, and no changes may be made to the manufacturing batch records except for administrative information, location, and language translation, where needed. For level 3 changes in manufacturing site, the SUPAC guidances recommend one batch with 3 months' accelerated stability data reported in a supplement and one batch on long-term stability data reported in an annual report if a significant body of information is available. If a significant body of information is not available, up to three batches with 3 months' accelerated stability data reported in a supplement and up to three batches on long-term stability data reported in an annual report are required to support the changes.

12.1.3   Batch Size

Postapproval change in the size of a batch from the pivotal or pilot biobatch material to a larger or a smaller production batch is referred to as scale-up or scale-down. As indicated in the SUPAC guidances, submission of additional information such as validation of scale-up change is required.

The SUPAC guidances define a level 1 change as a change in batch size up to and including a factor of 10 times the size of the pilot or biobatch, where: (a) the equipment used to produce the test batch(es) is of the same design and operating principles, (b) the batch(es) is (are) manufactured in full compliance with current good manufacturing practices, and (c) the same SOPs and controls, as well as the same formulation and manufacturing procedures, are used on the test batch(es) and on the full-scale production batch(es). The SUPAC guidances recommend that one batch on long-term stability testing be reported in an annual report to support the level 1 change in batch size. Level 2 changes in batch size are defined similarly with changes beyond a factor of 10 times the size of the pilot or biobatch. For a level 2 change, one batch with 3 months' accelerated stability testing and one batch on long-term stability testing are required to justify the change.

12.1.4   Manufacturing Equipment

Level 1 changes in manufacturing equipment consist of two categories. The first category applies to changes from nonautomated or nonmechanical equipment to automated or mechanical equipment to move ingredients, while the second category is changes to alternative equipment of the same design and operating principles of the same or of a different capacity. The SUPAC guidances suggest that one batch on long-term stability testing be conducted to support the change. A level 2 change in manufacturing equipment is a change in equipment to a different design and different operating principles. For level 2 changes in manufacturing equipment, SUPAC guidances recommend that one batch with 3 months' accelerated stability data be reported in a supplement and one batch on long-term stability data be reported in an annual report if a significant body of information is available. If a significant body of information is not available, up to three batches with 3 months' accelerated stability data reported in a supplement and up to three batches on long-term stability data reported in an annual report are required to support the changes.

12.1.5   Manufacturing Process

The SUPAC guidances define a level 1 change in manufacturing process as changes in mixing times and operating speeds *within* application or validation ranges of a manufacturing process. A level 2 change in manufacturing process is a change in mixing times and operating speeds *outside* application or validation ranges of a manufacturing process. For a level 1 change, no additional stability testing is required. For a level 2 change, one batch on long-term stability testing is recommended. Level 3 changes include changes in the type of process used in the manufacture of the product such as a change from wet granulation to direct compression of dry powder. For level 3 changes, the SUPAC guidances recommend one batch with 3 months' accelerated stability data reported in a supplement and one batch on long-term stability data reported in an annual report if a significant body of information is available. If a significant body of information is not available, up to three batches with 3 months' accelerated stability data reported in a supplement and up to three batches on long-term stability data reported in an annual report are required to support the changes.

12.1.6   Remarks

Table 12.1 provides a summary of regulatory requirements of stability testing with respect to different levels of changes in components and composition, manufacturing site, batch size (scale-up or scale-down), and manufacturing equipment and process.

For dissolution testing for postapproval changes, the SUPAC guidances classify *in vitro* dissolution testing into the following three cases:

- Case A: Dissolution of Q $= 85\%$ in 15 minutes in 900 mL of 0.1N hydrochloride (HCl), using the United States Pharmacopoeia (USP) $<711>$ Apparatus 1 at 100 revolutions per minute (rpm) or Apparatus 2 at 50 rpm.

- Case B: Multipoint dissolution profile in the application or compendial medium at 15, 30, 45, 60, and 120 minutes or unit that an asymptote is reached for the proposed and currently accepted formulation.

TABLE 12.1: Stability Testing Requirements for Scale-Up and Postapproval Changes

Post-Approval Change	Level 1	Level 2	Level 3 Information Available	Level 3 Information Not Available
Components and composition	a	a, b	a, b	c, d
Site	NA	NA	a, b	c, d
Batch Size	a	a, b	NA	NA
Equipment*	a	(a, b) or (c, d)	NA	NA
Process	NA	a	a, b	c, d

a: One batch on long-term stability data reported in annual report
b: One batch with 3 months' accelerated stability data in supplement
c: Up to three batches on long-term stability data reported in annual report
d: Up to three batches with three months' accelerated data reported in supplement
*: (a, b) If a significant body information is available; (c,d) if a significant body information is not available

- Case C: Multipoint dissolution profiles performed in water, 0.1N HCl, and USP buffer media at pH 4.5, 6.5, and 7.5 (five separate profiles) for the proposed and currently accepted formulations. Adequate sampling should be performed at 15, 30, 45, 60, and 120 minutes until either 90% of the drug from the drug product is dissolved or an asymptote is reached. A surfactant may be used with appropriate justification.

Table 12.2 provides a summary of regulatory requirements of dissolution testing with respect to different levels of changes in components and composition, manufacturing site, batch size (scale-up or scale-down), and manufacturing equipment and process.

TABLE 12.2: Dissolution Testing Requirements for Scale-Up and Postapproval Changes[a]

Postapproval Change	Level 1	Level 2	Level 3
Components and Composition	NA	A*, B*, C*	B
Site	NA	NA	B
Batch Size	NA	B	NA
Equipment	NA	C	NA
Process	NA	B	B

[a]Cases A, B, and C are as defined in Section 12.1.
A*: Case A dissolution testing for high permeability/high solubility products
B*: Case B dissolution testing for low permeability/high solubility products
C*: Case C dissolution testing for high permeability/low solubility products

12.2   Storage Conditions in Different Climatic Zones

Haynes (1971) pointed out that changes in the field storage temperature could cause the reaction rate constant of some drug products to change according to the Arrhenius relationship. Since drug products stored in pharmacies and warehouses for extended periods of time are exposed to a range of temperatures, the exact determination of drug shelf-life becomes almost impossible (Kommanaboyina and Rhodes, 1999). Grimm (1985, 1986) indicated that if the test batches are stored under incorrect conditions, the results from the tests are also incorrect, and false correlations could be drawn. The degradation curve of a drug product may not be consistent at different times under different environmental conditions. This will definitely have an impact on stability testing of the drug product. Some pharmaceutical companies consider cyclic testing, in which temperature and humidity are alternately increased and decreased in a pattern. Haynes (1971) established the mean kinetic temperature for a defined period. The mean kinetic temperature is a single derived temperature that affords the same thermal challenge to a drug substance or drug product as would be experienced over a range of both higher and lower temperatures for an equivalent defined period. Based on the mean kinetic temperature, the world is divided into four zones that are distinguished by their characteristic prevalent annual climatic conditions (Grimm, 1985, 1986). At different climatic zones, slightly different requirements for storage conditions in accelerated testing and long-term stability testing are imposed.

12.2.1   Mean Kinetic Temperature

Haynes (1971) suggested obtaining a single equivalent temperature by substituting the average rate constants over a defined time period (e.g., a week, a month, or a year) into the Arrhenius equation. Haynes referred to this single equivalent temperature as virtual temperature. As indicated in Chapter 2, the loss rate constant in any time according to the Arrhenius relationship is given by

$$k_i = Ae^{-\Delta H/RT_i}, \quad i = 1, \ldots, n, \tag{12.1}$$

where A is a constant for a given reaction, ΔH is heat of activation, R is the universal gas constant, and T_i is the absolute temperature in the ith time. It can be verified that the average of the k_i's is proportional to the average of $e^{-\Delta H/RT_i}$, $i = 1, \ldots, n$. The T_i values can be obtained easily for most locations and, with a suitable value of ΔH, used to calculate the average of the $e^{-\Delta H/RT_i}$. This numerical value is defined to be equal to $e^{-\Delta H/RT_K}$, where T_K is the virtual temperature. The solution of this for T_K in absolute degrees is given by (Haynes, 1971)

$$T_K = \left[\frac{\Delta H/R}{-\ln\left(\dfrac{e^{x_1} + e^{x_2} + \cdots + e^{x_n}}{n}\right)} \right], \tag{12.2}$$

where T_K is the mean kinetic temperature and

$$x_i = -\Delta H/RT_i.$$

With Equation 12.2, one can determine, from the individual temperatures for a day, a month, or even a year, the respective mean kinetic temperature that corresponds to the actual (thermal) stress at different or varying temperatures.

As indicated by Grimm (1985, 1986), a climate is characterized by the temperature and the partial pressure of water vapor (humidity). Both temperature and humidity decisively influence stability. The connection between the vapor pressure of water and the absolute temperature can be expressed by

$$\ln P = -\frac{\Delta H_v}{R}\frac{1}{T} + \text{constant}, \tag{12.3}$$

where ΔH_v is the molar (latent) heat of evaporation. This can be determined from the slope of a plot of $\ln P$ against $1/T$. Simple algebra gives the following equation

$$\ln \frac{P_2}{P_1} = \frac{\Delta H_v(T_2 - T_1)}{R T_1 T_2}, \tag{12.4}$$

where P_1 and P_2 are the vapor pressures for T_1 and T_2. At a given temperature, the partial pressure of water vapor in moist air can become maximally equal to the vapor pressure of water at the same temperature, that is, the saturation pressure P_S. Unsaturated air contains water vapor whose partial pressure, P_D, is lower than the saturation pressure P_S at the given temperature. This gives the relative humidity:

$$\varphi = \frac{P_D}{P_S} \times 100\%. \tag{12.5}$$

Grimm (1985, 1986) suggested using Equations 12.2 and 12.4 to derive relevant storage conditions for the various climatic regions. By this means, an exact determination of drug shelf-life at different climatic regions is possible.

12.2.2 Classification of Climatic Zones

As indicated by Grimm (1985, 1986), the earth can be divided into four climatic zones to which individual countries can be assigned. The four climatic zones are characterized as follows:

Climatic Zone	Characteristics
I	Temperate climate
II	Mediterranean-like and subtropical climates
III	Hot dry climate, dry regions
IV	Hot, humid climate, tropics

Grimm (1985, 1986) noted that if storage conditions are to be derived for a world-wide stability test for the individual climatic zones, then it must be clearly stated what requirements they must fulfill. Grimm provided the following points to consider when deriving storage conditions for the individual climatic zones:

- The storage condition should deal with climatic influences that apply to the respective climatic zones in which the drug is to be subsequently used.

- The storage condition must not represent an unrealistic stress, because otherwise inappropriate expiration dates are derived.

- When determining the mean value for the temperature, Haynes's formula Equation 12.2 must be used to calculate the mean kinetic temperature.

- The relative humidity must be considered in addition to variations in temperature.

- The storage conditions for the individual climatic zones should be logically interconnected so that an overall assessment for all climatic zones can be made.

- The individual storage conditions for the climatic zones are to be clearly separate from those of stress studies and, where applicable, from those of the follow-up investigations.

- The determined storage conditions must be definitively specified and appropriately supervised.

Table 12.3 summarizes storage conditions in terms of temperature and relative humidity for the individual climatic zones as suggested by Grimm (1985, 1986). An approach for classification of countries according to climatic zones I, II, III, and IV can be found in the literature (Dietz et al., 1993; Grimm, 1998).

The ICH guidance Q1A (R2) *Stability Testing of New Drug Substances and Products*, which is referred to as the parent guidance, outlines the stability data package for a new drug substance or drug product that is considered sufficient for a registration application in territories in climatic zones III and IV (Grimm, 1985 and 1986; Schumacher, 1974). The parent guidance can be followed to generate stability data packages for registration applications in other countries or regions in zones I and II. For territories in climatic zones III and IV, the data package as described in the parent guidance can be considered applicable except for certain storage conditions (see, e.g., ICH Q1F (2004) *Stability Data Package for Registration in Climatic Zones III and IV*).

TABLE 12.3:   Storage Conditions for Individual Climatic Zones

Climatic Zones	Temperature	Relative Humidity
I	21°C ± 2°C	45% ± 5%
II	25°C ± 2°C	60% ± 5%
III	31°C ± 2°C	40% ± 5%
IV	31°C ± 2°C	70% ± 5%

Source: Grimm (1985, 1986). *Drugs Made in Germany*, 28, 196–202 & 29, 39–47.

TABLE 12.4: Storage Conditions for Climatic Zones I and II

Study	Temperature/Relative Humidity (RH)	Minimum Time Period Covered by Data at Submission
Long-term[a]	$\pm25°C \pm 2°C$ / $60\% \pm 5\%$ or $30°C \pm 2°C$ / $65\% \pm 5\%$	12 months
Intermediate[b]	$\pm30°C \pm 2°C$ / $65\% \pm 5\%\pm$	6 months
Accelerated	$\pm40°C \pm 2°C$ / $75\% \pm 5\%\pm$	6 months

[a]It is up to the applicant to decide whether long-term stability studies are performed at $25°C \pm2°C$/ 60% RH ± 5% RH or $30°C \pm 2°C$ /65% RH ± 5%RH.

[b]If $30°C \pm2°C$ /65% RH ± 5% RH is the long-term condition, there is no intermediate condtion.

Source: ICH Q1A (R2) *Stability Testing of New Drug Substances and Products.*

12.2.2.1 Climatic Zones I and II

The climate in zone I is characterized by cold winters with high relative humidity and warm summers with low relative humidity. Marked differences occur in both temperature and relative humidity between 7 a.m. and 2 p.m., whereas the pressure varies much less. Zone II is a subtropical and mediterranean-like climate, which includes a large number of countries with different climatic conditions. The ICH tripartite regions, the European Union (EU), Japan, and the United States, are all in climatic zones I and II.

As indicated in the ICH Q1A (R2) (2003) guideline for stability, a drug substance should be evaluated under storage conditions (with appropriate tolerances) that test its thermal stability and, if applicable, its sensitivity to moisture. The storage conditions and the lengths of studies chosen should be sufficient to cover storage, shipment, and subsequent use. Long-term, accelerated, and intermediate stability testing as suggested in the ICH Q1A (R2) guideline are summarized in Table 12.4. The ICH Q1A (R2) guideline indicates that if long-term studies are conducted at $25°C \pm 2°C$/60% ± 5% relative humidity and significant change occurs at any time during the 6 months of testing at the accelerated storage condition, additional testing at the intermediate storage condition should be conducted and evaluated against significant criteria. Testing at the intermediate storage condition should include all tests, unless otherwise justified. The initial application should include a minimum of 6 months' data from a 12-month study at the intermediate storage condition. Note that the ICH Q1A (R2) guideline defines significant change of a drug product as follows:

- A 5% change in assay from its initial value or failure to meet the acceptance criteria for potency when using a biological or immunological procedure.

- Any degradation product's exceeding its acceptance criterion.

- Failure to meet the acceptance criteria for appearance, physical attributes, and functionality (e.g., color, phase separation, resuspendibility, caking, hardness, dose delivery per actuation). However, some changes in physical attributes (e.g., softening of suppositories, melting of creams) may be expected under accelerated conditions.

TABLE 12.5: Storage Conditions for Drug Substances Intended for Storage in Refrigerators and Freezers

Refrigerator Study	Temperature/Relative Humidity (RH)	Minimum Time Period Covered by Data at Submission
Long-term	5°C ± 3°C	12 months
Accelerated	25°C ± 2°C/65% RH ± 5%RH	6 months
Freezer Long-term	−20°C ± 5°C	12 months

Source: ICH Q1A (R2) *Stability Testing of New Drug Substances and Products.*

- Failure to meet the acceptance criterion for pH.

- Failure to meet the acceptance criteria for dissolution for 12 dosage units.

For drug substances and products intended for storage in a refrigerator or a freezer, storage conditions suggested by the ICH Q1A (R2) guideline are given in Table 12.5. For drug substances and products intended for storage in a refrigerator, if significant change occurs between 3 to 6 months of testing at the accelerated storage condition, the proposed retest period should be based on the real-time data available at the long-term storage condition. However, if significant change occurs within the first 3 months of testing at the accelerated storage condition, a discussion should be provided to address the effect of short-term excursions outside the label storage condition, for example, during shipping or handling. This discussion can be supported, if appropriate, by further testing on a single batch of the drug substance for a period shorter than 3 months but with more frequent testing than usual. It is considered unnecessary to continue to test a drug substance through 6 months when a significant change has occurred within the first 3 months. For drug substances intended for storage in a freezer, the retest period should be based on the real-time data obtained at the long-term storage condition. In the absence of an accelerated storage condition for drug substances intended to be stored in a freezer, testing on a single batch at an elevated temperature (e.g., 5°C ± 3°C or 25°C ± 2°C) for an appropriate time period should be conducted to address the effect of short-term excursions outside the proposed label storage condition, for example, during shipping or handling. For drug substances intended for storage below −20°C, the ICH Q1A (R2) guideline suggests it should be treated on a case-by-case basis.

12.2.2.2 Climatic Zones III and IV

The climate in zone III is hot and dry. A typical example is Baghdad. The climate in zone IV is hot and humid, which is a typical tropical climate. Manila is a typical example of climatic zone IV. For the general case, the recommended long-term and accelerated storage conditions for climatic zones III and IV are given in Table 12.6,

TABLE 12.6: Storage Conditions for Climatic Zones III and IV

Study	Temperature/RH	Minimum Time Period Covered by Data at Submission
Long-term	$30°C \pm 2°C/\ 65\% \pm 5\%$	12 months
Accelerated	$40°C \pm 2°C/\ 75\% \pm 5\%$	6 months

Source: ICH Q1F *Stability Data Package for Registration Application in Climatic Zones III and IV.*

which shows that no intermediate storage condition for stability studies is recommended for climatic zones III and IV. Therefore, the intermediate storage condition is not relevant when the principles of retest period or shelf-life extrapolation described in the ICH Q1E (2004) guideline for *Evaluation of Stability Data* are applied.

12.3 Optimal Designs in Stability Studies

As indicated in Chapter 4, many criteria for selection of an appropriate design have been proposed in the literature. These criteria mainly focus on the power of detecting factor effect. However, as indicated by Ju and Chow (1995), the primary goal of stability studies, is to estimate the shelf-life. As a result, Ju and Chow proposed a criterion such that for a fixed sample size, the design with the best precision for shelf-life estimation is the best design. For a fixed desired precision of shelf-life estimation, the design with the smallest sample size is the best design. Hedayat et al. (2006) provided a mathematical expression for Ju and Chow's criterion as follows:

$$d_t^* = \left\{ d : \min_{d:n(d)=n} x'(t)(X'X)^- x(t) \right\}, \tag{12.6}$$

where X is the design matrix, $(X'X)^-$ is the generalized inverse of $X'X$, and $n(d)$ is the sample size for design d. As indicated by Hedayat et al. (2006), the optimal design d_t^* depends on the time point t because it minimizes the variance of the estimated shelf-life at time point t. In other words, the design d_t^*, which is optimal at time point t, may not be optimal at every time point. Thus, it is difficult to compare designs according to this optimality criterion. To overcome this problem, Hedayat et al. (2006) proposed the following criteria:

Criterion 1: Among designs with the same sample size, the min–max optimal design d_t^* at time t is given by

$$d_t^* = \left\{ d : \min_{d:n(d)=n} \max_{all-level-combination} x'(t)(X'X)^- x(t) \right\}. \tag{12.7}$$

The min–max optimal design d_t^* depends on time t. If one needs to compare designs throughout the whole test period, then criterion 2 should be used.

Criterion 2: Among designs with the same sample size, d^* is uniformly optimal if d^* is min–max optimal design at every time point t. The uniformly optimal design does not always exist. In such a case, to compare designs throughout the whole test period, criterion 3 is useful.

Criterion 3: Among designs with the same sample size, the min–max optimal design d^* is

$$d^* = \left\{ d : \min_{d:n(d)=n} \max_{t:t \in dT} \max_{all-level-combination} x'(t)(X'X)^- x(t) \right\}, \qquad (12.8)$$

where T is the time vector containing all time points during the test period. This criterion can be used for comparing designs with the same sample size. If designs with different sample sizes are compared, this criterion has to be adjusted to the same sample size. Hedayat et al. (2006) suggested criterion 4.

Criterion 4: The sample-size-adjusted uniformly optimal design d^* at time t is

$$d_t^* = \left\{ d : \min_{d} \max_{all-level-combination} x'(t)(X'X)^- x(t) \times n(d) \right\}. \qquad (12.9)$$

Criterion 5: d^* is sample-size-adjusted uniformly optimal design if d^* is the sample-size-adjusted min–max optimal design at every time point t.

Criterion 6: The sample-size-adjusted min–max–max optimal design d^* is

$$d^* = \left\{ d : \min_{d} \max_{t:t \in T} \max_{all-level-combination} x'(t)(X'X)^- x(t) \times n(d) \right\}. \qquad (12.10)$$

To illustrate the use of the above criteria, Hedayat et al. (2006) compared several designs for a 2-year long-term stability study. According to the FDA stability guidelines, for a 2-year long-term stability study, every selected level combination should be tested at 0, 12, and 24 months and at least at one additional time point within the first year. As a result, Hedayat et al. (2006) considered various designs with all possible time vectors (Table 12.7). These designs include: (a) a balanced design (design B), (b) a complete one-half design (design C1/2), (c) a complete one-third design (design C1/3), (d) a complete two-thirds design (design C2/3), (e) a fractional factorial design (design F), and (f) a uniform design (design U). For every type of design, B, C1/2, C1/3, C2/3, F, and U, Hedayat et al. (2006) suggested the following steps for searching optimal designs:

- Step 1: List all possible time sets.

- Step 2: List all possible time allocations of time vectors.

- Step 3: For every design, calculate the maximum value of $x'(t)(X'X)^- x(t)$ at time t over all level combinations.

TABLE 12.7: Time Vectors

Tu0 = {0, 3, 6, 9, 12, 18, 24}
Tu1 = {0, 3, 6, 9, 12, 24}
Tu2 = {0, 3, 6, 12, 18, 24}
Tu3 = {0, 3, 9, 12, 18, 24}
Tu4 = {0, 6, 9, 12, 18, 24}
Tu5 = {0, 3, 6, 12, 24}
Tu6 = {0, 3, 9, 12, 24}
Tu7 = {0, 3, 12, 18, 24}
Tu8 = {0, 6, 9, 12, 24}
Tu9 = {0, 6, 12, 18, 24}
Tu10 = {0, 9, 12, 18, 24}
Tu11 = {0, 3, 12, 24}
Tu12 = {0, 6, 12, 24}
Tu13 = {0, 9, 12, 24}

- Step 4: Among the designs with the same time set, compare the value of

$$\max_{all-level-combination} x'(t)(X'X)^- x(t)$$

at time t. Then, the design with the minimum value is the min–max optimal design at time t, and its corresponding allocation of time vectors is optimal at time t. If such a design is min–max optimal throughout the whole test period, its corresponding allocation of time vectors is uniformly optimal.

- Step 5: Among the same type of designs with the same sample size, compare the min–max optimal designs with different time sets at time t. The design with the minimum values is optimal, and its corresponding time vector is optimal.

Following the above steps, Hedayat et al. (2006) provided search results for optimal designs under the model with and without factor interactions. Table 12.8 and Table 12.9 provide min–max $x'(t)(X'X)^- x(t)$ under the model without and with factor interactions, respectively. The trace values from the maximum to the minimum among designs with the same sample sizes are given in Table 12.10. Table 12.11a and Table 12.11b provide min–max $x'(t)(X'X)^- x(t) \times n(d)$ under the model with and without factor interactions, respectively. Table 12.12 and Table 12.13 provide summaries of designs with the same sample size by sorting the values of

$$\max_{all-level-combination} x'(t)(X'X)^- x(t)$$

at each time point from the smallest to the largest under the model without and with factor interactions, respectively. While Table 12.14 and Table 12.15 provide summaries of designs with the same sample size by sorting the values of

$$\max_{all-level-combination} x'(t)(X'X)^- x(t) \times n(d)$$

at each time point from the smallest to the largest under the model without and with factor interactions, respectively.

TABLE 12.8: Min–Max $x'(t)(X'X)^{-}x(t)$ Under the Model Without Factor Interactions

Type	Design	# of Obs.	Time	Month 0	3	6	9	12	18	24	Max Month
Balanced	B01	54	Tu1, Tu2, Tu3	0.24	0.17	0.12	0.10	0.10	0.19	0.39	0.39
Balanced	B02	54	Tu1, Tu2, Tu4	0.26	0.18	0.13	0.10	0.10	0.19	0.39	0.39
Balanced	B03	54	Tu1, Tu3, Tu4	0.27	0.18	0.13	0.10	0.10	0.19	0.39	0.39
Balanced	B04	54	Tu2, Tu3, Tu4	0.27	0.19	0.13	0.10	0.09	0.16	0.33	0.33
Balanced	B05	45	Tu5, Tu6, Tu7	0.27	0.19	0.14	0.11	0.12	0.22	0.42	0.42
Balanced	B06	45	Tu5, Tu6, Tu8	0.27	0.19	0.14	0.11	0.12	0.23	0.45	0.45
Balanced	B07	45	Tu5, Tu6, Tu9	0.29	0.21	0.15	0.12	0.12	0.22	0.43	0.43
Balanced	B08	45	Tu5, Tu6, Tu10	0.31	0.22	0.16	0.12	0.12	0.22	0.42	0.42
Balanced	B09	45	Tu5, Tu7, Tu8	0.28	0.20	0.14	0.12	0.12	0.22	0.43	0.43
Balanced	B10	45	Tu5, Tu7, Tu9	0.30	0.21	0.15	0.12	0.12	0.20	0.40	0.40
Balanced	B11	45	Tu5, Tu7, Tu10	0.32	0.23	0.16	0.12	0.12	0.20	0.40	0.40
Balanced	B12	45	Tu5, Tu8, Tu9	0.30	0.21	0.15	0.12	0.12	0.22	0.43	0.43
Balanced	B13	45	Tu5, Tu8, Tu10	0.32	0.23	0.16	0.12	0.12	0.21	0.42	0.42
Balanced	B14	45	Tu5, Tu9, Tu10	0.33	0.23	0.17	0.12	0.12	0.20	0.40	0.40
Balanced	B15	45	Tu6, Tu7, Tu8	0.28	0.20	0.15	0.12	0.12	0.21	0.42	0.42
Balanced	B16	45	Tu6, Tu7, Tu9	0.31	0.22	0.16	0.12	0.11	0.20	0.39	0.39
Balanced	B17	45	Tu6, Tu7, Tu10	0.33	0.23	0.16	0.12	0.11	0.20	0.39	0.39
Balanced	B18	45	Tu6, Tu8, Tu9	0.31	0.22	0.15	0.12	0.12	0.21	0.42	0.42
Balanced	B19	45	Tu6, Tu8, Tu10	0.33	0.23	0.16	0.12	0.12	0.21	0.42	0.42
Balanced	B20	45	Tu6, Tu9, Tu10	0.34	0.24	0.17	0.13	0.11	0.19	0.39	0.39
Balanced	B21	45	Tu7, Tu8, Tu9	0.32	0.22	0.16	0.12	0.11	0.19	0.39	0.39
Balanced	B22	45	Tu7, Tu8, Tu10	0.34	0.24	0.17	0.12	0.11	0.19	0.39	0.39

(Continued)

TABLE 12.8: Min–Max $x'(t)(X'X)^{-}x(t)$ Under the Model Without Factor Interactions (Continued)

Type	Design	# of Obs.	Time	Month							Max Month
				0	3	6	9	12	18	24	
Balanced	B23	45	Tu7, Tu9, Tu10	0.35	0.25	0.17	0.13	0.11	0.17	0.33	0.35
Balanced	B24	45	Tu8, Tu9, Tu10	0.35	0.24	0.17	0.13	0.11	0.19	0.39	0.39
Balanced	B25	36	Tu11, Tu12, Tu13	0.35	0.25	0.18	0.14	0.14	0.24	0.46	0.46
Complete–one-half	C1/2_1	45	Tu6, Tu9	0.31	0.22	0.15	0.12	0.12	0.21	0.41	0.41
Complete–one-half	C1/2_2	45	Tu5, Tu10	0.33	0.23	0.16	0.12	0.12	0.21	0.42	0.42
Complete–one-half	C1/2_3	45	Tu7, Tu8	0.29	0.21	0.15	0.12	0.12	0.21	0.41	0.41
Complete–one-half	C1/2_4	45	Tu6, Tu9	0.31	0.22	0.16	0.12	0.12	0.20	0.40	0.40
Complete–one-half	C1/2_5	45	Tu5, Tu10	0.34	0.24	0.17	0.17	0.12	0.21	0.41	0.41
Complete–one-half	C1/2_6	45	Tu7, Tu8	0.29	0.21	0.15	0.12	0.11	0.20	0.40	0.40
Complete–one-third	C1/3_1	39	Tu7, Tu12,Tu13	0.35	0.25	0.18	0.14	0.14	0.22	0.43	0.43
Complete–one-third	C1/3_2	39	Tu9, Tu11,Tu13	0.35	0.25	0.18	0.14	0.14	0.22	0.43	0.43
Complete–one-third	C1/3_3	39	Tu10, Tu11,Tu12	0.35	0.25	0.18	0.14	0.14	0.22	0.43	0.43
Complete–two-third	C2/3_1	51	Tu2, Tu3, Tu8	0.27	0.19	0.13	0.11	0.11	0.19	0.39	0.39
Complete–two-third	C2/3_2	51	Tu2, Tu4, Tu6	0.27	0.19	0.13	0.11	0.11	0.19	0.39	0.39
Complete–two-third	C2/3_3	51	Tu3, Tu4, Tu5	0.27	0.19	0.13	0.10	0.11	0.20	0.39	0.39
Fractional	F1	42	Tu0, Tu0, Tu0	0.26	0.18	0.12	0.10	0.10	0.19	0.39	0.39
Fractional	F2	36	Tu1, Tu2, Tu3	0.30	0.21	0.15	0.12	0.12	0.24	0.48	0.48

(*Continued*)

TABLE 12.8: Min–Max $x'(t)(X'X)^{-}x(t)$ Under the Model Without Factor Interactions (Continued)

Type	Design	# of Obs.	Time	Month 0	3	6	9	12	18	24	Max Month
Fractional	F3	36	Tu1, Tu2, Tu4	0.32	0.22	0.15	0.12	0.12	0.24	0.48	0.48
Fractional	F4	36	Tu1, Tu3, Tu4	0.33	0.22	0.16	0.12	0.12	0.24	0.48	0.48
Fractional	F5	36	Tu2, Tu3, Tu4	0.33	0.23	0.16	0.12	0.11	0.20	0.39	0.39
Uniform	U0	63	Tu0	0.22	0.15	0.10	0.08	0.08	0.16	0.32	0.32
Uniform	U01	54	Tu1	0.22	0.15	0.11	0.09	0.11	0.22	0.44	0.44
Uniform	U02	54	Tu2	0.24	0.17	0.12	0.10	0.10	0.17	0.33	0.33
Uniform	U03	54	Tu3	0.26	0.18	0.13	0.10	0.09	0.16	0.32	0.32
Uniform	U04	54	Tu4	0.29	0.20	0.14	0.10	0.09	0.16	0.33	0.33
Uniform	U05	45	Tu5	0.24	0.17	0.13	0.11	0.13	0.24	0.46	0.46
Uniform	U06	45	Tu6	0.26	0.18	0.13	0.11	0.12	0.22	0.44	0.44
Uniform	U07	45	Tu7	0.29	0.21	0.15	0.12	0.11	0.17	0.33	0.33
Uniform	U08	45	Tu8	0.29	0.20	0.14	0.11	0.12	0.22	0.45	0.45
Uniform	U09	45	Tu9	0.33	0.24	0.17	0.13	0.11	0.17	0.33	0.33
Uniform	U10	45	Tu10	0.38	0.27	0.18	0.13	0.11	0.16	0.33	0.38
Uniform	U11	36	Tu11	0.29	0.21	0.16	0.14	0.15	0.25	0.46	0.46
Uniform	U12	36	Tu12	0.33	0.24	0.17	0.14	0.14	0.24	0.46	0.46
Uniform	U13	36	Tu13	0.38	0.27	0.19	0.15	0.14	0.22	0.45	0.45

Source: Headayat, A.S, Yan, X., and Lin, L. (2006). *Journal of Biopharmaceutical Statistics,* 16, 35–59.

TABLE 12.9: Min–Max $x'(t)(X'X)^- x(t)$ Under the Model with Factor Interactions

Type	Design	# of Obs.	Time	Month 0	3	6	9	12	18	24	Max Month
Balanced	B01	54	Tu1, Tu2, Tu3	0.46	0.32	0.23	0.18	0.19	0.39	0.79	0.79
Balanced	B02	54	Tu1, Tu2, Tu4	0.53	0.36	0.25	0.18	0.19	0.39	0.79	0.79
Balanced	B03	54	Tu1, Tu3, Tu4	0.53	0.36	0.25	0.18	0.19	0.39	0.79	0.79
Balanced	B04	54	Tu2, Tu3, Tu4	0.53	0.36	0.25	0.18	0.17	0.30	0.59	0.59
Balanced	B05	45	Tu5, Tu6, Tu7	0.52	0.38	0.27	0.21	0.23	0.43	0.83	0.83
Balanced	B06	45	Tu5, Tu6, Tu8	0.53	0.36	0.26	0.20	0.23	0.43	0.83	0.83
Balanced	B07	45	Tu5, Tu6, Tu9	0.60	0.43	0.30	0.23	0.23	0.43	0.83	0.83
Balanced	B08	45	Tu5, Tu6, Tu10	0.68	0.48	0.33	0.24	0.23	0.43	0.83	0.83
Balanced	B09	45	Tu5, Tu7, Tu8	0.53	0.38	0.27	0.21	0.23	0.43	0.83	0.83
Balanced	B10	45	Tu5, Tu7, Tu9	0.60	0.42	0.30	0.23	0.23	0.43	0.83	0.83
Balanced	B11	45	Tu5, Tu7, Tu10	0.68	0.48	0.33	0.24	0.23	0.43	0.83	0.83
Balanced	B12	45	Tu5, Tu8, Tu9	0.60	0.43	0.30	0.23	0.23	0.43	0.83	0.83
Balanced	B13	45	Tu5, Tu8, Tu10	0.68	0.48	0.33	0.24	0.23	0.43	0.83	0.83
Balanced	B14	45	Tu5, Tu9, Tu10	0.68	0.48	0.33	0.24	0.23	0.43	0.83	0.83
Balanced	B15	45	Tu6, Tu7, Tu8	0.53	0.38	0.27	0.21	0.22	0.40	0.80	0.80
Balanced	B16	45	Tu6, Tu7, Tu9	0.60	0.43	0.30	0.23	0.22	0.40	0.79	0.79
Balanced	B17	45	Tu6, Tu7, Tu10	0.68	0.48	0.33	0.24	0.22	0.40	0.79	0.79
Balanced	B18	45	Tu6, Tu8, Tu9	0.60	0.43	0.30	0.23	0.22	0.40	0.80	0.80
Balanced	B19	45	Tu6, Tu8, Tu10	0.68	0.48	0.33	0.24	0.22	0.40	0.80	0.80
Balanced	B20	45	Tu6, Tu9, Tu10	0.68	0.48	0.33	0.24	0.22	0.40	0.79	0.79
Balanced	B21	45	Tu7, Tu8, Tu9	0.60	0.43	0.30	0.23	0.21	0.39	0.80	0.80
Balanced	B22	45	Tu7, Tu8, Tu10	0.68	0.48	0.33	0.24	0.21	0.39	0.80	0.80

(Continued)

TABLE 12.9: Min–Max $x'(t)(X'X)^{-} x(t)$ Under the Model with Factor Interactions (Continued)

Type	Design	# of Obs.	Time	Month							Max Month
				0	3	6	9	12	18	24	
Balanced	B23	45	Tu7, Tu9, Tu10	0.68	0.48	0.33	0.24	0.20	0.31	0.60	0.68
Balanced	B24	45	Tu8, Tu9, Tu10	0.68	0.48	0.33	0.24	0.21	0.39	0.80	0.80
Balanced	B25	36	Tu11, Tu12, Tu13	0.68	0.48	0.34	0.27	0.26	0.45	0.83	0.83
Complete–one-half	C1/2_1	45	Tu6, Tu9	0.60	0.42	0.30	0.22	0.22	0.40	0.79	0.79
Complete–one-half	C1/2_2	45	Tu5, Tu10	0.68	0.48	0.33	0.24	0.23	0.43	0.83	0.83
Complete–one-half	C1/2_3	45	Tu7, Tu8	0.53	0.37	0.27	0.21	0.21	0.39	0.80	0.80
Complete–one-half	C1/2_4	45	Tu6, Tu9	0.60	0.43	0.30	0.23	0.22	0.40	0.79	0.79
Complete–one-half	C1/2_5	45	Tu5, Tu10	0.68	0.48	0.33	0.24	0.23	0.43	0.83	0.83
Complete–one-half	C1/2_6	45	Tu7, Tu8	0.53	0.38	0.27	0.21	0.21	0.39	0.80	0.80
Complete–one-third	C1/3_1	39	Tu7, Tu12, Tu13	0.68	0.48	0.34	0.27	0.26	0.43	0.83	0.83
Complete–one-third	C1/3_2	39	Tu9, Tu11, Tu13	0.68	0.48	0.34	0.27	0.26	0.45	0.83	0.83
Complete–one-third	C1/3_3	39	Tu10, Tu11, Tu12	0.68	0.48	0.33	0.26	0.26	0.45	0.83	0.83
Complete–two-third	C2/3_1	51	Tu2, Tu3, Tu8	0.53	0.36	0.26	0.20	0.21	0.39	0.80	0.80
Complete–two-third	C2/3_2	51	Tu2, Tu4, Tu6	0.53	0.36	0.25	0.20	0.22	0.40	0.79	0.79
Complete–two-third	C2/3_3	51	Tu3, Tu4, Tu5	0.53	0.36	0.25	0.20	0.23	0.43	0.83	0.83
Fractional	F1	42	Tu0, Tu0, Tu0	0.39	0.27	0.19	0.15	0.15	0.28	0.58	0.58
Fractional	F2	36	Tu1, Tu2, Tu3	0.46	0.32	0.23	0.18	0.19	0.39	0.79	0.79
Fractional	F3	36	Tu1, Tu2, Tu4	0.53	0.36	0.25	0.18	0.19	0.39	0.79	0.79
Fractional	F4	36	Tu1, Tu3, Tu4	0.53	0.36	0.25	0.18	0.19	0.39	0.79	0.79
Fractional	F5	36	Tu2, Tu3, Tu4	0.53	0.36	0.25	0.18	0.17	0.30	0.59	0.59
Uniform	U0	63	Tu0	0.39	0.27	0.19	0.15	0.15	0.28	0.58	0.58

(Continued)

TABLE 12.9: Min–Max $x'(t)(X'X)^{-}x(t)$ Under the Model with Factor Interactions (Continued)

Type	Design	# of Obs.	Time	Month							Max Month
				0	3	6	9	12	18	24	
Uniform	U01	54	Tu1	0.39	0.27	0.19	0.17	0.19	0.39	0.79	0.79
Uniform	U02	54	Tu2	0.42	0.30	0.21	0.17	0.17	0.30	0.59	0.59
Uniform	U03	54	Tu3	0.46	0.32	0.23	0.18	0.17	0.29	0.58	0.58
Uniform	U04	54	Tu4	0.53	0.36	0.25	0.18	0.17	0.28	0.59	0.59
Uniform	U05	45	Tu5	0.43	0.30	0.23	0.20	0.23	0.43	0.83	0.83
Uniform	U06	45	Tu6	0.46	0.32	0.24	0.20	0.22	0.40	0.79	0.79
Uniform	U07	45	Tu7	0.52	0.38	0.27	0.21	0.20	0.31	0.59	0.59
Uniform	U08	45	Tu8	0.53	0.36	0.26	0.20	0.21	0.39	0.80	0.80
Uniform	U09	45	Tu9	0.60	0.43	0.30	0.23	0.20	0.30	0.60	0.60
Uniform	U10	45	Tu10	0.68	0.48	0.33	0.24	0.20	0.29	0.59	0.68
Uniform	U11	36	Tu11	0.52	0.38	0.29	0.25	0.26	0.45	0.83	0.83
Uniform	U12	36	Tu12	0.60	0.43	0.31	0.26	0.26	0.43	0.83	0.83
Uniform	U13	36	Tu13	0.68	0.48	0.34	0.27	0.25	0.40	0.80	0.80

Source: Hedayat, A.S., Yan, X., and Lin, L. (2006). Journal of Biopharmaceutical Statistics, 16, 35–59.

TABLE 12.10: $Tr(C_d)$ Are Sorted from the Maximum to the Minimum Value Among Designs with the Same Sample Size

# of Observations	Design	Time	Trace
63	U0	Tu0	7020.00
54	U04	Tu4	6966.00
54	U03	Tu3	6804.00
54	B04	Tu2, Tu3, Tu4	6765.65
54	U02	Tu2	6534.00
54	B03	Tu1, Tu3, Tu4	6223.77
54	B02	Tu1, Tu2, Tu4	6139.20
54	B01	Tu1, Tu2, Tu3	6091.07
54	U01	Tu1	5076.00
51	C2/3_1	Tu2, Tu3, Tu8	6069.76
51	C2/3_2	Tu2, Tu4, Tu6	6052.61
51	C2/3_3	Tu3, Tu4, Tu5	6024.02
45	U10	Tu10	6750.00
45	B23	Tu7, Tu9, Tu10	6513.56
45	U09	Tu9	6480.00
45	U07	Tu7	6318.00
45	B24	Tu8, Tu9, Tu10	6036.66
45	B22	Tu7, Tu8, Tu10	5985.30
45	B20	Tu6, Tu9, Tu10	5972.24
45	B17	Tu6, Tu7, Tu10	5921.30
45	B21	Tu7, Tu8, Tu9	5904.16
45	B14	Tu5, Tu9, Tu10	5862.27
45	B16	Tu6, Tu7, Tu9	5840.92
45	B11	Tu5, Tu7, Tu10	5812.02
45	B10	Tu5, Tu7, Tu9	5732.89
45	B19	Tu6, Tu8, Tu10	5478.02
45	B18	Tu6, Tu8, Tu9	5405.35
45	B13	Tu5, Tu8, Tu10	5374.16
45	B15	Tu6, Tu7, Tu8	5360.58
45	B08	Tu5, Tu6, Tu10	5314.50
45	B12	Tu5, Tu8, Tu9	5303.05
45	B09	Tu5, Tu7, Tu8	5259.23
45	B07	Tu5, Tu6, Tu9	5244.41
45	B05	Tu5, Tu6, Tu7	5201.20
45	U08	Tu8	5022.00
45	U06	Tu6	4860.00
45	B06	Tu5, Tu6, Tu8	4820.71
45	U05	Tu5	4590.00
39	C1/3_1	Tu7, Tu12, Tu13	5161.10
39	C1/3_2	Tu9, Tu11, Tu13	5140.99
39	C1/3_3	Tu10, Tu11, Tu12	5107.47
36	U13	Tu13	4806.00
36	B25	Tu11, Tu12, Tu13	4568.53
36	U12	Tu12	4536.00
36	U11	Tu11	4374.00

Source: Hedayat, A.S., Yan, X., and Lin, L. (2006). *Journal of Biopharmaceutical Statistics*, 16, 35–59.

TABLE 12.11a: Min–Max $x'(t)(X'X)^- x(t) \times n(d)$ Under the Model With Factor Interactions

Type	Design	# of Obs.	Time	Month							Max Month
				0	3	6	9	12	18	24	
Balanced	B01	54	Tu1, Tu2, Tu3	25.01	17.47	12.31	9.53	10.35	21.15	42.75	42.75
Balanced	B02	54	Tu1, Tu2, Tu4	28.43	19.62	13.44	9.92	10.35	21.15	42.75	42.75
Balanced	B03	54	Tu1, Tu3, Tu4	28.43	19.62	13.44	9.92	10.35	21.15	42.75	42.75
Balanced	B04	54	Tu2, Tu3, Tu4	28.43	19.62	13.44	9.92	9.28	16.11	32.02	32.02
Balanced	B05	45	Tu5, Tu6, Tu7	23.50	16.87	12.25	9.64	10.13	19.13	37.13	37.13
Balanced	B06	45	Tu5, Tu6, Tu8	23.78	16.36	11.51	9.20	10.13	19.13	37.13	37.13
Balanced	B07	45	Tu5, Tu6, Tu9	27.00	19.13	13.50	10.13	10.13	19.13	37.13	37.13
Balanced	B08	45	Tu5, Tu6, Tu10	30.57	21.52	14.92	10.76	10.13	19.13	37.13	37.13
Balanced	B09	45	Tu5, Tu7, Tu8	23.78	16.88	12.25	9.64	10.13	19.13	37.13	37.13
Balanced	B10	45	Tu5, Tu7, Tu9	27.00	19.12	13.50	10.13	10.13	19.13	37.13	37.13
Balanced	B11	45	Tu5, Tu7, Tu10	30.57	21.52	14.92	10.76	10.13	19.13	37.13	37.13
Balanced	B12	45	Tu5, Tu8, Tu9	27.00	19.13	13.50	10.13	10.13	19.13	37.13	37.13
Balanced	B13	45	Tu5, Tu8, Tu10	30.57	21.52	14.92	10.76	10.13	19.13	37.13	37.13
Balanced	B14	45	Tu5, Tu9, Tu10	30.57	21.52	14.92	10.76	10.13	19.13	37.13	37.13
Balanced	B15	45	Tu6, Tu7, Tu8	23.78	16.88	12.25	9.64	9.74	18.09	36.05	36.05
Balanced	B16	45	Tu6, Tu7, Tu9	27.00	19.13	13.50	10.13	9.74	18.09	36.05	36.05
Balanced	B17	45	Tu6, Tu7, Tu10	30.57	21.52	14.92	10.76	9.74	18.09	35.72	35.72
Balanced	B18	45	Tu6, Tu8, Tu9	27.00	19.13	13.50	10.13	9.74	18.09	36.05	36.05
Balanced	B19	45	Tu6, Tu8, Tu10	30.57	21.52	14.92	10.76	9.74	18.09	36.05	36.05
Balanced	B20	45	Tu6, Tu9, Tu10	30.57	21.52	14.92	10.76	9.74	18.09	35.72	35.72
Balanced	B21	45	Tu7, Tu8, Tu9	27.00	19.13	13.50	10.13	9.46	17.64	36.05	36.05
Balanced	B22	45	Tu7, Tu8, Tu10	30.57	21.52	14.92	10.76	9.46	17.64	36.05	36.05

(Continued)

TABLE 12.11a: Min–Max $x'(t)(X'X)^{-} x(t) \times n(d)$ Under the Model With Factor Interactions (Continued)

Type	Design	# of Obs.	Time	Month							Max Month
				0	3	6	9	12	18	24	
Balanced	B23	45	Tu7, Tu9, Tu10	30.57	21.52	14.92	10.76	9.05	13.86	27.00	30.57
Balanced	B24	45	Tu8, Tu9, Tu10	30.57	21.52	14.92	10.76	9.46	17.64	36.05	36.05
Balanced	B25	36	Tu11, Tu12, Tu13	24.46	17.31	12.37	9.62	9.52	16.03	29.96	29.96
Complete-one-half	C1/2_1	45	Tu6, Tu9	27.00	19.12	13.50	10.12	9.74	18.09	35.72	35.72
Complete-one-half	C1/2_2	45	Tu5, Tu10	30.57	21.52	14.92	10.76	10.13	19.13	37.13	37.13
Complete-one-half	C1/2_3	45	Tu7, Tu8	23.78	16.87	12.25	9.64	9.46	17.64	36.05	36.05
Complete-one-half	C1/2_4	45	Tu6, Tu9	27.00	19.13	13.50	10.13	9.74	18.09	35.72	35.72
Complete-one-half	C1/2_5	45	Tu5, Tu10	30.57	21.52	14.92	10.76	10.13	19.13	37.13	37.13
Complete-one-half	C1/2_6	45	Tu7, Tu8	23.78	16.88	12.25	9.64	9.46	17.64	36.05	36.05
Complete-one-third	C1/3_1	39	Tu7, Tu12, Tu13	26.50	18.76	13.40	10.42	10.03	16.71	32.31	32.31
Complete-one-third	C1/3_2	39	Tu9, Tu11, Tu13	26.50	18.76	13.40	10.42	10.32	17.36	32.46	32.46
Complete-one-third	C1/3_3	39	Tu10, Tu11, Tu12	26.49	18.65	12.93	10.03	10.32	17.36	32.46	32.46
Complete-two-third	C2/3_1	51	Tu2, Tu3, Tu8	26.95	18.55	13.04	10.43	10.72	19.99	40.86	40.86
Complete-two-third	C2/3_2	51	Tu2, Tu4, Tu6	26.85	18.53	12.70	10.25	11.04	20.51	40.48	40.48
Complete-two-third	C2/3_3	51	Tu3, Tu4, Tu5	26.85	18.53	12.70	10.20	11.48	21.68	42.08	42.08
Fractional	F1	42	Tu0, Tu0, Tu0	16.35	11.19	7.80	6.16	6.29	11.82	24.40	24.40
Fractional	F2	36	Tu1, Tu2, Tu3	16.68	11.65	8.21	6.35	6.90	14.10	28.50	28.50
Fractional	F3	36	Tu1, Tu2, Tu4	18.96	13.08	8.96	6.61	6.90	14.10	28.50	28.50
Fractional	F4	36	Tu1, Tu3, Tu4	18.96	13.08	8.96	6.61	6.90	14.10	28.50	28.50

(*Continued*)

TABLE 12.11a: Min–Max $x'(t)(X'X)^{-}x(t) \times n(d)$ Under the Model With Factor Interactions (Continued)

Type	Design	# of Obs.	Time	Month								Max Month
				0	3	6	9	12	18	24	
Fractional	F5	36	Tu2, Tu3, Tu4	18.96	13.08	8.96	6.61	6.19	10.74	21.35	21.35
Uniform	U0	63	Tu0	24.52	16.79	11.69	9.24	9.43	17.73	36.59	36.59
Uniform	U01	54	Tu1	21.15	14.40	10.35	9.00	10.35	21.15	42.75	42.75
Uniform	U02	54	Tu2	22.93	16.11	11.56	9.28	9.28	16.11	32.02	32.02
Uniform	U03	54	Tu3	25.01	17.47	12.31	9.53	9.13	15.49	31.37	31.37
Uniform	U04	54	Tu4	28.43	19.62	13.44	9.92	9.04	15.21	31.96	31.96
Uniform	U05	45	Tu5	19.13	13.50	10.13	9.00	10.13	19.13	37.13	37.13
Uniform	U06	45	Tu6	20.88	14.61	10.67	9.05	9.74	18.09	35.72	35.72
Uniform	U07	45	Tu7	23.50	16.88	12.25	9.64	9.04	13.86	26.72	26.72
Uniform	U08	45	Tu8	23.78	16.36	11.51	9.20	9.46	17.64	36.05	36.05
Uniform	U09	45	Tu9	27.00	19.13	13.50	10.13	9.00	13.50	27.00	27.00
Uniform	U10	45	Tu10	30.57	21.52	14.92	10.76	9.05	12.96	26.66	30.57
Uniform	U11	36	Tu11	18.81	13.70	10.45	9.06	9.52	16.03	29.96	29.96
Uniform	U12	36	Tu12	21.60	15.43	11.31	9.26	9.26	15.43	29.83	29.83
Uniform	U13	36	Tu13	24.46	17.31	12.37	9.62	9.07	14.56	28.85	28.85

TABLE 12.11b: Min–Max $x'(t)(X'X)^- x(t) \times n(d)$ Under the Model Without Factor Interactions

Type	Design	# of Obs.	Time	Month								Max Month
				0	3	6	9	12	18	24	
Balanced	B01	54	Tu1, Tu2, Tu3	13.17	9.17	6.51	5.18	5.42	10.35	20.86	20.86
Balanced	B02	54	Tu1, Tu2, Tu4	14.16	9.78	6.82	5.27	5.39	10.31	20.92	20.92
Balanced	B03	54	Tu1, Tu3, Tu4	14.44	9.97	6.92	5.31	5.36	10.22	20.80	20.80
Balanced	B04	54	Tu2, Tu3, Tu4	14.70	10.20	7.09	5.38	5.11	8.77	17.68	17.68
Balanced	B05	45	Tu5, Tu6, Tu7	12.18	8.68	6.34	5.16	5.41	9.83	19.12	19.12
Balanced	B06	45	Tu5, Tu6, Tu8	12.28	8.52	6.11	5.06	5.50	10.33	20.30	20.30
Balanced	B07	45	Tu5, Tu6, Tu9	13.16	9.30	6.67	5.26	5.39	9.78	19.13	19.13
Balanced	B08	45	Tu5, Tu6, Tu10	14.05	9.88	7.00	5.39	5.37	9.69	19.04	19.04
Balanced	B09	45	Tu5, Tu7, Tu8	12.58	8.89	6.45	5.20	5.38	9.77	19.13	19.13
Balanced	B10	45	Tu5, Tu7, Tu9	13.54	9.61	6.89	5.37	5.25	9.17	17.87	17.87
Balanced	B11	45	Tu5, Tu7, Tu10	14.48	10.24	7.25	5.52	5.23	9.09	17.78	17.78
Balanced	B12	45	Tu5, Tu8, Tu9	13.56	9.55	6.80	5.31	5.35	9.72	19.13	19.13
Balanced	B13	45	Tu5, Tu8, Tu10	14.50	10.17	7.15	5.44	5.34	9.63	19.04	19.04
Balanced	B14	45	Tu5, Tu9, Tu10	14.98	10.57	7.44	5.60	5.21	9.04	17.78	17.78
Balanced	B15	45	Tu6, Tu7, Tu8	12.81	9.04	6.53	5.22	5.28	9.47	18.80	18.80
Balanced	B16	45	Tu6, Tu7, Tu9	13.80	9.78	6.98	5.40	5.17	8.90	17.54	17.54
Balanced	B17	45	Tu6, Tu7, Tu10	14.77	10.43	7.35	5.55	5.15	8.82	17.46	17.46
Balanced	B18	45	Tu6, Tu8, Tu9	13.81	9.71	6.89	5.34	5.26	9.42	18.85	18.85
Balanced	B19	45	Tu6, Tu8, Tu10	14.78	10.36	7.26	5.48	5.24	9.34	18.79	18.79
Balanced	B20	45	Tu6, Tu9, Tu10	15.27	10.75	7.54	5.63	5.13	8.77	17.47	17.47
Balanced	B21	45	Tu7, Tu8, Tu9	14.18	10.02	7.11	5.44	5.11	8.76	17.64	17.64

(Continued)

TABLE 12.11b: Min–Max $x'(t)(X'X)^- x(t) \times n(d)$ Under the Model Without Factor Interactions (Continued)

Type	Design	# of Obs.	Time	Month							Max Month
				0	3	6	9	12	18	24	
Balanced	B22	45	Tu7, Tu8, Tu10	15.20	10.70	7.50	5.60	5.09	8.69	17.58	17.58
Balanced	B23	45	Tu7, Tu9, Tu10	15.66	11.08	7.78	5.75	5.01	7.55	14.92	15.66
Balanced	B24	45	Tu8, Tu9, Tu10	15.69	11.02	7.68	5.68	5.08	8.64	17.61	17.61
Balanced	B25	36	Tu11, Tu12, Tu13	12.53	8.93	6.49	5.21	5.20	8.66	16.47	16.47
Complete–one-half	C1/2_1	45	Tu6, Tu9	13.82	9.76	6.95	5.39	5.26	9.37	18.54	18.54
Complete–one-half	C1/2_2	45	Tu5, Tu10	14.63	10.31	7.28	5.54	5.36	9.66	18.93	18.93
Complete–one-half	C1/2_3	45	Tu7, Tu8	13.13	9.27	6.67	5.28	5.18	9.23	18.64	18.64
Complete–one-half	C1/2_4	45	Tu6, Tu9	14.10	9.97	7.08	5.43	5.23	9.16	18.14	18.14
Complete–one-half	C1/2_5	45	Tu5, Tu10	15.11	10.64	7.47	5.60	5.32	9.39	18.47	18.47
Complete–one-half	C1/2_6	45	Tu7, Tu8	13.12	9.30	6.71	5.30	5.16	9.06	18.22	18.22
Complete–one-third	C1/3_1	39	Tu7, Tu12,Tu13	13.57	9.65	6.96	5.49	5.27	8.57	16.65	16.65
Complete–one-third	C1/3_2	39	Tu9, Tu11,Tu13	13.57	9.66	6.97	5.50	5.32	8.70	16.64	16.64
Complete–one-third	C1/3_3	39	Tu10, Tu11,Tu12	13.57	9.63	6.84	5.44	5.34	8.75	16.76	16.76
Complete–two-third	C2/3_1	51	Tu2, Tu3, Tu8	13.90	9.65	6.82	5.42	5.44	9.78	19.84	19.84
Complete–two-third	C2/3_2	51	Tu2, Tu4, Tu6	13.89	9.64	6.74	5.37	5.50	9.88	19.77	19.77
Complete–two-third	C2/3_3	51	Tu3, Tu4, Tu5	13.89	9.64	6.74	5.33	5.56	10.07	20.00	20.00
Fractional	F1	42	Tu0, Tu0, Tu0	10.90	7.46	5.20	4.11	4.19	7.88	16.26	16.26
Fractional	F2	36	Tu1, Tu2, Tu3	10.63	7.40	5.24	4.15	4.39	8.55	17.26	17.26
Fractional	F3	36	Tu1, Tu2, Tu4	11.60	8.00	5.55	4.25	4.37	8.52	17.30	17.30
Fractional	F4	36	Tu1, Tu3, Tu4	11.71	8.07	5.59	4.26	4.36	8.48	17.25	17.25
Fractional	F5	36	Tu2, Tu3, Tu4	11.93	8.27	5.73	4.32	4.09	7.05	14.18	14.18
Uniform	U0	63	Tu0	13.62	9.33	6.50	5.13	5.24	9.85	20.33	20.33

(Continued)

TABLE 12.11b: Min–Max $x'(t)(X'X)^- x(t) \times n(d)$ Under the Model Without Factor Interactions (Continued)

Type	Design	# of Obs.	Time	Month							Max Month
				0	3	6	9	12	18	24	
Uniform	U01	54	Tu1	11.75	8.00	5.75	5.00	5.75	11.75	23.75	23.75
Uniform	U02	54	Tu2	12.74	8.95	6.42	5.16	5.16	8.95	17.79	17.79
Uniform	U03	54	Tu3	13.90	9.71	6.84	5.29	5.07	8.60	17.43	17.43
Uniform	U04	54	Tu4	15.80	10.90	7.47	5.51	5.02	8.45	17.76	17.76
Uniform	U05	45	Tu5	10.63	7.50	5.63	5.00	5.63	10.63	20.63	20.63
Uniform	U06	45	Tu6	11.60	8.12	5.93	5.03	5.41	10.05	19.85	19.85
Uniform	U07	45	Tu7	13.06	9.38	6.81	5.36	5.02	7.70	14.84	14.84
Uniform	U08	45	Tu8	13.21	9.09	6.39	5.11	5.26	9.80	20.03	20.03
Uniform	U09	45	Tu9	15.00	10.63	7.50	5.63	5.00	7.50	15.00	15.00
Uniform	U10	45	Tu10	16.98	11.96	8.29	5.98	5.03	7.20	14.81	16.98
Uniform	U11	36	Tu11	10.45	7.61	5.81	5.03	5.29	8.90	16.65	16.65
Uniform	U12	36	Tu12	12.00	8.57	6.29	5.14	5.14	8.57	16.57	16.57
Uniform	U13	36	Tu13	13.59	9.62	6.87	5.34	5.04	8.09	16.03	16.03

Source: Hedayat, A.S., Yan, X., and Lin, L. (2006). *Journal of Biopharmaceutical Statistics*, 16, 35–59.

TABLE 12.12: Designs with the Same Sample Size are Sorted from the Smallest Value to the Largest Value of Max *all-level-combination* $x'(t)(X'X)^- x(t)$ at Each Time Point

# of Obs.	Month 0	3	6	9	12	18	24	Max Month
63	U0	U0	U0	U0	U0	U0	U0	U0
54	U01	U01	U01	U01	U04	U04	U03	U03
54	U02	U02	U02	U02	U03	U03	U04	U04
54	B01	B01	B01	B01	B04	B04	B04	B04
54	U03	U03	B02	B02	U02	U02	U02	U02
54	B02	B02	U03	U03	B03	B03	B03	B03
54	B03	B03	B03	B03	B02	B02	B01	B01
54	B04	B04	B04	B04	B01	B01	B02	B02
54	U04	U04	U04	U04	U01	U01	U01	U01
51	C2/3_3	C2/3_2	C2/3_2	C2/3_3	C2/3_1	C2/3_1	C2/3_2	C2/3_2
51	C2/3_2	C2/3_3	C2/3_3	C2/3_2	C2/3_2	C2/3_2	C2/3_1	C2/3_1
51	C2/3_1	C2/3_1	C2/3_1	C2/3_1	C2/3_3	C2/3_3	C2/3_3	C2/3_3
45	U05	U05	U05	U05	U09	U10	U10	U07
45	U06	U06	U06	U06	B23	U09	U07	U09
45	B05	B06	B06	B06	U07	B23	B23	B23
45	B06	B05	B05	U08	U10	U07	U09	U10
45	B09	B09	U08	B05	B24	B24	B17	B17
45	B15	B15	B09	B09	B22	B22	B20	B20
45	U07	U08	B15	B15	B21	B21	B16	B16
45	C1/2_6	C1/2_3	B07	B07	B20	B20	B22	B22
45	C1/2_3	C1/2_6	C1/2_3	C1/2_3	B17	B17	B24	B24
45	B07	B07	C1/2_6	C1/2_6	C1/2_6	B16	B21	B21
45	U08	U07	B12	B12	B16	B14	B11	B11
45	B10	B12	U07	B18	C1/2_3	C1/2_6	B14	B14
45	B12	B10	B18	U07	B14	B11	B10	B10
45	B16	B18	B10	B10	C1/2_4	C1/2_4	C1/2_4	C1/2_4
45	B18	C1/2_1	C1/2_1	B08	B11	B10	C1/2_6	C1/2_6
45	C1/2_1	B16	B16	C1/2_1	B19	C1/2_3	C1/2_5	C1/2_5
45	B08	B08	B08	B16	B10	B19	C1/2_1	C1/2_1
45	C1/2_4	C1/2_4	C1/2_4	C1/2_4	U08	C1/2_1	C1/2_3	C1/2_3
45	B21	B21	B21	B21	B18	C1/2_5	B19	B19
45	B11	B13	B13	B13	C1/2_1	B18	B15	B15
45	B13	B11	B11	B19	B15	B15	B18	B18
45	C1/2_2	C1/2_2	B19	B11	C1/2_5	B13	C1/2_2	C1/2_2
45	B17	B19	C1/2_2	C1/2_2	B13	C1/2_2	B08	B08
45	B19	B17	B17	B17	B12	B08	B13	B13
45	B14	B14	B14	B14	C1/2_2	B12	B05	B05
45	U09	U09	C1/2_5	C1/2_5	B08	B09	B07	B07
45	C1/2_5	C1/2_5	U09	B22	B09	B07	B09	B09
45	B22	B22	B22	U09	B07	U08	B12	B12
45	B20	B20	B20	B20	U06	B05	U06	U06
45	B23	B24	B24	B24	B05	U06	U08	U08

(Continued)

TABLE 12.12: Designs with the Same Sample Size are Sorted from the Smallest Value to the Largest Value of Max *all-level-combination* $x'(t)(X'X)^- x(t)$ at Each Time Point (Continued)

# of Obs.	Month							Max Month
	0	3	6	9	12	18	24	
45	B24	B23	B23	B23	B06	B06	B06	B06
45	U10	U10	U10	U10	U05	U05	U05	U05
42	F1	F1	F1	F1	F1	F1	F1	F1
39	C1/3_2	C1/3_3	C1/3_3	C1/3_3	C1/3_1	C1/3_1	C1/3_2	C1/3_2
39	C1/3_1	C1/3_1	C1/3_1	C1/3_1	C1/3_2	C1/3_2	C1/3_1	C1/3_1
39	C1/3_3	C1/3_2	C1/3_2	C1/3_2	C1/3_3	C1/3_3	C1/3_3	C1/3_3
36	U11	F2	F2	F2	F5	F5	F5	F5
36	F2	U11	F3	F3	F4	U13	U13	U13
36	F3	F3	F4	F4	F3	F4	B25	B25
36	F4	F4	F5	F5	F2	F3	U12	U12
36	F5	F5	U11	U11	U13	F2	U11	U11
36	U12	U12	U12	U12	U12	U12	F4	F4
36	B25	B25	B25	B25	B25	B25	F2	F2
36	U13	U13	U13	U13	U11	U11	F3	F3

Source: Hedayat, A.S., Yan, X., and Lin, L. (2006). *Journal of Biopharmaceutical Statistics*, 16, 35–59.

Under the model without factor interactions, Tables 12.8, 12.11b, 12.12 and 12.14 lead to the following conclusions:

- Among all types of designs, the complete design U0 is uniformly optimal.

- Among every group of designs with the same sample size, there are no uniformly optimal designs.

- The min–max–max optimal designs can be found for the designs with the same sample size. For example, for a design with a sample size of 54, the uniform design U03 is the min–max–max optimal design.

- The sample-size-adjusted min–max–max optimal designs can be found when comparing different designs with $x'(t)(X'X)^- x(t) \times n(d)$. The fractional design F5 is the sample-size-adjusted min–max–max optimal design among all designs.

Under the model with factor interactions, from Tables 12.9, 12.11a, 12.13, and 12.15, the following conclusions can be made:

- For each type of designs, the designs with the same time vectors but different time vector allocations are equivalent to each other under the min–max optimality criterion. In other words, for every design with the same time vectors and the same type, regardless of the allocation of the time vectors, the value of $\max_{all-level-combination} x'(t)(X'X)^- x(t)$ is the same at time t.

- The fractional design F1 is equivalent to the complete design U0 under the uniform optimality criterion. Designs F1 and U0 are the uniformly optimal designs of all types of designs.

TABLE 12.13: Designs with the Same Sample Size Are Sorted from the Smallest Value to the Largest Value of $\text{Max}_{all-level-combination}\, x'(t)(X'X)^- x(t)$ at Each Time Point

	Month							
# of Obs.	0	3	6	9	12	18	24	Max Month
63	U0	U0	U0	U0	U0	U0	U0	U0
54	U01	U01	U01	U01	U04	U04	U03	U03
54	U02	U02	U02	U02	U03	U03	U04	U04
54	B01	B01	B01	B01	B04	B04	B04	B04
54	U03	U03	U03	U03	U02	U02	U02	U02
54	B02	B02	B03	B03	B02	B02	B02	B02
54	B03	B03	B02	B02	B01	B01	B01	B01
54	B04	B04	B04	B04	B03	B03	B03	B03
54	U04	U04	U04	U04	U01	U01	U01	U01
51	C2/3_3	C2/3_3	C2/3_3	C2/3_3	C2/3_1	C2/3_1	C2/3_2	C2/3_2
51	C2/3_2	C2/3_2	C2/3_2	C2/3_2	C2/3_2	C2/3_2	C2/3_1	C2/3_1
51	C2/3_1	C2/3_1	C2/3_1	C2/3_1	C2/3_3	C2/3_3	C2/3_3	C2/3_3
45	U05	U05	U05	U05	U09	U10	U10	U07
45	U06	U06	U06	U06	U07	U09	U07	U09
45	B05	B06	B06	B06	B23	B23	B23	B23
45	U07	U08	U08	U08	U10	U07	U09	U10
45	C1/2_6	C1/2_3	B05	C1/2_3	C1/2_6	C1/2_6	C1/2_4	C1/2_4
45	B06	B05	C1/2_3	B05	C1/2_3	C1/2_3	C1/2_1	C1/2_1
45	B09	B15	B15	B15	B24	B22	B16	B16
45	B15	C1/2_6	C1/2_6	C1/2_6	B21	B24	B20	B20
45	C1/2_3	B09	B09	B09	B22	B21	B17	B17
45	U08	U07	U07	U07	U08	U08	U06	U06
45	C1/2_1	C1/2_1	C1/2_1	C1/2_1	C1/2_4	C1/2_4	C1/2_6	C1/2_6
45	B18	B10	B18	B18	C1/2_1	C1/2_1	B15	B15
45	B07	C1/2_4	C1/2_4	C1/2_4	B20	B15	C1/2_3	C1/2_3
45	B12	B18	B10	B10	B16	B16	B18	B18
45	B10	B21	B16	B16	B19	B18	B19	B19
45	C1/2_4	B16	B07	B07	B18	B20	B22	B22
45	B21	B07	B21	B21	B15	B19	B24	B24
45	B16	B12	B12	B12	B17	B17	B21	B21
45	U09	U09	U09	U09	U06	U06	U08	U08
45	C1/2_2	B22	C1/2_2	C1/2_2	C1/2_5	B05	C1/2_5	C1/2_5
45	B14	C1/2_2	C1/2_5	B19	B05	C1/2_5	C1/2_2	C1/2_2
45	B13	C1/2_5	B19	C1/2_5	C1/2_2	C1/2_2	B05	B05
45	B19	B24	B24	B24	B06	B13	B13	B13
45	B23	B11	B13	B20	B10	B10	B10	B10
45	B20	B19	B11	B13	B13	B06	B06	B06
45	B08	B17	B20	B11	B12	B12	B12	B12
45	B22	B20	B17	B23	B08	B08	B08	B08
45	B11	B13	B22	B17	B09	B11	B14	B14
45	C1/2_5	B23	B23	B22	B07	B14	B11	B11
45	B24	B14	B14	B08	B14	B07	B07	B07

(Continued)

TABLE 12.13: Designs with the Same Sample Size Are Sorted from the Smallest Value to the Largest Value of $\mathrm{Max}_{all-level-combination}\, x'(t)(X'X)^{-}x(t)$ at Each Time Point

	Month							
# of Obs.	0	3	6	9	12	18	24	Max Month
45	B17	B08	B08	B14	B11	B09	B09	B09
45	U10	U10	U10	U10	U05	U05	U05	U05
42	F1	F1	F1	F1	F1	F1	F1	F1
39	C1/3_3	C1/3_3	C1/3_3	C1/3_3	C1/3_1	C1/3_1	C1/3_1	C1/3_1
39	C1/3_2	C1/3_2	C1/3_1	C1/3_1	C1/3_2	C1/3_2	C1/3_2	C1/3_2
39	C1/3_1	C1/3_1	C1/3_2	C1/3_2	C1/3_3	C1/3_3	C1/3_3	C1/3_3
36	F2	F2	F2	F2	F5	F5	F5	F5
36	U11	F4	F4	F5	F3	F3	F3	F3
36	F4	F5	F5	F4	F2	F2	F2	F2
36	F5	F3	F3	F3	F4	F4	F4	F4
36	F3	U11	U11	U11	U13	U13	U13	U13
36	U12	U12	U12	U12	U12	U12	U12	U12
36	B25	B25	B25	B25	B25	B25	B25	B25
36	U13	U13	U13	U13	U11	U11	U11	U11

Source: Hedayat, A.S., Yan. X. and Lin, L. (2006). *Journal of Biopharmaceutical Statistics*, 16, 35–59.

- Among all types of designs, the sample-size-adjusted uniformly optimal design does not exist.

- Among every group of designs with the same sample size, there are no uniformly optimal designs.

- The min–max–max optimal designs can be found for the designs with the same sample size. For example, for a design with a sample size of 54, the uniform design U03 is the min–max–max optimal design.

- The sample-size-adjusted min–max–max optimal designs can be found when comparing different designs with $x'(t)(X'X)^{-}x(t) \times n(d)$. The fractional design F5 is the sample-size-adjusted min–max–max optimal design among all designs.

12.3.1 Remarks

Hedayat et al. (2006) focused on searching for optimal design among designs with two factors (three levels in each factor) based on the proposed optimality criteria under a model with or without factor interactions. In practice, it is of interest to study the general case where one factor has m levels and the other factor has n levels. In such a case, the problem becomes how to allocate k time vectors T_1, T_2, $\cdots$, T_k to the $m \times n$ tables and how to select the optimal time vector. It is also of interest to study the case when there are three factors. Furthermore, as indicated in Chapter 6, batch factor is often considered as random effect. As a result, the search for an optimal design under a mixed effects model should be explored.

TABLE 12.14: Designs Are Sorted from the Smallest Value to the Largest Value of Max$_{all-level-combination}$ $x'(t) (X' X)^- x(t) \times n(d)$ at Each Time Point

			Month				
0	**3**	**6**	**9**	**12**	**18**	**24**	**Max Month**
U11	F2	F1	F1	F5	F5	F5	F5
U05	F1	F2	F2	F1	U10	U10	U07
F2	U05	F3	F3	F4	U09	U07	U09
F1	U11	F4	F4	F3	B23	B23	B23
U06	U01	U05	F5	F2	U07	U09	U13
F3	F3	F5	U05	U09	F1	U13	F1
F4	F4	U01	U01	B23	U13	F1	B25
U01	U06	U11	U06	U04	U04	B25	U12
F5	F5	U06	U11	U07	F4	U12	C1/3_2
U12	B06	B06	B06	U10	F3	C1/3_2	U11
B05	U12	U12	U08	U13	F2	U11	C1/3_1
B06	B05	B05	U0	U03	U12	C1/3_1	C1/3_3
B25	B09	U08	U12	B24	C1/3_1	C1/3_3	U10
B09	B25	U02	U02	B22	U03	F4	F4
U02	U02	B09	B05	B04	B24	F2	F2
B15	B15	B25	B01	B21	B25	F3	F3
U07	U08	U0	B09	B20	B22	U03	U03
C1/2_6	B01	B01	B25	U12	C1/3_2	B17	B17
C1/2_3	C1/2_3	B15	B15	B17	C1/3_3	B20	B20
B07	C1/2_6	B07	B07	U02	B21	B16	B16
B01	B07	C1/2_3	B02	C1/2_6	B04	B22	B22
U08	U0	C1/2_6	C1/2_3	B16	B20	B24	B24
B10	U07	C2/3_2	U03	C1/2_3	B17	B21	B21
B12	B12	C2/3_3	C1/2_6	B25	B16	B04	B04
C1/3_2	B10	B12	B03	B14	U11	U04	U04
C1/3_1	U13	U07	B12	C1/2_4	U02	B11	B11
C1/3_3	C1/3_3	B02	C2/3_3	B11	B14	B14	B14
U13	C2/3_2	C2/3_1	B18	U0	C1/2_6	U02	U02
U0	C2/3_3	C2/3_3	U13	B19	B11	B10	B10
B16	C2/3_1	U03	U07	B10	C1/2_4	C1/2_4	C1/2_4
B18	C1/3_1	U13	C2/3_2	U08	B10	C1/2_6	C1/2_6
C1/2_1	C1/3_2	B18	B10	B18	C1/2_3	C1/2_5	C1/2_5
C2/3_3	U03	B10	B04	C1/2_1	B19	C1/2_1	C1/2_1
C2/3_2	B18	B03	B08	C1/3_1	C1/2_1	C1/2_3	C1/2_3
U03	C1/2_1	C1/2_1	C1/2_1	B15	C1/2_5	B19	B19
C2/3_1	B02	C1/3_1	B16	U11	B18	B15	B15
B08	B16	C1/3_2	C2/3_1	C1/2_5	B15	B18	B18
C1/2_4	B08	B16	C1/2_4	C1/3_2	B13	C1/2_2	C1/2_2
B02	C1/2_4	B08	C1/3_3	B13	C1/2_2	B08	B08
B21	B03	C1/2_4	B21	C1/3_3	B08	B13	B13
B03	B21	B04	B13	B12	B12	B05	B05
B11	B13	B21	B19	B03	B09	B07	B07
B13	B04	B13	C1/3_1	C1/2_2	B07	B09	B09

(*Continued*)

TABLE 12.14: Designs Are Sorted from the Smallest Value to the Largest Value of $\text{Max}_{all-level-combination}$ $x'(t)\,(X'\,X)^{-}\,x(t) \times n(d)$ at Each Time Point (Continued)

			Month				
0	**3**	**6**	**9**	**12**	**18**	**24**	**Max Month**
C1/2_2	B11	B11	C1/3_2	B08	C2/3_1	B12	B12
B04	C1/2_2	B19	U04	B09	U08	C2/3_2	C2/3_2
B17	B19	C1/2_2	B11	B02	B05	C2/3_1	C2/3_1
B19	B17	B17	C1/2_2	B07	U0	U06	U06
B14	B14	B14	B17	U06	C2/3_2	C2/3_3	C2/3_3
U09	U09	U04	B14	B05	U06	U08	U08
C1/2_5	C1/2_5	C1/2_5	C1/2_5	B01	C2/3_3	B06	B06
B22	B22	U09	B22	C2/3_1	B03	U0	U0
B20	B20	B22	U09	B06	B02	U05	U05
B23	U04	B20	B20	C2/3_2	B06	B03	B03
B24	B24	B24	B24	C2/3_3	B01	B01	B01
U04	B23	B23	B23	U05	U05	B02	B02
U10	U10	U10	U10	U01	U01	U01	U01

Source: Hedayat, A.S., Yan, X., and Lin, L. (2006). *Journal of Biopharmaceutical Statistics*, 16, 35–59.

Hedayat et al. (2006) proposed using an information matrix C_d to investigate whether the optimal allocations of time vectors have good properties in terms of detecting slope differences between levels of one factor under the following model with common intercept and no factor interactions:

$$y_{ijk} = \alpha_0 + (\beta_{A_i} + \beta_{B_j})t_{ijk} + \varepsilon_{ijk}$$

where ε_{ijk} are independent and identically distributed normal with mean 0 and variance σ^2, $i = 1, 2, 3$, $j = 1, 2, 3$, $k = 1, 2, \cdots, n_{ij}$. If one is interested in testing the hypothesis

$$H_0 : \beta_{A_1} = \beta_{A_2} = \beta_{A_3},$$

the information matrix for β_{A_i} is given by

$$C_d = X_1' X_1 - X_1' X_2 (X_2' X_2)^{-} X_2' X_1,$$

where

$$X_1 = (X_{\beta_{A_1}}, X_{\beta_{A_2}}, X_{\beta_{A_3}})$$

$$= (I_3 \otimes J_3) \otimes \begin{pmatrix} J_{n_1} \\ \vdots \\ J_{n_9} \end{pmatrix} \odot T,$$

and

$$X_2 = (X_{\beta_{B_1}}, X_{\beta_{B_2}}, X_{\beta_{B_3}})$$

$$= (J_3 \otimes I_3) \otimes \begin{pmatrix} J_{n_1} \\ \vdots \\ J_{n_9} \end{pmatrix} \odot T.$$

TABLE 12.15: Designs Are Sorted from the Smallest Value to the Largest Value of $\text{Max}_{all-level-combination}$ x' (t) $(x'$ $x)^-$ $x(t) \times n(d)$ at Each Time Point

			Month				
0	3	6	9	12	18	24	Max Month
F1	F1	F1	F1	F5	F5	F5	F5
F2	F2	F2	F2	F1	F1	F1	F1
U11	F4	F4	F5	F3	U10	U10	U07
F4	F5	F5	F4	F2	U09	U07	U09
F5	F3	F3	F3	F4	B23	B23	F3
F3	U05	U05	U05	U09	U07	U09	F2
U05	U11	U01	U01	U04	F3	F3	F4
U06	U01	U11	U06	U07	F2	F2	U13
U01	U06	U06	U11	B23	F4	F4	U12
U12	U12	U12	B06	U10	U13	U13	B25
U02	U02	B06	U08	U13	U04	U12	U11
B05	B06	U08	U0	U03	U12	B25	B23
U07	U08	U02	U12	U12	U03	U11	U10
C1/2_6	U0	U0	U02	B04	B25	U03	U03
B06	C1/2_3	B05	B01	U02	U11	U04	U04
B09	B05	C1/2_3	U03	U0	B04	B04	B04
B15	B15	B15	B25	C1/2_6	U02	U02	U02
C1/2_3	C1/2_6	C1/2_6	U13	C1/2_3	C1/3_1	C1/3_1	C1/3_1
U08	B09	B09	C1/2_3	B24	C1/3_2	C1/3_2	C1/3_2
B25	U07	U07	B05	B21	C1/3_3	C1/3_3	C1/3_3
U13	B25	B01	B15	B22	C1/2_6	C1/2_4	C1/2_4
U0	U13	U03	C1/2_6	U08	C1/2_3	C1/2_1	C1/2_1
B01	B01	B25	B09	B25	B22	B16	B16
U03	U03	U13	U07	U11	B21	B20	B20
C1/3_3	C2/3_3	C2/3_3	B03	C1/2_4	B24	B17	B17
C1/3_2	C2/3_2	C2/3_2	B02	C1/2_1	U08	U06	U06
C1/3_1	C2/3_1	C1/3_3	B04	B20	U0	C1/2_6	C1/2_6
C2/3_3	C1/3_3	C2/3_1	U04	B16	C1/2_4	B15	B15
C2/3_2	C1/3_2	C1/3_1	C1/3_3	B19	C1/2_1	C1/2_3	C1/2_3
C2/3_1	C1/3_1	C1/3_2	C1/2_1	B18	B15	B18	B18
C1/2_1	C1/2_1	B03	B18	B15	B16	B19	B19
B18	B10	B02	C1/2_4	B17	B18	B22	B22
B07	C1/2_4	B04	B10	U06	B20	B24	B24
B12	B18	U04	B16	C1/3_1	B19	B21	B21
B10	B21	C1/2_1	B07	C1/2_5	B17	U08	U08
C1/2_4	B16	B18	B21	B05	U06	U0	U0
B21	B07	C1/2_4	B12	C1/2_2	B05	C1/2_5	C1/2_5
B16	B12	B10	U09	B06	C1/2_5	C1/2_2	C1/2_2
U09	U09	B16	C2/3_3	B10	C1/2_2	B05	B05
B02	B02	B07	C2/3_2	B13	B13	B13	B13
B03	B03	B21	C1/3_1	B12	B10	B10	B10
B04	B04	B12	C1/3_2	B08	B06	B06	B06
U04	U04	U09	C2/3_1	B09	B12	B12	B12

(Continued)

TABLE 12.15: Designs Are Sorted from the Smallest Value to the Largest Value of $\text{Max}_{all-level-combination}\ x'(t)\ (x'\ x)^-\ x(t) \times n(d)$ at Each Time Point

Month							
0	**3**	**6**	**9**	**12**	**18**	**24**	**Max Month**
C1/2_2	B22	C1/2_2	C1/2_2	B07	B08	B08	B08
B14	C1/2_2	C1/2_5	B19	B14	B11	B14	B14
B13	C12_5	B19	C1/2_5	B11	B14	B07	B07
B19	B24	B24	B24	U05	B07	B11	B11
B23	B11	B13	B20	C1/3_2	B09	B09	B09
B20	B19	B11	B13	C1/3_3	U05	U05	U05
B08	B17	B20	B11	B02	C2/3_1	C2/3_2	C2/3_2
B22	B20	B17	B23	B01	C2/3_2	C2/3_1	C2/3_1
B11	B13	B22	B17	B03	B02	C2/3_3	C2/3_3
C1/2_5	B23	B23	B22	U01	B01	B02	B02
B24	B14	B14	B08	C2/3_1	B03	B01	B01
B17	B08	B08	B14	C2/3_2	U01	B03	B03
U10	U10	U10	U10	C2/3_3	C2/3_3	U01	U01

Source: Hedayat, A.S., Yan, X., and Lin, L. (2006). *Journal of Biopharmaceutical Statistics*, 16, 35–59.

in which T is the time vector, $\otimes$ represents the Kronckknecker multiplication, and $\odot$ is the Hadamard product (i.e., entry-wise multiplication). To investigate the property of the information of C_d, Hedayat et al. (2006) suggested Kiefer's theorem (see Hedayat, 2001) be used, which is stated as follows:

Theorem 12.1 *Let $d^* \in D$ such that*

(a) d^* maximizes $Tr(C_d)$.

(b) C_{d^*} is completely symmetric, that is, C_{d^*} is in the form of $aI + bJ$.

Then d^* is uniformly optimal in the class of the design with $C_d 1 = 0$.

12.4 Current Issues in Stability Analysis

As discussed in the previous chapters, statistical methods for analysis of stability data collected from either short-term or long-term stability studies are well established in the literature. However, some issues are of particular interest to pharmaceutical researchers and scientists and biostatisticians (see, e.g., Tsong, 2003). These issues include the use of overage to account for stability loss prior to expiration dates, the impact of assay sensitivity on drug shelf-life estimations, stability analysis with multiple drug characteristics, the use of the 0.25 significance level for other design factors rather than batch, the impact of random batch effects on statistical inference in complicated stability studies, the impact of the deviation from linearity on the validity and reliability of estimated drug shelf-life, and sample size justification based on the precision of the estimated drug shelf-life. These practical issues, which may or may not have been partially or completely resolved in the literature, are briefly described below.

12.4.1 Overage

In the pharmaceutical industry it is common practice to add an amount of active ingredients to account for a possible stability loss over a desired expiration period, provided that the resulting initial value is within the USP/NF upper specification limit. The additional amount of active ingredient is called an *overage* of the drug product. For example, suppose the desired shelf-life of a drug product is 36 months and the expected stability loss over 36 months is 15% of label claim, and the USP/NF specification limits are (90%,110%) of label claim. Then, the pharmaceutical company may consider adding an overage of 5% of the label claim to the drug product to ensure that the drug product will remain within the (90%,110%) specification limits at the end of shelf-life. The overage of 5% may result in an initial average potency of 105%. Adding an overage of 5% may increase the manufacturing cost and increase the chance of a potential safety problem due to occasional instances of superpotency. If we take into consideration possible expected and unexpected sources of variation, the true strength at the time the drug product is manufactured may exceed the upper approved specification limit of 110%. Thus, when an overage is to be added to account for the possible stability loss over the desired shelf-life, it is suggested that all possible sources of variability be taken into account for determining drug shelf-life.

12.4.2 Assay Sensitivity

As discussed in Chapters 3 to 6, it is ideal to have a small σ^2 in order to enjoy the good statistical property of a small error asymptotic. However, for some drug products, σ^2 may be large owing to the analytical method used when the drug product was developed. A large σ^2 will definitely have an negative impact on stability analysis for estimating an accurate and reliable drug shelf-life because the assay may not be sensitive enough to detect a samll degradation in a short period of time. As a result, a stability analysis based on assay results with poor sensitivity may provide a biased estimate of the drug shelf-life. Consequently, the drug product may fail to meet the approved specification limit prior to its expiration date. Thus, it is suggested that the assay sensitivity be taken into consideration when performing a stability analysis for drug products with poor assay sensitivity, especially for drug products approved by the FDA prior to 1962.

To account for the relative nature of the data variability due to assay sensitivity, Pogany (2006) suggested that the following capability index be employed:

$$C_p = \frac{UCL - LCL}{6\hat{\sigma}},$$

where (LCL, UCL) and $\hat{\sigma}$ are the acceptance limits and sample standard deviation of the assay results, respectivley. If $C_p > 2.5$, we conclude that there is little or no data variability (see also Bar, 2003).

12.4.3   Multiple Drug Characteristics

Stability analyses with multiple drug characteristics can be classified into three categories. The first category includes multivariate analyses based on a primary drug characteristic such as potency (strength) and a secondary drug characteristic such as color or odor. As indicated in Chapter 8, the primary drug characteristic is a continuous variable, and the secondary drug characteristics may be a discrete response. As a result, it is a challenge to determine drug shelf-life based on a multivariate analysis with one continuous variable and one discrete response. The second category is referred to as stability analysis of potency (percent of label claim) and dissolution (not only USP/NF dissolution testing, but also dissolution profile testing) assuming there is a correlation between potency and dissolution over time. As indicated in Chapter 1, stability (potency) and dissolution were the top 3 and 5 reasons for drug recalls in the fiscal year of 2004. One could consider incorporting both drug characteristics for an overall assessment of the drug expiration dating period. The third category includes drug products with multiple components, as discussed in Chapter 9. Chow and Shao (2007) proposed a method for determining shelf-life for drug products with multiple components under certain assumptions. These assumptions, however, cannot be verified because the pharmacological activities of these components and the interactions among them are usually unknown (Chow, Pong, and Chang, 2006).

Stability analysis with multiple characteristics as described above is a challenge to pharmaceutical researchers and scientists and biostatisticians. Further research is needed.

12.4.4   Batch Similarity

As indicated in the 1987 FDA stability guideline, a 0.25 level of significance is recommended for testing batch similarity. The 0.25 level of significance has been criticized by many sponsors as being too conservative for future production batches. Lin and Tsong (1990) pointed out that the level of significance required for a given minimum relative efficiency of the estimate based on results of the pooling test (in comparison to simply using the worst batch estimate) depends on sample size, time points measured, mean slope of all batches, and tightness of the stability data. In stability analyses the poolability of batches is not limited to batches of the same strength or package type. In practice, we may have several strengths that are packed in different types of packages. As a result, it is not clear whether the same 0.25 level of significance should be applied to other design factors when pooling batches for an overall assessment of drug shelf-life. Current practice is to use the 0.25 level of significance for the design factor of batch and use 0.05 for the design factors of strength and package type. In practice, it is of interest to study the impact on shelf-life estimation using different levels of significance on different design factors. Statistical justifications for selecting significance levels for design factors need to be provided in a full-scale stability analysis with all design factors.

In Chapter 6 we introduced several methods for determining drug shelf-life under the random effects model. These methods may be applied to some complicated situations such as the two-phase analysis for determining drug shelf-life for frozen products (Chapter 10). This requires further research.

12.4.5 Deviation from Linearity

Throughout this book estimation of an expiration dating period for a drug product is based on fitting a linear regression model to the percent of label claim on the original scale. In other words, the assumption of a linear degradation is made for the entire range of time points selected in the stability study. However, as pointed out in Chapter 2, the true degradation function is generally not a linear function rather than a simple exponential first order function. Morris (1992) compares the differences in estimated shelf-life between a linear regression and a simple exponential model. The results indicate that if the total amount of degradation over the entire range of time points is less than 15% of the label claim, the difference in estimated shelf-life between the two models is quite trivial. If the loss of strength is expected to be greater than 15%, the exponential model should be used for estimating the expiration dating period. Under the current FDA stability guidelines, the acceptable lower specification limit specified in the USP/NF is usually set to be 90%. This indicates that only a maximum amount of degradation of 10% is allowed. As a result, the current method for fitting a linear regression model to the percent of label claim on the original scale is adequate. An estimated shelf-life based on extrapolation from the estimated regression line over the range of the time intervals observed may be biased because it is not known whether an empirical linear relationship between the strength and time still holds from the last observed time interval to the estimated shelf-life. It is suggested that the use of extrapolation beyond the range of observed time points be carefully evaluated. Thus, the 1987 FDA stability guideline requires that an expiration dating period granted on the basis of extrapolation be verified by stability data up to the granted expiration time as soon as these data become available.

12.4.6 Time-Dependent Degradation

In the previous chapters, for determination of drug shelf-life, we assume that the primary drug characteristic such as potency for stability testing is expected to decrease linearly over time. The degradation rate is a fixed constant. In this case statistical methods suggested in the FDA stability guidelines can be used to determine the drug shelf-life based on the collected stability data (FDA, 1987, 1998). However, the primary drug characteristic of some drug products may not decrease linearly over time, and the degradation rate may be time dependent. A typical example is drug products containing levothyroxine sodium, which is the sodium salt of the levo isomer of the thyroid hormone thyroxine. Thyroid hormones affect protein, lipid, and carbohydrate metabolism, growth, and development. As indicated in Issue No. 157 of *Federal Register*, there is evidence showing significant stability and potency problems with orally administered levothyroxine sodium products (*Federal Register*, 1997). These products fail to maintain potency through the expiration date. In addition, tablets of the same dosage strength from the same manufacturer vary from lot to lot in the amount of active ingredient present. This lack of stability and consistent potency has the potential to cause serious health consequences to the public.

In a stability study Won (1992) reported that levothyroxine sodium exhibits a biphasic first-order degradation profile with an initial fast degradation rate followed by a slower rate. This observation suggests a time-dependent degradation for drug products containing levothyroxine sodium. In this case the usual approach for determining drug shelf-life is not appropriate. As a result, future research for appropriate statistical methods for determining drug shelf-life under the assumption of time-dependent degradation is needed. In addition, it is of interest in studying the impact of assay sensitivity on the possible inconsistent potency over time for drug products approved prior to 1962, such as drug products containing levothyroxine sodium. There was no standard for the use of a stability-indicating assay for quality control until 1982 (Garnick et al., 1982). The accuracy and reliability of the stability-indicating assay based on high-performance liquid chromatography may not be sensitive enough to provide an accurate and reliable estimate of drug shelf-life under the assumption of time-dependent degradation.

12.4.7 Sample Size Justification

Both the FDA stability guidelines and the ICH guidelines for stability recommend that stability testing be done at 3-month intervals during the first year, 6-month intervals during the second year, and annually thereafter. The 1987 FDA stability guideline also encourages testing an increasing number of replicates at later sampling time points. However, no statistical justification for these recommendations is provided. As indicated by Chow, Shao, and Wang (2003), sample size justification can be made based on either power analysis or precision assessment. Ju and Chow (1995) pointed out that the primary objective of a stability study is to estimate the shelf-life of a drug product. Thus, the precision of the estimated shelf-life should be assured. It is therefore suggested that sample size justification be provided based on the precision of an estimated drug shelf-life. The sample size justification should include knowledge of assay sensitivity, safety margin (or tolerable limit of the estimated shelf-life), and design factors. For example, the following statement can be provide to justify the selected sample size: The selected sample size (e.g., number of batches for each combination of strength and package type, number of replicates at prespecified sampling time intervals or additional sampling time intervals) will provide a 95% assurance that the true shelf-life is within the safety margin of 2 months or 24 months assuming that the assay sensitivity is less than 5% of label claim.

12.5 SAS Programs for Stability Analysis

To assist the sponsors in stability analysis, the Division of Biometrics, Center for Drug Evaluation and Research (CDER) at the FDA and has developed a SAS/PC program system STAB for estimating the expiration dating period of a drug product based on linear regression analysis. The SAS/PC program system STAB was released on March 23, 1992, and updated on July 13, 2005. The statistical foundation of the SAS program was documented in the 1987 FDA stability guideline *Guideline for Submitting Documentation for Human Drugs and Biologics.* SAS/PC version 6.03 is required to run the program. DOS version 3 or later is assumed in these instructions, but not required.

As indicated on the FDA website (http://www.fda.gov/cder/sas/index.htm), to install the program, one has to make sure that the CONFIG.SYS file contains the lines

```
DEVICE=C:\DOS\ANSI.SYS
```

```
FILES=50
```

Then, reboot the system after the changes are made in the CONFIG.SYS file. If the DOS operating system is not in the directory C:\DOS, create an appropriate directory. Note that without ANSI.SYS, one cannot leave the SAS display manager by the "X" command. Without sufficient file handles, the stability program will not run. One can load the software (presumably from floppy drive A, to the current default drive and SAS directory) with the following commands or equivalent procedure for loading:

```
MKDIR\SAS\STAB
```

```
XCOPY A:*.* \SAS\STAB
```

The STAB system contains a data file, six SAS macro files, and an optional SAS/GRAPH file:

STAB Files		
SAS Macro Files	**Control Files**	**Help Files**
ANALYS1.SAS	EXAMPLE.DAT	HELPSTAB.BAT
ANALYS2.SAS	STAB.DAT	README.TXT
ANALYS3.SAS	STABNC.DAT	
ANALYS4.SAS		
ANALYS5.SAS		
ANALYS6.SAS		
STABGRAF.SAS		

Stab.dat is a SAS control file, to which the user adds current variable values, titles, and the data. STABNC.DAT is an alternate SAS control file with a few comments. ANALYS1.SAS to ANALYS6.SAS are the SAS macro programs, while STABGRAF.SAS is an optional program for SAS/GRAPH. Readme.bat and Helpstab.bat are optional files to display the document. To apply the STAB system, one should follow the following steps:

- Step 1: Make a duplicate of the STAB.DAT file (e.g., STAB1.DAT) to avoid changing the original.

- Step 2: Start SAS (presumably using the display manager with \SAS as the default directory.)

- Step 3: Load STAB1.DAT by entering INCLUDE 'STAB/STAB1.DAT' on the command line.

- Step 4: When STAB1.DAT is loaded, enter the data and other modifications, as per the instructions given in comments in STAB1.DAT.

- Step 5: After completing data and information entry, enter Submit on the command line to run the STAB system.

- Step 6: Check the log listing for errors, and if there are none, save the output listing (lst) as detailed in the comments given in STAB.DAT.

- Step 7: If you want high resolution SAS graphs, enter Clear on the command line of the program window if a program is visible.

- Step 8: The STABGRAF.SAS file uses a permanent dataset created by STAB.DAT, so STAB.DAT needs to be run before STABGRAF.SAS. Load STABGRAF.SAS by entering on the command line

```
INCLUDE 'STAB\STABGRAF.SAS'
```

- Step 9: Make any modifications needed (check graphics device type and the axis scales) and run by entering Submit on the command line.

The program may run through a series of graphs. After viewing the current graph, press the space bar once. The primary SAS programs are listed in Appendix B. For further information regarding the SAS/PC program system STAB, the readers should contact Atiar Rahman at rahmanat@cder.fda.gov.

Appendix A

Guidance for Industry[1]

Q1A(R2) Stability Testing of New Drug Substances and Products

> This guidance represents the Food and Drug Administration's (FDA's) current thinking on this topic. It does not create or confer any rights for or on any person and does not operate to bind FDA or the public. You can use an alternative approach if the approach satisfies the requirements of the applicable statutes and regulations. If you want to discuss an alternative approach, contact the FDA staff responsible for implementing this guidance. If you cannot identify the appropriate FDA staff, call the appropriate number listed on the title page of this guidance.

A.1 Introduction[2]

This guidance is the second revision of *Q1A Stability Testing of New Drug Substances and Products,* which was first published in September 1994 and revised in August 2001. The purpose of this revision is to harmonize the intermediate storage condition for zones I and II with the long-term condition for zones III and IV recommended in the ICH guidance *Q1F Stability Data Package for Registration Applications in Climatic Zones III and IV.* The changes made in this second revision are listed in the attachment to this guidance.

A.1.1 Objectives of the Guidance

This guidance is intended to define what stability data package for a new drug substance or drug product is sufficient for a registration application within the three

---

[1] This guidance was developed within the Expert Working Group (Quality) of the International Conference on Harmonisation of Technical Requirements for Registration of Pharmaceuticals for Human Use (ICH) and has been subject to consultation by the regulatory parties, in accordance with the ICH process. This document was endorsed by the ICH Steering Committee at *Step 4* of the ICH process, February 2003. At *Step 4* of the process, the final draft is recommended for adoption to the regulatory bodies of the European Union, Japan, and the United States.

[2] Arabic numbers reflect the organizational breakdown in the document endorsed by the ICH Steering Committee at Step 4 of the ICH process.

regions of the European Union (EU), Japan, and the United States. It does not seek to address the testing for registration in or export to other areas of the world. The guidance exemplifies the core stability data package for new drug substances and products, but leaves sufficient flexibility to encompass the variety of different practical situations that may be encountered due to specific scientific considerations and characteristics of the materials being evaluated. Alternative approaches can be used when there are scientifically justifiable reasons.

A.1.2 Scope of the Guidance

The guidance addresses the information to be submitted in registration applications for new molecular entities and associated drug products. This guidance does not currently seek to cover the information to be submitted for abbreviated or abridged applications, variations, or clinical trial applications.

Specific details of the sampling and testing for particular dosage forms in their proposed container closures are not covered in this guidance.

Further guidance on new dosage forms and on biotechnological/biological products can be found in ICH guidances *Q1C Stability Testing for New Dosage Forms and Q5C Quality of Biotechnological Products: Stability Testing of Biotechnological/Biological Products*, respectively.

A.1.3 General Principles

The purpose of stability testing is to provide evidence on how the quality of a drug substance or drug product varies with time under the influence of a variety of environmental factors, such as temperature, humidity, and light, and to establish a retest period for the drug substance or a shelf life for the drug product and recommended storage conditions.

The choice of test conditions defined in this guidance is based on an analysis of the effects of climatic conditions in the three regions of the EU, Japan, and the United States. The mean kinetic temperature in any part of the world can be derived from climatic data, and the world can be divided into four climatic zones, I-IV. This guidance addresses climatic zones I and II. The principle has been established that stability information generated in any one of the three regions of the EU, Japan, and the United States would be mutually acceptable to the other two regions, provided the information is consistent with this guidance and the labeling is in accord with national/regional requirements.

FDA's guidance documents, including this guidance, do not establish legally enforceable responsibilities. Instead, guidances describe the Agency's current thinking on a topic and should be viewed only as recommendations, unless specific regulatory or statutory requirements are cited. The use of the word *should* in Agency guidances means that something is suggested or recommended, but not required.

A.2 Guidance

A.2.1 Drug Substance

A.2.1.1 General

Information on the stability of the drug substance is an integral part of the systematic approach to stability evaluation.

A.2.1.2 Stress Testing

Stress testing of the drug substance can help identify the likely degradation products, which can in turn help establish the degradation pathways and the intrinsic stability of the molecule and validate the stability indicating power of the analytical procedures used. The nature of the stress testing will depend on the individual drug substance and the type of drug product involved.

Stress testing is likely to be carried out on a single batch of the drug substance. The testing should include the effect of temperatures (in 10°C increments (e.g., 50°C, 60°C) above that for accelerated testing), humidity (e.g., 75 percent relative humidity or greater) where appropriate, oxidation, and photolysis on the drug substance. The testing should also evaluate the susceptibility of the drug substance to hydrolysis across a wide range of pH values when in solution or suspension. Photostability testing should be an integral part of stress testing. The standard conditions for photostability testing are described in ICH *Q1B Photostability Testing of New Drug Substances and Products*.

Examining degradation products under stress conditions is useful in establishing degradation pathways and developing and validating suitable analytical procedures. However, such examination may not be necessary for certain degradation products if it has been demonstrated that they are not formed under accelerated or long-term storage conditions.

Results from these studies will form an integral part of the information provided to regulatory authorities.

A.2.1.3 Selection of Batches

Data from formal stability studies should be provided on at least three primary batches of the drug substance. The batches should be manufactured to a minimum of pilot scale by the same synthetic route as production batches and using a method of manufacture and procedure that simulates the final process to be used for production batches. The overall quality of the batches of drug substance placed on formal stability studies should be representative of the quality of the material to be made on a production scale.

Other supporting data can be provided.

A.2.1.4   Container Closure System

The stability studies should be conducted on the drug substance packaged in a container closure system that is the same as or simulates the packaging proposed for storage and distribution.

A.2.1.5   Specification

Specification, which is a list of tests, references to analytical procedures, and proposed acceptance criteria, is addressed in ICH *Q6A Specifications: Test Procedures and Acceptance Criteria for New Drug Substances and New Drug Products: Chemical Substances* and *Q6B Specifications: Test Procedures and Acceptance Criteria for New Drug Substances and New Drug Products: Biotechnological/Biological Products*. In addition, specification for degradation products in a drug substance is discussed in ICH *Q3A Impurities in New Drug Substances*.

Stability studies should include testing of those attributes of the drug substance that are susceptible to change during storage and are likely to influence quality, safety, and/or efficacy. The testing should cover, as appropriate, the physical, chemical, biological, and microbiological attributes. Validated stability-indicating analytical procedures should be applied. Whether and to what extent replication should be performed should depend on the results from validation studies.

A.2.1.6   Testing Frequency

For long-term studies, frequency of testing should be sufficient to establish the stability profile of the drug substance. For drug substances with a proposed retest period of at least 12 months, the frequency of testing at the long-term storage condition should normally be every 3 months over the first year, every 6 months over the second year, and annually thereafter through the proposed retest period.

At the accelerated storage condition, a minimum of three time points, including the initial and final time points (e.g., 0, 3, and 6 months), from a 6-month study is recommended. Where an expectation (based on development experience) exists that the results from accelerated studies are likely to approach significant change criteria, increased testing should be conducted either by adding samples at the final time point or including a fourth time point in the study design.

When testing at the intermediate storage condition is called for as a result of significant change at the accelerated storage condition, a minimum of four time points, including the initial and final time points (e.g., 0, 6, 9, 12 months), from a 12-month study is recommended.

A.2.1.7   Storage Conditions

In general, a drug substance should be evaluated under storage conditions (with appropriate tolerances) that test its thermal stability and, if applicable, its sensitivity to moisture. The storage conditions and the lengths of studies chosen should be sufficient to cover storage, shipment, and subsequent use.

The long-term testing should cover a minimum of 12 months' duration on at least three primary batches at the time of submission and should be continued for a period of time sufficient to cover the proposed retest period. Additional data accumulated during the assessment period of the registration application should be submitted to the authorities if requested. Data from the accelerated storage condition and, if appropriate, from the intermediate storage condition can be used to evaluate the effect of short-term excursions outside the label storage conditions (such as might occur during shipping).

Long-term, accelerated, and, where appropriate, intermediate storage conditions for drug substances are detailed in the sections below. The general case should apply if the drug substance is not specifically covered by a subsequent section. Alternative storage conditions can be used if justified.

A.2.1.7.1   General Case

Study	Storage Condition	Minimum Time Period Covered by Data at Submission
Long-term*	$25°C \pm 2°C/60\%$ RH $\pm$ 5% RH or $30°C \pm 2°C/65\%$ RH $\pm$ 5% RH	12 months
Intermediate**	$30°C \pm 2°C/65\%$ RH $\pm$ 5% RH	6 months
Accelerated	$40°C \pm 2°C/75\%$ RH $\pm$ 5% RH	6 months

*It is up to the applicant to decide whether long-term stability studies are performed at $25°C \pm 2°C/60\%$ RH $\pm$ 5% RH or $30°C \pm 2°C/65\%$ RH $\pm$ 5% RH.
** If $30°C \pm 2°C/65\%$ RH $\pm$ 5% RH is the long-term condition, there is no intermediate condition.

If long-term studies are conducted at $25°C \pm 2°C/60\%$ RH $\pm$ 5% RH and *significant change* occurs at any time during 6 months' testing at the accelerated storage condition, additional testing at the intermediate storage condition should be conducted and evaluated against significant change criteria. Testing at the intermediate storage condition should include all tests, unless otherwise justified. The initial application should include a minimum of 6 months' data from a 12-month study at the intermediate storage condition.

Significant change for a drug substance is defined as failure to meet its specification.

A.2.1.7.2   Drug Substances Intended for Storage in a Refrigerator

Study	Storage Condition	Minimum Time Period Covered by Data at Submission
Long-term	$5°C \pm 3°C$	12 months
Accelerated	$25°C \pm 2°C/60\%$ RH $\pm$ 5% RH	6 months

Data from refrigerated storage should be assessed according to the evaluation section of this guidance, except where explicitly noted below.

If significant change occurs between 3 and 6 months' testing at the accelerated storage condition, the proposed retest period should be based on the real time data available at the long-term storage condition.

If significant change occurs within the first 3 months' testing at the accelerated storage condition, a discussion should be provided to address the effect of short-term excursions outside the label storage condition (e.g., during shipping or handling). This discussion can be supported, if appropriate, by further testing on a single batch of the drug substance for a period shorter than 3 months but with more frequent testing than usual. It is considered unnecessary to continue to test a drug substance through 6 months when a significant change has occurred within the first 3 months.

A.2.1.7.3   Drug Substances Intended for Storage in a Freezer

Study	Storage Condition	Minimum Time Period Covered by Data at Submission
Long-term	$-20°C \pm 5°C$	12 months

For drug substances intended for storage in a freezer, the retest period should be based on the real time data obtained at the long-term storage condition. In the absence of an accelerated storage condition for drug substances intended to be stored in a freezer, testing on a single batch at an elevated temperature (e.g., $5°C \pm 3°C$ or $25°C \pm 2°C$) for an appropriate time period should be conducted to address the effect of short-term excursions outside the proposed label storage condition (e.g., during shipping or handling).

A.2.1.7.4   Drug Substances Intended for Storage Below $-20°C$

Drug substances intended for storage below $-20°C$ should be treated on a case-by-case basis.

A.2.1.8   Stability Commitment

When available long-term stability data on primary batches do not cover the proposed retest period granted at the time of approval, a commitment should be made to continue the stability studies postapproval to firmly establish the retest period.

Where the submission includes long-term stability data on three production batches covering the proposed retest period, a postapproval commitment is considered unnecessary. Otherwise, one of the following commitments should be made:

- If the submission includes data from stability studies on at least three production batches, a commitment should be made to continue these studies through the proposed retest period.

- If the submission includes data from stability studies on fewer than three production batches, a commitment should be made to continue these studies through the proposed retest period and to place additional production batches, to a total of at least three, on long-term stability studies through the proposed retest period.

- If the submission does not include stability data on production batches, a commitment should be made to place the first three production batches on long-term stability studies through the proposed retest period.

The stability protocol used for long-term studies for the stability commitment should be the same as that for the primary batches, unless otherwise scientifically justified.

A.2.1.9 Evaluation

The purpose of the stability study is to establish, based on testing a minimum of three batches of the drug substance and evaluating the stability information (including, as appropriate, results of the physical, chemical, biological, and microbiological tests), a retest period applicable to all future batches of the drug substance manufactured under similar circumstances. The degree of variability of individual batches affects the confidence that a future production batch will remain within specification throughout the assigned retest period.

The data may show so little degradation and so little variability that it is apparent from looking at the data that the requested retest period will be granted. Under these circumstances, it is normally unnecessary to go through the formal statistical analysis; providing a justification for the omission should be sufficient.

An approach for analyzing the data on a quantitative attribute that is expected to change with time is to determine the time at which the 95 percent, one-sided confidence limit for the mean curve intersects the acceptance criterion. If analysis shows that the batch-to-batch variability is small, it is advantageous to combine the data into one overall estimate. This can be done by first applying appropriate statistical tests (e.g., p values for level of significance of rejection of more than 0.25) to the slopes of the regression lines and zero time intercepts for the individual batches. If it is inappropriate to combine data from several batches, the overall retest period should be based on the minimum time a batch can be expected to remain within acceptance criteria.

The nature of any degradation relationship will determine whether the data should be transformed for linear regression analysis. Usually the relationship can be represented by a linear, quadratic, or cubic function on an arithmetic or logarithmic scale. Statistical methods should be employed to test the goodness of fit of the data on all batches and combined batches (where appropriate) to the assumed degradation line or curve.

Limited extrapolation of the real time data from the long-term storage condition beyond the observed range to extend the retest period can be undertaken at approval time if justified. This justification should be based, for example, on what is known about the mechanism of degradation, the results of testing under accelerated conditions, the goodness of fit of any mathematical model, batch size, and/or existence of supporting stability data. However, this extrapolation assumes that the same degradation relationship will continue to apply beyond the observed data.

Any evaluation should cover not only the assay, but also the levels of degradation products and other appropriate attributes.

A.2.1.10   Statements/Labeling

A storage statement should be established for the labeling in accordance with relevant national/regional requirements. The statement should be based on the stability evaluation of the drug substance. Where applicable, specific instructions should be provided, particularly for drug substances that cannot tolerate freezing. Terms such as *ambient conditions* or *room temperature* should be avoided.

A retest period should be derived from the stability information, and a retest date should be displayed on the container label if appropriate.

A.2.2   Drug Product

A.2.2.1   General

The design of the formal stability studies for the drug product should be based on knowledge of the behavior and properties of the drug substance, results from stability studies on the drug substance, and experience gained from clinical formulation studies. The likely changes on storage and the rationale for the selection of attributes to be tested in the formal stability studies should be stated.

A.2.2.2   Photostability Testing

Photostability testing should be conducted on at least one primary batch of the drug product if appropriate. The standard conditions for photostability testing are described in ICH Q1B.

A.2.2.3   Selection of Batches

Data from stability studies should be provided on at least three primary batches of the drug product. The primary batches should be of the same formulation and packaged in the same container closure system as proposed for marketing. The manufacturing process used for primary batches should simulate that to be applied to production batches and should provide product of the same quality and meet the same specification as that intended for marketing. Two of the three batches should be at least pilot scale batches, and the third one can be smaller if justified. Where possible, batches of the drug product should be manufactured by using different batches of the drug substance.

Stability studies should be performed on each individual strength and container size of the drug product unless bracketing or matrixing is applied.

Other supporting data can be provided.

A.2.2.4   Container Closure System

Stability testing should be conducted on the dosage form packaged in the container closure system proposed for marketing (including, as appropriate, any secondary packaging and container label). Any available studies carried out on the drug product outside its immediate container or in other packaging materials can form a useful part of the stress testing of the dosage form or can be considered as supporting information, respectively.

A.2.2.5 Specification

Specification, which is a list of tests, references to analytical procedures, and proposed acceptance criteria, including the concept of different acceptance criteria for release and shelf life specifications, is addressed in ICH Q6A and Q6B. In addition, specification for degradation products in a drug product is addressed in ICH *Q3B Impurities in New Drug Products.*

Stability studies should include testing of those attributes of the drug product that are susceptible to change during storage and are likely to influence quality, safety, and/or efficacy. The testing should cover, as appropriate, the physical, chemical, biological, and microbiological attributes, preservative content (e.g., antioxidant, antimicrobial preservative), and functionality tests (e.g., for a dose delivery system). Analytical procedures should be fully validated and stability indicating. Whether and to what extent replication should be performed will depend on the results of validation studies.

Shelf life acceptance criteria should be derived from consideration of all available stability information. It may be appropriate to have justifiable differences between the shelf life and release acceptance criteria based on the stability evaluation and the changes observed on storage. Any differences between the release and shelf life acceptance criteria for antimicrobial preservative content should be supported by a validated correlation of chemical content and preservative effectiveness demonstrated during drug development on the product in its final formulation (except for preservative concentration) intended for marketing. A single primary stability batch of the drug product should be tested for antimicrobial preservative effectiveness (in addition to preservative content) at the proposed shelf life for verification purposes, regardless of whether there is a difference between the release and shelf life acceptance criteria for preservative content.

A.2.2.6 Testing Frequency

For long-term studies, frequency of testing should be sufficient to establish the stability profile of the drug product. For products with a proposed shelf life of at least 12 months, the frequency of testing at the long-term storage condition should normally be every 3 months over the first year, every 6 months over the second year, and annually thereafter through the proposed shelf life.

At the accelerated storage condition, a minimum of three time points, including the initial and final time points (e.g., 0, 3, and 6 months), from a 6-month study is recommended. Where an expectation (based on development experience) exists that results from accelerated testing are likely to approach significant change criteria, increased testing should be conducted either by adding samples at the final time point or by including a fourth time point in the study design.

When testing at the intermediate storage condition is called for as a result of significant change at the accelerated storage condition, a minimum of four time points, including the initial and final time points (e.g., 0, 6, 9, 12 months), from a 12-month study is recommended.

Reduced designs (i.e., matrixing or bracketing), where the testing frequency is reduced or certain factor combinations are not tested at all, can be applied if justified.

A.2.2.7   Storage Conditions

In general, a drug product should be evaluated under storage conditions (with appropriate tolerances) that test its thermal stability and, if applicable, its sensitivity to moisture or potential for solvent loss. The storage conditions and the lengths of studies chosen should be sufficient to cover storage, shipment, and subsequent use.

Stability testing of the drug product after constitution or dilution, if applicable, should be conducted to provide information for the labeling on the preparation, storage condition, and in-use period of the constituted or diluted product. This testing should be performed on the constituted or diluted product through the proposed in-use period on primary batches as part of the formal stability studies at initial and final time points, and if full shelf life, long-term data will not be available before submission, at 12 months or the last time point for which data will be available. In general, this testing need not be repeated on commitment batches.

The long-term testing should cover a minimum of 12 months' duration on at least three primary batches at the time of submission and should be continued for a period of time sufficient to cover the proposed shelf life. Additional data accumulated during the assessment period of the registration application should be submitted to the authorities if requested. Data from the accelerated storage condition and, if appropriate, from the intermediate storage condition can be used to evaluate the effect of short-term excursions outside the label storage conditions (such as might occur during shipping).

Long-term, accelerated; and, where appropriate, intermediate storage conditions for drug products are detailed in the sections below. The general case should apply if the drug product is not specifically covered by a subsequent section. Alternative storage conditions can be used if justified.

A.2.2.7.1   General Case

Study	Storage Condition	Minimum Time Period Covered by Data at Submission
Long-term*	25°C ± 2°C/60% RH ± 5% RH or 30°C ± 2°C/65% RH ± 5% RH	12 months
Intermediate**	30°C ± 2°C/65% RH ± 5% RH	6 months
Accelerated	40°C ± 2°C/75% RH ± 5% RH	6 months

*It is up to the applicant to decide whether long-term stability studies are performed at 25°C ± 2°C/60% RH ± 5% RH or 30°C ± 2°C/65% RH ± 5% RH.
**If 30°C ± 2°C/65% RH ± 5% RH is the long-term condition, there is no intermediate condition.

If long-term studies are conducted at 25°C ± 2°C/60% RH ± 5% RH and *significant change* occurs at any time during 6 months' testing at the accelerated storage condition, additional testing at the intermediate storage condition should be conducted and evaluated against significant change criteria. The initial application should include a minimum of 6 months' data from a 12-month study at the intermediate storage condition.

In general, *significant change* for a drug product is defined as one or more of the following (as appropriate for the dosage form):

- A 5 percent change in assay from its initial value, or failure to meet the acceptance criteria for potency when using biological or immunological procedures

- Any product's degradation exceeding its acceptance criterion

- Failure to meet the acceptance criteria for appearance, physical attributes, and functionality test (e.g., color, phase separation, resuspendibility, caking, hardness, dose delivery per actuation). However, some changes in physical attributes (e.g., softening of suppositories, melting of creams) may be expected under accelerated conditions

- Failure to meet the acceptance criterion for pH

- Failure to meet the acceptance criteria for dissolution for 12 dosage units

A.2.2.7.2 Drug Products Packaged in Impermeable Containers

Sensitivity to moisture or potential for solvent loss is not a concern for drug products packaged in impermeable containers that provide a permanent barrier to passage of moisture or solvent. Thus, stability studies for products stored in impermeable containers can be conducted under any controlled or ambient humidity condition.

A.2.2.7.3 Drug Products Packaged in Semipermeable Containers

Aqueous-based products packaged in semipermeable containers should be evaluated for potential water loss in addition to physical, chemical, biological, and microbiological stability. This evaluation can be carried out under conditions of low relative humidity, as discussed below. Ultimately, it should be demonstrated that aqueous-based drug products stored in semipermeable containers can withstand low relative humidity environments. Other comparable approaches can be developed and reported for nonaqueous, solvent-based products.

Study	Storage Condition	Minimum Time Period Covered by Data at Submission
Long-term*	$25°C \pm 2°C/40\%$ RH $\pm 5\%$ RH or $30°C \pm 2°C/35\%$ RH $\pm 5\%$ RH	12 months
Intermediate**	$30°C \pm 2°C/65\%$ RH $\pm 5\%$ RH	6 months
Accelerated	$40°C \pm 2°C/$not more than (NMT) 25% RH	6 months

*It is up to the applicant to decide whether long-term stability studies are performed at $25°C \pm 2°C/40\%$ RH $\pm 5\%$ RH or $30°C \pm 2°C/35\%$ RH $\pm 5\%$ RH.

**If $30°C \pm 2°C/35\%$ RH $\pm 5\%$ RH is the long-term condition, there is no intermediate condition.

When long-term studies are conducted at 25°C ± 2°C/40% RH ± 5% RH and significant change other than water loss occurs during the 6 months' testing at the accelerated storage condition, additional testing at the intermediate storage condition should be performed, as described under the general case, to evaluate the temperature effect at 30°C. A significant change in water loss alone at the accelerated storage condition does not necessitate testing at the intermediate storage condition. However, data should be provided to demonstrate that the drug product will not have significant water loss throughout the proposed shelf life if stored at 25°C and the reference relative humidity of 40 percent RH.

A 5 percent loss in water from its initial value is considered a significant change for a product packaged in a semipermeable container after an equivalent of 3 months' storage at 40°C/NMT 25 percent RH. However, for small containers (1 mL or less) or unit-dose products, a water loss of 5 percent or more after an equivalent of 3 months' storage at 40°C/NMT 25 percent RH may be appropriate if justified.

An alternative approach to studying at the reference relative humidity as recommended in the table above (for either long-term or accelerated testing) is performing the stability studies under higher relative humidity and deriving the water loss at the reference relative humidity through calculation. This can be achieved by experimentally determining the permeation coefficient for the container closure system or, as shown in the example below, using the calculated ratio of water loss rates between the two humidity conditions at the same temperature. The permeation coefficient for a container closure system can be experimentally determined by using the worst case scenario (e.g., the most diluted of a series of concentrations) for the proposed drug product.

Example of an approach for determining water loss: For a product in a given container closure system, container size, and fill, an appropriate approach for deriving the water loss rate at the reference relative humidity is to multiply the water loss rate measured at an alternative relative humidity at the same temperature by a water loss rate ratio shown in the table below. A linear water loss rate at the alternative relative humidity over the storage period should be demonstrated.

For example, at a given temperature (e.g., 40°C), the calculated water loss rate during storage at NMT 25 percent RH is the water loss rate measured at 75 percent RH multiplied by 3.0, the corresponding water loss rate ratio.

Alternative Relative Humidity	Reference Relative Humidity	Ratio of Water Loss Rates at a Given Temperature
60% RH	25% RH	1.9
60% RH	40% RH	1.5
65% RH	35% RH	1.9
75% RH	25% RH	3.0

Valid water loss rate ratios at relative humidity conditions other than those shown in the table above can also be used.

A.2.2.7.4   Drug Products Intended for Storage in a Refrigerator

Study	Storage Condition	Minimum Time Period Covered by Data at Submission
Long-term	5°C ± 3°C	12 months
Accelerated	25°C ± 2°C/60% RH ± 5%RH	6 months

If the drug product is packaged in a semipermeable container, appropriate information should be provided to assess the extent of water loss.

Data from refrigerated storage should be assessed according to the evaluation section of this guidance, except where explicitly noted below.

If significant change occurs between 3 and 6 months' testing at the accelerated storage condition, the proposed shelf life should be based on the real time data available from the long-term storage condition.

If significant change occurs within the first 3 months' testing at the accelerated storage condition, a discussion should be provided to address the effect of short-term excursions outside the label storage condition (e.g., during shipment and handling). This discussion can be supported, if appropriate, by further testing on a single batch of the drug product for a period shorter than 3 months but with more frequent testing than usual. It is considered unnecessary to continue to test a product through 6 months when a significant change has occurred within the first 3 months.

A.2.2.7.5   Drug Products Intended for Storage in a Freezer

Study	Storage Condition	Minimum Time Period Covered by Data at Submission
Long-term	−20°C ± 5°C	12 months

For drug products intended for storage in a freezer, the shelf life should be based on the real time data obtained at the long-term storage condition. In the absence of an accelerated storage condition for drug products intended to be stored in a freezer, testing on a single batch at an elevated temperature (e.g., 5°C ± 3°C or 25°C ± 2°C) for an appropriate time period should be conducted to address the effect of short-term excursions outside the proposed label storage condition.

A.2.2.7.6   Drug Products Intended for Storage Below −20°C

Drug products intended for storage below −20°C should be treated on a case-by-case basis.

A.2.2.8 Stability Commitment

When available long-term stability data on primary batches do not cover the proposed shelf life granted at the time of approval, a commitment should be made to continue the stability studies postapproval to firmly establish the shelf life.

Where the submission includes long-term stability data from three production batches covering the proposed shelf life, a postapproval commitment is considered unnecessary. Otherwise, one of the following commitments should be made:

- If the submission includes data from stability studies on at least three production batches, a commitment should be made to continue the long-term studies through the proposed shelf life and the accelerated studies for 6 months.

- If the submission includes data from stability studies on fewer than three production batches, a commitment should be made to continue the long-term studies through the proposed shelf life and the accelerated studies for 6 months, and to place additional production batches, to a total of at least three, on long-term stability studies through the proposed shelf life and on accelerated studies for 6 months.

- If the submission does not include stability data on production batches, a commitment should be made to place the first three production batches on long-term stability studies through the proposed shelf life and on accelerated studies for 6 months.

The stability protocol used for studies on commitment batches should be the same as that for the primary batches, unless otherwise scientifically justified.

Where intermediate testing is called for by a significant change at the accelerated storage condition for the primary batches, testing on the commitment batches can be conducted at either the intermediate or the accelerated storage condition. However, if significant change occurs at the accelerated storage condition on the commitment batches, testing at the intermediate storage condition should also be conducted.

A.2.2.9 Evaluation

A systematic approach should be adopted in the presentation and evaluation of the stability information, which should include, as appropriate, results from the physical, chemical, biological, and microbiological tests, including particular attributes of the dosage form (e.g., dissolution rate for solid oral dosage forms).

The purpose of the stability study is to establish, based on testing a minimum of three batches of the drug product, a shelf life and label storage instructions applicable to all future batches of the drug product manufactured and packaged under similar circumstances. The degree of variability of individual batches affects the confidence that a future production batch will remain within specification throughout its shelf life.

Where the data show so little degradation and so little variability that it is apparent from looking at the data that the requested shelf life will be granted, it is normally unnecessary to go through the formal statistical analysis; providing a justification for the omission should be sufficient.

An approach for analyzing data of a quantitative attribute that is expected to change with time is to determine the time at which the 95 percent one-sided confidence limit for the mean curve intersects the acceptance criterion. If analysis shows that the batch-to-batch variability is small, it is advantageous to combine the data into one overall estimate. This can be done by first applying appropriate statistical tests (e.g., p values for level of significance of rejection of more than 0.25) to the slopes of the regression lines and zero time intercepts for the individual batches. If it is inappropriate to combine data from several batches, the overall shelf life should be based on the minimum time a batch can be expected to remain within acceptance criteria.

The nature of the degradation relationship will determine whether the data should be transformed for linear regression analysis. Usually the relationship can be represented by a linear, quadratic, or cubic function on an arithmetic or logarithmic scale. Statistical methods should be employed to test the goodness of fit on all batches and combined batches (where appropriate) to the assumed degradation line or curve.

Limited extrapolation of the real time data from the long-term storage condition beyond the observed range to extend the shelf life can be undertaken at approval time if justified. This justification should be based, for example, on what is known about the mechanisms of degradation, the results of testing under accelerated conditions, the goodness of fit of any mathematical model, batch size, and/or existence of supporting stability data. However, this extrapolation assumes that the same degradation relationship will continue to apply beyond the observed data.

Any evaluation should consider not only the assay but also the degradation products and other appropriate attributes. Where appropriate, attention should be paid to reviewing the adequacy of the mass balance and different stability and degradation performance.

A.2.2.10 Statements/Labeling

A storage statement should be established for the labeling in accordance with relevant national/regional requirements. The statement should be based on the stability evaluation of the drug product. Where applicable, specific instruction should be provided, particularly for drug products that cannot tolerate freezing. Terms such as *ambient conditions* or *room temperature* should be avoided.

There should be a direct link between the label storage statement and the demonstrated stability of the drug product. An expiration date should be displayed on the container label.

A.3 Glossary

The following definitions are provided to facilitate interpretation of the guidance.

Accelerated testing: Studies designed to increase the rate of chemical degradation or physical change of a drug substance or drug product by using exaggerated storage conditions as part of the formal stability studies. Data from these studies, in addition to long-term stability studies, can be used to assess longer term chemical effects at nonaccelerated conditions and to evaluate the effect of short-term excursions outside the label storage conditions such as might occur during shipping. Results from accelerated testing studies are not always predictive of physical changes.

Bracketing: The design of a stability schedule such that only samples on the extremes of certain design factors (e.g., strength, package size) are tested at all time points as in a full design. The design assumes that the stability of any intermediate levels is represented by the stability of the extremes tested. Where a range of strengths is to be tested, bracketing is applicable if the strengths are identical or very closely related in composition (e.g., for a tablet range made with different compression weights of a similar basic granulation, or a capsule range made by filling different plug fill weights of the same basic composition into different size capsule shells). Bracketing can be applied to different container sizes or different fills in the same container closure system.

Climatic zones: The four zones in the world that are distinguished by their characteristic, prevalent annual climatic conditions. This is based on the concept described by W. Grimm (*Drugs Made in Germany*, 28:196–202, 1985 and 29:39–47, 1986).

Commitment batches: Production batches of a drug substance or drug product for which the stability studies are initiated or completed postapproval through a commitment made in the registration application.

Container closure system: The sum of packaging components that together contain and protect the dosage form. This includes primary packaging components and secondary packaging components if the latter are intended to provide additional protection to the drug product. A packaging system is equivalent to a container closure system.

Dosage form: A pharmaceutical product type (e.g., tablet, capsule, solution, cream) that contains a drug substance generally, but not necessarily, in association with excipients.

Drug product: The dosage form in the final immediate packaging intended for marketing.

Drug substance: The unformulated drug substance that may subsequently be formulated with excipients to produce the dosage form.

Excipient: Anything other than the drug substance in the dosage form.

Expiration date: The date placed on the container label of a drug product designating the time prior to which a batch of the product is expected to remain within the approved shelf life specification, if stored under defined conditions, and after which it must not be used.

Formal stability studies: Long-term and accelerated (and intermediate) studies undertaken on primary and/or commitment batches according to a prescribed stability protocol to establish or confirm the retest period of a drug substance or the shelf life of a drug product.

Impermeable containers: Containers that provide a permanent barrier to the passage of gases or solvents (e.g., sealed aluminum tubes for semi-solids, sealed glass ampoules for solutions).

Intermediate testing: Studies conducted at 30°C/65% RH and designed to moderately increase the rate of chemical degradation or physical changes for a drug substance or drug product intended to be stored long-term at 25°C.

Long-term testing: Stability studies under the recommended storage condition for the retest period or shelf life proposed (or approved) for labeling.

Mass balance: The process of adding together the assay value and levels of degradation products to see how closely these add up to 100 percent of the initial value, with due consideration of the margin of analytical error.

Matrixing: The design of a stability schedule such that a selected subset of the total number of possible samples for all factor combinations is tested at a specified time point. At a subsequent time point, another subset of samples for all factor combinations is tested. The design assumes that the stability of each subset of samples tested represents the stability of all samples at a given time point. The differences in the samples for the same drug product should be identified as, for example, covering different batches, different strengths, different sizes of the same container closure system, and, possibly in some cases, different container closure systems.

Mean kinetic temperature: A single derived temperature that, if maintained over a defined period of time, affords the same thermal challenge to a drug substance or drug product as would be experienced over a range of both higher and lower temperatures for an equivalent defined period. The mean kinetic temperature is higher than the arithmetic mean temperature and takes into account the Arrhenius equation.

When establishing the mean kinetic temperature for a defined period, the formula of J. D. Haynes (*J. Pharm. Sci.,* 60:927–929, 1971) can be used.

New molecular entity: An active pharmaceutical substance not previously contained in any drug product registered with the national or regional authority concerned. A new salt, ester, or noncovalent bond derivative of an approved drug substance is considered a new molecular entity for the purpose of stability testing under this guidance.

Pilot scale batch: A batch of a drug substance or drug product manufactured by a procedure fully representative of and simulating that to be applied to a full production scale batch. For solid oral dosage forms, a pilot scale is generally, at a minimum, one-tenth that of a full production scale or 100,000 tablets or capsules, whichever is larger.

Primary batch: A batch of a drug substance or drug product used in a formal stability study, from which stability data are submitted in a registration application for the purpose of establishing a retest period or shelf life, respectively. A primary batch of a drug substance should be at least a pilot scale batch. For a drug product, two of the three batches should be at least pilot scale batch, and the third batch can be smaller if it is representative with regard to the critical manufacturing steps. However, a primary batch may be a production batch.

Production batch: A batch of a drug substance or drug product manufactured at production scale by using production equipment in a production facility as specified in the application.

Retest date: The date after which samples of the drug substance should be examined to ensure that the material is still in compliance with the specification and thus suitable for use in the manufacture of a given drug product.

Retest period: The period of time during which the drug substance is expected to remain within its specification and, therefore, can be used in the manufacture of a given drug product, provided that the drug substance has been stored under the defined conditions. After this period, a batch of drug substance destined for use in the manufacture of a drug product should be retested for compliance with the specification and then used immediately. A batch of drug substance can be retested multiple times and a different portion of the batch used after each retest, as long as it continues to comply with the specification. For most biotechnological/biological substances known to be labile, it is more appropriate to establish a shelf life than a retest period. The same may be true for certain antibiotics.

Semipermeable containers: Containers that allow the passage of solvent, usually water, while preventing solute loss. The mechanism for solvent transport occurs by absorption into one container surface, diffusion through the bulk of the container material, and desorption from the other surface. Transport is driven by a partial pressure gradient. Examples of semipermeable containers include plastic bags and semirigid, low-density polyethylene (LDPE) pouches for large volume parenterals (LVPs), and LDPE ampoules, bottles, and vials.

Shelf life (also referred to as expiration dating period): The time period during which a drug product is expected to remain within the approved shelf life specification, provided that it is stored under the conditions defined on the container label.

Specification: See ICH Q6A and Q6B.

Specification, Release: The combination of physical, chemical, biological, and microbiological tests and acceptance criteria that determine the suitability of a drug product at the time of its release.

Specification, Shelf life: The combination of physical, chemical, biological, and microbiological tests and acceptance criteria that determine the suitability of a drug substance throughout its retest period, or that a drug product should meet throughout its shelf life.

Storage condition tolerances: The acceptable variations in temperature and relative humidity of storage facilities for formal stability studies. The equipment should be capable of controlling the storage condition within the ranges defined in this guidance. The actual temperature and humidity (when controlled) should be monitored during stability storage. Short-term spikes due to opening of doors of the storage facility are accepted as unavoidable. The effect of excursions due to equipment failure should be addressed and reported if judged to affect stability results. Excursions that exceed the defined tolerances for more than 24 hours should be described in the study report and their effect assessed.

Stress testing (drug substance): Studies undertaken to elucidate the intrinsic stability of the drug substance. Such testing is part of the development strategy and is normally carried out under more severe conditions than those used for accelerated testing.

Stress testing (drug product): Studies undertaken to assess the effect of severe conditions on the drug product. Such studies include photostability testing (see ICH Q1B) and specific testing of certain products (e.g., metered dose inhalers, creams, emulsions, refrigerated aqueous liquid products).

Supporting data: Data, other than those from formal stability studies, that support the analytical procedures, the proposed retest period or shelf life, and the label storage statements. Such data include; (1) stability data on early synthetic route batches of drug substance, small-scale batches of materials, investigational formulations not proposed for marketing, related formulations, and product presented in containers and closures other than those proposed for marketing; (2) information regarding test results on containers; and (3) other scientific rationales.

References[3]

ICH Q1B Photostability Testing of New Drug Substances and Products

ICH Q1C Stability Testing for New Dosage Forms

ICH Q3A Impurities in New Drug Substances

---

[3] We update guidances periodically. To make sure you have the most recent version of a guidance, check the CDER guidance page at http://www.fda.gov/cder/guidance/index.htm

ICH Q3B Impurities in New Drug Products

ICH Q5C Quality of Biotechnological Products: Stability Testing of Biotechnological/Biological Products

ICH Q6A Specifications: Test Procedures and Acceptance Criteria for New Drug Substances and New Drug Products: Chemical Substances

ICH Q6B Specifications: Test Procedures and Acceptance Criteria for New Drug Substances and New Drug Products: Biotechnological/Biological Products

---

Attachment

List of Revision 2 Changes

The revisions to this *Q1A* guidance result from adoption of the ICH guidance *Q1F Stability Data Package for Registration Applications in Climatic Zones III and IV*. The following changes were made.

1. The intermediate storage condition has been changed from 30°C ± 2°C/60% RH ± 5% RH to 30°C ± 2°C/65% RH ± 5% RH in the following sections:
   - II.A.7.a (2.1.7.1) Drug Substance - Storage Conditions - General case
   - II.B.7.a (2.2.7.1) Drug Product - Storage Conditions - General case
   - II.B.7.c (2.2.7.3) Drug products packaged in semipermeable containers
   - Glossary (3) *Intermediate testing*

2. 30°C ± 2°C/65% RH ± 5% RH has been added as a suitable alternative long-term storage condition to 25°C ± 2°C/60% RH ± 5% in the following sections:
   - II.A.7.a (2.1.7.1) Drug Substance - Storage Conditions - General case
   - II.B.7.a (2.2.7.1) Drug Product - Storage Conditions - General case

3. 30°C ± 2°C/35% RH ± 5% RH has been added as a suitable alternative long-term storage condition to 25°C ± 2°C/40% RH ± 5% and the corresponding example for the ratio of water-loss rates has been included in the following section:
   - II.B.7.c (2.2.7.3) Drug products packaged in semipermeable containers

Midstream switch of the intermediate storage condition from 30°C ± 2°C/60% RH ± 5% RH to 30°C ± 2°C/65% RH ± 5% RH can be appropriate provided that the respective storage conditions and the date of the switch are clearly documented and stated in the registration application.

It is recommended that registration applications contain data from complete studies at the intermediate storage condition 30°C ± 2°C/65% RH ± 5% RH, if applicable, by three years after the date of publication of this revised guideline in the respective ICH tripartite region.

Appendix B

SAS Macro Files for STAB System for Stability Analysis

ANALYS1.SAS

```
*==============================================================;
* First Macro File of the Stability Analysis Program          ;
*==============================================================;

OPTIONS MPRINT NOSOURCE;                    /* Debugging line */

*%%%%%%%%%%%%%%%%%%%%%%%%%%%%%%%%%%%%%%%%%%%%%%;
*   MTITLE                                               %;
*   Title Macro                                          %;
*%%%%%%%%%%%%%%%%%%%%%%%%%%%%%%%%%%%%%%%%%%%%%%;

%MACRO MTITLE;
    %LET NT=1;
    %DO I=1 %TO 6;
    %IF %QUOTE(&&TT&I) NE   %THEN %DO;
    %LET NT=%EVAL(&NT+1);
       TITLE&NT "&&TT&I";   %END; %END;
%MEND MTITLE;

******************************************************************;
* Program for Reading Input Data and Printing Hard Copy         *;
******************************************************************;

DATA STABLE; SET LIB.&SSDNAME;

IF "&UNIT"="WEEK" THEN TIME=TIME*(7/30);

PROC SORT DATA=STABLE; BY BATCH TIME;

* Print data in a compact format;

DATA STABLE1; SET STABLE; BY BATCH TIME;
RETAIN REP;
IF FIRST.TIME THEN REP=0;
```

```
  REP=REP+1;

PROC SORT DATA=STABLE1; BY TIME REP;

* Macro variable NBATCH = Number of batch;

DATA TEMP; SET STABLE; BY BATCH;
  IF LAST.BATCH;
DATA TEMP; SET TEMP;
  CALL SYMPUT('NBATCH',_N_);

DATA STABLE1; SET STABLE1; BY TIME REP;
PROC TRANSPOSE DATA=STABLE1 OUT=STABLE2;
ID BATCH;
BY TIME REP;
VAR LEVEL;
PROC PRINT DATA=STABLE2(DROP=REP _NAME_) NOOBS;
%MTITLE
RUN;
```

ANALYS2.SAS

```
*==============================================================;
* Second Macro File of the Stability Analysis Program         ;
*==============================================================;

* Tests for equalities of intercepts and slopes;

*************************************************************.
* OUTSTAT= option in the PROC GLM statement produced an output  *;
*       data set that contains:                                 *;
* _SOURCE_: contains the name of the model effect or contrast    *;
*       label from which the corresponding statistics are        *;
*       generated; * e.g. ERROR, INTERCEPT ...;                 *;
* -TYPE_: contains the values SS1,SS2,SS3,SS4 or CONTRAST,;      *;
*       corresponding the various types of sums of squares       *;
* _NAME_: contains the name of one of the dependent variables;   *;
*       in the data set;                                         *;
* also contains  SS: sum square                                 *;
*             DF: degree of freedom                             *;
*             F: F values                                       *;
*          and PROB: probabilities                              *;
*     respectively, for each model generated in the analysis;    *;
*************************************************************.
```

```
TITLE;
PROC GLM DATA=STABLE OUTSTAT=SSTABLE NOPRINT;
CLASS BATCH;
MODEL LEVEL=TIME BATCH TIME*BATCH/INT SS1 SOLUTION;
RUN;
OPTION MISSING=' ';
PROC SORT DATA=SSTABLE; BY _NAME_;

*****************************************************************.
* The following program generates statistical analysis         *;
* from OUSTSTAT of the following form                          *;
* _SOURCE_        _TYPE_      |       SOURCE                    *;
* BATCH           SS1         |       B                        *;
* TIME*BATCH      SS1         |       C                        *;
* ERROR           ERROR       |       D                        *;
* INTERCEPT       SS1         |       E                        *;
*****************************************************************.
DATA SSTABLE; SET SSTABLE; BY _NAME_;
RETAIN SSE DFE SSA DFA 0 DFD SSD;
KEEP SOURCE SS DF MS F P;
P=PROB;
IF _TYPE_ = 'ERROR' THEN DO; SOURCE = 'D';
  MS=SS/DF; DFD=DF; SSD=SS; F=.; P=.; OUTPUT; END;
ELSE DO;
  SSE=SSE+SS;
  DFE=DFE+DF;
  IF _SOURCE_='BATCH' THEN DO;
    SOURCE='B';
    SSA=SSA+SS;  DFA=DFA+DF;
    MS=SS/DF; OUTPUT; END;
  IF _SOURCE_='TIME*BATCH' THEN DO;
    SOURCE='C';
    SSA=SSA+SS;  DFA=DFA+DF;
    MS=SS/DF; OUTPUT; END;
END;
IF LAST._NAME_ THEN DO;
  SOURCE='A'; SS=SSA; DF=DFA;
  MS=SS/DF; F=MS*DFD/SSD; P=1-PROBF(F,DF,DFD); OUTPUT;
F=.; P=.;
  SOURCE='E'; SS=SSE; DF=DFE;
  MS=SS/DF; OUTPUT;
END;
RUN;
PROC SORT DATA=SSTABLE; BY SOURCE;
```

```
PROC PRINT DATA=SSTABLE;
   VAR SS DF MS F P;
   ID SOURCE;

%MTITLE
OPTIONS PS=35;
FOOTNOTE1 '***************************************************';
FOOTNOTE2 '* Statistical Analysis:                         *';
FOOTNOTE3 '*       Key to sources of variation             *';
FOOTNOTE4 '* A = sep. intercep, sep slope | com intercep, com slope   *';
FOOTNOTE5 '* B = sep. intercep, com slope | com intercep, com slope    *';
FOOTNOTE6 '* C = sep. intercep, sep slope | sep intercep, com slope    *';
FOOTNOTE7 '* D = Residual                                  *';
FOOTNOTE8 '* E = Full Model                                *';
FOOTNOTE9 '***************************************************';
RUN;
```

ANALYS3.SAS

```
*=============================================================;
* Third Macro File of the Stability Analysis Program          ;
*=============================================================;

*************************************************************.
* Model selection                                          *;
*    (Tests for poolability of stability batch data)       *;
*    Based on the P-values P_A P_B P_C                     *;
*    MODEL1: common intercept and common slope             *;
*        Single Common Regression                          *;
*    P_B >= 0.25  P_C >= 0.25                              *;
*    MODEL2: separate intercepts and common slope          *;
*        Analysis of Covariance model                      *;
*    P_B < 0.25  P_C >= 0.25                               *;
*    MODEL3: separate intercepts and separate slope        *;
*        Individual Regressions                            *;
*    P_B  < 0.25  P_C < 0.25                               *;
*************************************************************.
OPTIONS PS=58;
DATA MODEL(KEEP=P_A P_B P_C) DF(KEEP=DF_D);
  SET SSTABLE(KEEP=SOURCE P DF);
FOOTNOTE;
RETAIN P_A P_B P_C DF_B DF_C DF_TEMP 0;
  IF SOURCE='A' THEN P_A=P;
```

```
 IF SOURCE='B' THEN DO;
  P_B=P; DF_B=DF; END;
 IF SOURCE='C' THEN DO;
  P_C=P; DF_C=DF;
   IF (P_B GE .25) AND (P_C GE .25) THEN DO;
    CALL SYMPUT ('M', 'MODEL1'); DF_TEMP=DF_B+DF_C; END;
 ELSE IF (P_B LT .25) AND (P_C GE .25) THEN DO;
    CALL SYMPUT ('M', 'MODEL2'); DF_TEMP=DF_B; END;
 ELSE DO;   CALL SYMPUT ('M', 'MODEL3'); DF_TEMP=0; END;
   OUTPUT MODEL; END;
 IF SOURCE='D' THEN DO;
  DF_D=DF+DF_TEMP; OUTPUT DF; END;
FOOTNOTE;
* Create data points for fitting the curve;
PROC MEANS MAX DATA=TEMP NOPRINT;
VAR TIME;
OUTPUT OUT=MAXTIME MAX=MAXT;
RUN;

DATA STABLE3; MERGE TEMP (KEEP=BATCH) MAXTIME (KEEP=MAXT);
KEEP BATCH TIME LEVEL MAXT UTIME;
RETAIN MAXTEMP;
IF _N_=1 THEN MAXTEMP=MAXT;
ELSE MAXT=MAXTEMP;
IF MAXT GE 24 THEN UTIME=MIN(96+(MAXT-24)*3, 84);
ELSE UTIME=4*MAXT;
DO TIME=0 TO UTIME;
  LEVEL=.; OUTPUT; END;
CALL SYMPUT ('MCTIME', UTIME);
DATA STABLNEW;
  MERGE STABLE3 STABLE; BY BATCH TIME;
RUN;
```

ANALYS4.SAS

```
*===============================================================;
* Fourth Macro File of the Stability Analysis Program          ;
*===============================================================;

*%%%%%%%%%%%%%%%%%%%%%%%%%%%%%%%%%%%%%%%%%%%%%%%;
*           MMODEL1                           %;
* MODEL 1 - COMMON INTERCEPT & COMMON SLOPE   %;
* STDP=standard error of the mean predicted value  %;
*%%%%%%%%%%%%%%%%%%%%%%%%%%%%%%%%%%%%%%%%%%%%%%%;
```

```
%MACRO MMODEL1;
   PROC SORT DATA=STABLNEW; BY TIME LEVEL;
   DATA MODELX; SET STABLNEW;
      BY TIME;
   * IF FIRST.TIME NE 1 THEN DO;
   IF NOT (FIRST.TIME) THEN DO;
   IF LEVEL > .;
   END;
   PROC REG DATA=MODELX OUTEST=FITTED NOPRINT;
   *  PROC GLM DATA=MODELX NOPRINT;
   MODEL LEVEL = TIME;
   OUTPUT OUT=LIB.MODELXP PREDICTED=PREDICT STDP=STD_ERR;
   %MODELX
DATA LIB.MODELXP; SET LIB.MODELXP;
   BATCH='All';
   proc sort data=lib.modelxp; by batch time;
   * %LET NBATCH=1;
%MEND MMODEL1;

*%%%%%%%%%%%%%%%%%%%%%%%%%%%%%%%%%%%%%%%%%%%%%%%;
*               MMODEL2                         %;
* MODEL 2 - SEPARATE INTERCEPTS & COMMON SLOPE  %;
* STDP=standard error of the mean predicted value  %;
*%%%%%%%%%%%%%%%%%%%%%%%%%%%%%%%%%%%%%%%%%%%%%%%;
%MACRO MMODEL2;
DATA STABLNEW TEMP2 (KEEP=BNO BATCH); SET STABLNEW; BY BATCH;
RETAIN BNO 0;
IF FIRST.BATCH THEN DO; BNO=BNO+1; OUTPUT TEMP2; END;

   %DO I = 1 %TO %EVAL(&NBATCH-1);
      IF BNO = &I THEN DUMMY&I=1;
      ELSE DUMMY&I=0;
   %END;
 OUTPUT STABLNEW;

   PROC REG DATA=STABLNEW OUTEST=FITTED NOPRINT;
   MODEL LEVEL = %DO I=1 %TO %EVAL(&NBATCH-1);
           DUMMY&I
           %END; TIME;
   OUTPUT OUT=LIB.MODELXP P=PREDICT STDP=STD_ERR;

   DATA FITTED; SET FITTED(RENAME=(INTERCEP=TEMP));
   KEEP INTERCEP TIME BNO;
   %DO I=1 %TO %EVAL(&NBATCH-1);
    BNO=&i;
    INTERCEP=TEMP+DUMMY&I;
```

```
     OUTPUT;
   %END;
    INTERCEP=TEMP;
    BNO=&NBATCH;
    OUTPUT;

DATA FITTED; MERGE FITTED TEMP2; BY BNO;
   %MODELX
%MEND MMODEL2;

*%%%%%%%%%%%%%%%%%%%%%%%%%%%%%%%%%%%%%%%%%%;
*            MMODEL3                               %;
* MODEL 3 - SEPARATE INTERCEPTS & SEPARATE SLOPE   %;
* STDP=standard error of the mean predicted value  %;
*%%%%%%%%%%%%%%%%%%%%%%%%%%%%%%%%%%%%%%%%%%;
%MACRO MMODEL3;
   PROC SORT DATA=STABLNEW; BY BATCH;
   PROC REG DATA=STABLNEW OUTEST=FITTED NOPRINT;
   MODEL LEVEL = TIME;
   BY BATCH;
   OUTPUT OUT=LIB.MODELXP PREDICTED=PREDICT STDP=STD_ERR;
    PROC FREQ DATA=STABLE; TABLE BATCH/NOPRINT OUT=DF;
    DATA DF; SET DF (KEEP=BATCH COUNT);
     DROP COUNT;
     DF_D=COUNT-2;
   %MODELX
%MEND MMODEL3;

*%%%%%%%%%%%%%%%%%%%%%%%%%%%%%%%%%%%%%%%%%%;
*            MODELX                                %;
*  Expiration Date Macro                           %;
*%%%%%%%%%%%%%%%%%%%%%%%%%%%%%%%%%%%%%%%%%%;
%MACRO MODELX;
DATA LIB.MODELXP; MERGE LIB.MODELXP DF;
   %IF &M=MODEL3 %THEN
   %DO;
     BY BATCH;
   %END;

RETAIN T_INV;
KEEP BATCH LEVEL TIME PREDICT STD_ERR
   EXP L_BOUND U_BOUND ALPHA2;
LABEL TIME='MONTH'
    LEVEL='PERCENT OF CLAIM';

IF  %IF &M=MODEL3 %THEN
```

```
%DO;
    FIRST.BATCH
%END;
%ELSE
%DO;
    _N_=1
%END;   THEN DO;
        ALPHA2=(1-&ALPHA)*100;
        CALL SYMPUT (LEFT('ALPHA3'),ALPHA2);
        IF &MTEST=1 THEN DO; ALPHA1=&ALPHA; END;
        ELSE DO; ALPHA1=0.5*&ALPHA; END;
        T_INV=TINV(1-ALPHA1,DF_D);
        END;
   L_BOUND = PREDICT - T_INV * STD_ERR;
   U_BOUND = PREDICT + T_INV * STD_ERR;

EXP=' ';
IF &MTEST=1 AND "&BOUND"="L" THEN DO;
  IF (L_BOUND LE &LL) THEN EXP='*';
 * IF L_BOUND GE 80;
  END;
ELSE IF &MTEST=1 AND ("&BOUND"="U") THEN DO;
  IF (U_BOUND GE &UL) THEN EXP='*';
  IF U_BOUND LE 120;
  END;
ELSE DO;
  IF (L_BOUND LE &LL) OR (U_BOUND GT &UL) THEN EXP='*';
 * IF L_BOUND GE 80 AND U_BOUND LE 120;
  END;
%MEND MODELX;
```

ANALYS5.SAS

```
*===============================================================;
* Fifth Macro File of the Stability Analysis Program           ;
*===============================================================;

*%%%%%%%%%%%%%%%%%%%%%%%%%%%%%%%%%%%%%%%%%%%%%%%;
*          MPLOT                                            %;
* Plot Macro                                                %;
*%%%%%%%%%%%%%%%%%%%%%%%%%%%%%%%%%%%%%%%%%%%%%%%;
%MACRO MPLOT;
PROC PLOT DATA=LIB.MODELXP
```

```
  /* FORMCHAR='B3C4DAC2BFC3C5B4C0C1D9'X */ ;
%IF &M NE MODEL1 %THEN
%DO;
  BY BATCH;
%END;
PLOT LEVEL*TIME='O'
   PREDICT*TIME='P'
   %IF &MTEST=1 %THEN
      %DO;
      %IF "&BOUND"="L" %THEN
      %DO;
        L_BOUND*TIME='L'
      %END;
      %ELSE
        %DO;
        U_BOUND*TIME='U'
      %END;
     %END;
     %ELSE
       %DO;
        L_BOUND*TIME='L'
        U_BOUND*TIME='U'
     %END;
     /OVERLAY
  VREF= 90 110  HAXIS=0 TO &MCTIME BY 3
  VAXIS=80 TO 120;
%MEND MPLOT;

*%%%%%%%%%%%%%%%%%%%%%%%%%%%%%%%%%%%%%%%%%%%%%%;
*   MACRO TO PRINT FITTED REGRESSION LINES              %;
*%%%%%%%%%%%%%%%%%%%%%%%%%%%%%%%%%%%%%%%%%%%%%%%;

%MACRO PRINTSUM ;
DATA _NULL_;
 SET LIB.MODELXP1;
  BY BATCH ;
 IF FIRST.BATCH THEN DO ;
 COUNT+1 ;
 CALL SYMPUT('START'||LEFT(COUNT),_N_) ;
 CALL SYMPUT('BATCH'||LEFT(COUNT),LEFT(BATCH));
 END;
 IF LAST.BATCH THEN DO;
  CALL SYMPUT('LAST'||LEFT(COUNT),_N_) ;
 END ;
 * IF EOF THEN CALL SYMPUT('TOTAL',COUNT) ;
```

```
RUN ;

DATA _NULL_ ; SET FITTED;
 COUNT+1 ;
 CALL SYMPUT('INTRCP'||LEFT(COUNT),INTERCEP);
 CALL SYMPUT('TIME'||LEFT(COUNT),TIME);
RUN ;

%DO K=1 %TO &NBATCH;
 TITLE7 "Batch &&BATCH&K" ;
 TITLE8 "Fitted Line :  Y = &&INTRCP&K + &&TIME&K  X " ;
PROC PRINT DATA=LIB.MODELXP (FIRSTOBS=&&START&K OBS
=&&LAST&K) NOOBS;
   VAR TIME L_BOUND PREDICT U_BOUND STD_ERR EXP;
 RUN;
%END;
%MEND PRINTSUM;

*%%%%%%%%%%%%%%%%%%%%%%%%%%%%%%%%%%%%%%%%%%;
*            STAB MACRO                                        %;
* Model Selection                                             %;
*%%%%%%%%%%%%%%%%%%%%%%%%%%%%%%%%%%%%%%%%%%;
%MACRO STAB;
 %IF &M=MODEL1 %THEN
 %DO;
   %MMODEL1
 %END;
 %ELSE
  %IF &M=MODEL2 %THEN
    %DO;
      %MMODEL2
    %END;
    %ELSE
       %DO;
          %MMODEL3
    %END;

PROC SORT DATA=LIB.MODELXP; BY BATCH;
TITLE7;
%MPLOT

PROC SORT DATA=LIB.MODELXP NODUPKEY OUT=LIB.MODELXP1;
 BY BATCH TIME;
RUN;
%PRINTSUM
RUN;
%MEND STAB;

%STAB
```

ANALYS6.SAS

```
*=============================================================;
* Sixth Macro File of the Stability Analysis Program          ;
*=============================================================;

*%%%%%%%%%%%%%%%%%%%%%%%%%%%%%%%%%%%%%%%%%%%%%%%%%;
*          TILMTEST                              %;
* Title of test                                 %;
*%%%%%%%%%%%%%%%%%%%%%%%%%%%%%%%%%%%%%%%%%%%%%%%%%;
%MACRO TILMTEST;
%IF &MTEST=1 %THEN
  %DO;
  %IF "&BOUND"="L" %THEN
  %DO;
    TITLE7 "&ALPHA3% One-Sided Lower Confidence Limit";
    %END;
  %ELSE
    %DO;
    TITLE7 "&ALPHA3% One-Sided Upper Confidence Limit";
    %END;
  %END;
    %ELSE
      %DO;
      TITLE7 "&ALPHA3% Two-Sided Confidence Limit";
      %END;
%MEND TILMTEST;

*%%%%%%%%%%%%%%%%%%%%%%%%%%%%%%%%%%%%%%%%%%%%%%%%%;
*          MTIMODEL                              %;
* Model title                                   %;
*%%%%%%%%%%%%%%%%%%%%%%%%%%%%%%%%%%%%%%%%%%%%%%%%%;
%MACRO MTIMODEL;
%IF &M=MODEL1 %THEN
%DO;
  TITLE9 "Common Intercept and Common Slope";
%END;
%ELSE
  %IF &M=MODEL2 %THEN
  %DO;
    TITLE9 "Separate Intercepts and Common Slope";
  %END;
    %ELSE
    %DO;
        TITLE9 "Separate Intercepts and Separate Slopes";
```

```
    %END;
%MEND MTIMODEL;

*****************************************************************;
* Summary Output Program                                       *;
*****************************************************************;
DATA SUMMARY; SET LIB.MODELXP(KEEP=TIME BATCH EXP);
    WHERE (EXP='');
RUN;

DATA SUMMARY; SET SUMMARY; BY BATCH;

    IF last.BATCH;
    %TILMTEST
    %MTIMODEL
    PROC PRINT DATA=SUMMARY  D SPLIT='*' NOOBS;
     LABEL TIME='ESTIMATED*DATING PERIOD*(MONTHS/WEEKS)';
        VAR BATCH TIME;

RUN;
TITLE;
RUN;
```

STABGRAF.SAS

```
*==============================================================;
* SAS/PC/STABGRAF.SAS (version 1)                              ;
* This program draws high resolution SAS graphs for stability analysis.  ;
*==============================================================;
* Before submitting this program, the user may need to modify  ;
* the libname, or statements which tell SAS the types of       ;
* graphics devices you are using.  See SAS Graphics Manual if  ;
* you have a device which is not listed here.                  ;
*==============================================================;

 LIBNAME LIB '\SAS\STAB';   /* Check that this is where STAB is */

/* Next line defines choice of graphics adapter for display on screen  */
  GOPTIONS DEVICE=ps2ega;

PROC SORT DATA=LIB.MODELXP; BY BATCH;

PROC GPLOT DATA=LIB.MODELXP;
  PLOT LEVEL*TIME=1 PREDICT*TIME=2
     L_BOUND*TIME=3 U_BOUND*TIME=4/FRAME
  OVERLAY AUTOHREF VREF=90 110 LHREF=34 LVREF=1
```

HAXIS=AXIS1 VAXIS=AXIS2 LEGEND=LEGEND;
BY BATCH;

/* More material which may need modification: */
/* Next statement defines the range of the horizontal axis and */
/* the number of marks   */
AXIS1 VALUE=(H=1.0 F=duplex) LABEL=(H=1.3 F=complex)
ORDER=0 TO 55 by 6;

/* Next statement defines the range of the vertical axis and   */
/* the number of marks   */
AXIS2 VALUE=(H=1.0 F=duplex) LABEL=(H=1.3 F=complex A=90 R=0)
ORDER=80 to 120 BY 10;

LEGEND VALUE=(H=1.3 F=COMPLEX);

SYMBOL1 V=STAR C=WHITE; SYMBOL2 I=JOIN L=20 C=RED ;
SYMBOL3 I=JOIN L=1 C=GREEN; SYMBOL4 I=JOIN L=1 C=GREEN;

RUN;

Example for Use

```
*=============================================================;
* DATE     : March 9, 1992                                    ;
*=============================================================;
OPTIONS FORMCHAR='B3C4DAC2BFC3C5B4C0C1D9'X;

%LET PATH=\SAS\STAB;                    /* line A */

LIBNAME LIB "&PATH";
OPTIONS MPRINT NODATE PS=50 MISSING='.';

%MACRO DEFAULT;
%GLOBAL NLINES SSDNAME MTEST ALPHA LL UL BOUND PRTSEL NUM
    UNIT TT1 TT2 TT3 TT4 TT5 TT6 NBATCH TIME INTERCEP;
%LET SSDNAME=STAB;
%LET ALPHA=0.05;    /* .05 is customarily required by FDA.  */

%LET LL=90;                    /* line B */
%LET UL=110;                   /* line B */
%LET BOUND=L;                  /* line C */
%LET MTEST=1;                  /* line C */
%LET UNIT=MONTH;               /* line D */
%LET TT1=Stability Analysis        ;  /* line E */
%LET TT2=                  ; /* line E */
%LET TT3=                  ; /* line E */
%LET TT4=                  ; /* line E */
%LET TT5=                  ; /* line E */
```

```
%LET TT6=                        ;   /* line E */
%MEND DEFAULT;
%DEFAULT

DATA LIB.&SSDNAME;
INPUT BATCH $ TIME LEVEL;
CARDS;

; /* end of input data */

%INCLUDE "&PATH\ANALYS1.SAS";
%INCLUDE "&PATH\ANALYS2.SAS";
%INCLUDE "&PATH\ANALYS3.SAS";
%INCLUDE "&PATH\ANALYS4.SAS";
%INCLUDE "&PATH\ANALYS5.SAS";
%INCLUDE "&PATH\ANALYS6.SAS";
RUN;
```

References

[1] Ahn, H., Chen, J.J., and Lin, T.Y.D. (1995). Classification of batches in a stability study. *Proceedings of the Biopharmaceutical Section of the American Statistical Association*, Alexandria, VA, pp. 256–261.

[2] Ahn, H., Chen, J., and Lin, T.D. (1997). A two-way analysis of covariance model for classification of stability data. *Biometrical Journal*, 39, 559–576.

[3] Altan, S. and Raghavarao. (2003). A note on kinetic modeling of stability data and implications on pooling. *Journal of Biopharmaceutical Statistics*, 13, 425–430.

[4] Anderson, R.L. (1982). Analysis of Variance Components. Unpublished lecture notes. University of Kentucky, Lexington.

[5] Bakshi, M. and Singh, S. (2002). Development of validated stability-indicating assay methods: critical review. *Journal of Pharmaceutical and Biomedical Analysis*, 28, 1011–1040.

[6] Bancroft, T.A. (1964). Analysis and inference for incompletely specified models involving the use of preliminary test(s) of significance. *Biometrics*, 20, 427–442.

[7] Bar, R. (2003). Statistical evaluation of stability data: criteria for change-over-time and data variability. *Journal of Pharmaceutical Science and Technology*, 57(5), 369–377.

[8] Barron, A.M. (1994). Use of fractional factorial (or matrix) designs in stability analysis. *Proceedings of the Biopharmaceutical Section of the American Statistical Association*, Alexandria, VA, 475–481.

[9] Bergum, J.S. (1990). Constructing acceptance limits for multiple stage tests. *Drug Development and Industrial Pharmacy*, 16, 2153–2166.

[10] Bohidar, N.R. and Peace, K. (1988). Pharamceutical formulation development. In *Biopharmaceutical Statistics for Drug Development*, Peace, K., Ed. Marcel Dekker, New York, pp. 149–229.

[11] Box, G.E.P. and Lucas, H.L. (1959). Design and experiments in non-linear situation. *Biometrics*, 46, 77–90.

[12] Brandt, A. and Collings, B.J. (1989). The development of estimators for parameters in a random effects model for shelf-life. *Proceedings of the Biopharmaceutical Section of the American Statistical Association*, Alexandria, VA, pp. 98–99.

[13] Carstensen, J.T. (1990). *Drug Stability*. Marcel Dekker, New York.

[14] Carstensen, J.T. and Rhodes, C. (2000). *Drug Stability: Principles and Practices*. Marcel Dekker, New York.

[15] Carstensen, J.T., Franchini, M., and Ertel, K. (1992). Statistical approaches to stability protocol design. *Journal of Pharmaceutical Sciences*, 81, 303–308.

[16] Carter, R.L. and Yang, M.C.K. (1986). Large sample inference in random co-efficient regression model. *Communications in Statistics: Theory and Methods*, 15, 2507–2525.

[17] CBP. (2001). Expiration dates: compliance guidelines. *The Script*, California Board of Pharmacy, July, 2001, p. 1.

[18] CDER. (2004). *CDER Report to the Nation: 2004*. Center for Drug Evaluation and Research, the U.S. Food and Drug Administration, Rockville, MD.

[19] Chambers, D. (1996). Matrixing/bracketing US industry views. *Proceedings of EFPIA Symposium: Advanced Topics in Pharmaceutical Stability Testing Building on the ICH Stability Guideline*. EFPIA, Brussels.

[20] Chen, C. (1996). US FDA's perspective of matrixing and bracketing. *Proceedings of EFPIA Symposium: Advanced Topics in Pharmaceutical Stability Testing Building on the ICH Stability Guideline*. EFPIA, Brussels.

[21] Chen, J.J., Ahn, H., and Tsong, Y. (1997). Shelf life estimation for multi-factor stability studies. *Drug Information Journal*, 31, 573–587.

[22] Chen, J.J., Hwang, J.S., and Tsong, Y. (1995). Estimation of the shelf-life of drugs with mixed effects models. *Journal of Biopharmaceutical Statistics*, 5, 131–140.

[23] Chen, W.J. and Tsong, Y. (2003). Significance levels for stability pooling test: a simulation study. *Journal of Biopharmaceutical Statistics*, 13, 355–374.

[24] Chen, Y.Q., Pong, A., and Xing, B. (2003). Rank regression in stability analysis. *Journal of Biopharmaceutical Statistics*, 13, 463–479.

[25] Chow, S.C. (1992). Statistical Design and Analysis of Stability Studies. Presented at the 48th Conference on Applied Statistics, Atlantic City, NJ.

[26] Chow, S.C. (1997). Pharmaceutical validation and process controls in drug development. *Drug Information Journal*, 31, 1195–1201.

[27] Chow, S.C. and Ki, F.Y.C. (1997). Statistical comparison between dissolution profiles of drug products. *Journal of Biopharmaceutical Statistics*, 7, 241–258.

[28] Chow, S.C. and Liu, J.P. (1995). *Statistical Design and Analysis in Pharmaceutical Science*. Marcel Dekker, New York.

[29] Chow, S.C. and Liu, J.P. (2000). *Design and Analysis of Bioavailability and Bioequivalence Studies. 2nd Edition*, Marcel Dekker, New York.

[30] Chow, S.C. and Pong, A. (1995). Current issues in regulatory requirements of drug stability. *Journal of Food and Drug Analysis*, 3, 75–85.

[31] Chow, S.C., Pong, A., and Chang, Y.W. (2006). On traditional Chinese medicine clinical trials. *Drug Information Journal*, 40, 395–406.

[32] Chow, S.C. and Shao, J. (1989). Test for batch-to-batch variation in stability analysis. *Statistics in Medicine*, 8, 883–890.

[33] Chow, S.C. and Shao, J. (1990a). Estimating drug shelf-life in NDA stability studies. *Proceedings of the Biopharmaceutical Section of the American Statistical Association*, Alexandria, VA, pp. 190–195.

[34] Chow, S.C. and Shao, J. (1990b). On the difference between the classical and inverse methods of calibration. *Journal of the Royal Statistical Society*, C, 39, No. 2, 219–228.

[35] Chow, S.C. and Shao, J. (1991). Estimating drug shelf-life with random batches. *Biometrics*, 47, 1071–1079.

[36] Chow, S.C. and Shao, J. (2002a). *Statistics in Drug Research*. Marcel Dekker New York, Chapter 4.

[37] Chow, S.C. and Shao, J. (2002b). On the assessment of similarity for dissolution profiles of two drug products. *Journal of Biopharmaceutical Statistics*, 12, 311–321.

[38] Chow, S.C. and Shao, J. (2003). Stability analysis with discrete responses. *Journal of Biopharmaceutical Statistics*, 13, 451–462.

[39] Chow, S.C. and Shao, J. (2007). Stability analysis for drugs with multiple ingredients. *Statistics in Medicine*. 26. In press.

[40] Chow, S.C., Shao, J., and Hu, O.Y.P. (2002a). Assessing sensitivity and similarity in bridging studies. *Journal of Biopharmaceutical Statistics*, 12, 385–400.

[41] Chow, S.C., Shao, J., and Wang, H. (2002b). Probability lower bounds for USP/NF tests. *Journal of Biopharmaceutical Statistics*, 12, 79–92.

[42] Chow, S.C., Shao, J., and Wang, H. (2003). *Sample Size Calculation in Clinical Research*. Marcel Dekker, New York.

[43] Chow, S.C. and Wang, S.G. (1994). On the estimation of variance components in stability analysis. *Communications in Statistics: Theory and Methods*, 23, 289–303.

[44] Connors, K.A., Amidon, G.L., and Stella, V.J. (1986). *Chemical Stability of Pharmaceuticals: A Handbook for Pharmacists*. John Wiley and Sons, New York.

[45] CPMP (1998). *Note for Guidance on Stability Testing of Existing Active Substances and Related Finished Products*. Committee for Proprietary Medicinal Products, EMEA, London.

[46] Davies, O.L. (1980). Note on regression with corrected responses. *Biometrics*, 36, 551–552.

[47] Davies, O.L. and Hudson, H.E. (1981). Stability of drugs: accelerated storage tests. In *Statistics in the Pharamceutical Industry*, Ed Buncher, C.R. and Tsay, J.Y., Eds. Marcel Dekker, New York, pp. 355–395.

[48] Davies, O.L. and Hudson, H.E. (1993). Stability of drugs. Part B: Accelerated storage tests. In *Statisics in the Pharmaceutical Industry*, 2nd ed. Buncher, C.R. and Tsay, J.Y., Eds., Marcel Dekker, New York, 445–480.

[49] Dawoodbhai, S., Suryanarayan, E., Woodruff, C., and Rhodes, C. (1991). Optimization of tablet formulation containing talc. *Drug Development and Industrial Pharmacy*, 17, 1343–1371.

[50] Dempster, A.P., Laird, N.M., Rubin, D.B. (1977). Maximum likelihood from incomplete data via the EM algorithm (with discussion). *Journal of the Royal Statistical Society*, B, 39, 1–38.

[51] Dempster, A.P., Rubin, D.B., and Tsitakawa, R.K. (1981). Estimation in covariance components models. *Journal of the American Statistical Association*, 76, 341–353

[52] DeWoody, K. and Raghavarao, D. (1997). Some optimal matrix designs in stability studies. *Journal of Biopharmaceutical Statistics*, 7, 205–213.

[53] Dietz, R., Feilner, K., Gerst, F., and Grimm, W. (1993). Drug stability testing: classification of countries according to climatic zone. *Drugs Made in Germany*, 36, 99–103.

[54] Draper, N.R. and Smith, H. (1981). *Applied Regression Analysis*, 2nd Ed., Wiley, New York.

[55] Easterling, R.G. (1969). Discrimination intervals for percentiles in regression. *Journal of the American Statistical Association*, 64, 1031–1041.

[56] Fahrmeir, L. and Tutz, G. (1994). *Multivariate Statistical Modelling Based on Generalized Linear Models*. Springer, New York.

[57] Fairweather, W.R. (2003). A quantitative assessment of factors influencing the probability of postmarketing out-of-specification observations. *Journal of Biopharmaceutical Statistics*, 13, 415–423.

[58] Fairweather, W.R. and Lin, T.D. (1999). Statistical and regulatory aspects of drug stability studies: an FDA perspective. In *International Stability Testing*, Mazzo, D., Ed. Interpharm Press, Buffalo Grove, pp. 107–132.

[59] Fairweather, W.R., Lin, T.D., and Kelly, R. (1995). Regulatory, design, and analysis aspects of complex stability stdies. *Journal of Pharmaceutical Sciences*, 84, 1322–1326.

[60] Fairweather, W.R., Mogg, R., Bennett, P.S., Zhong, J., Morrisey, C., and Schofield, T.L. (2003). Monitoring the stability of human vaccines. *Journal of Biopharmaceutical Statistics*, 13, 395–413.

[61] FDA. (1987). *Guideline for Submitting Documentation for the Stability of Human Drugs and Biologics*. Center for Drugs and Biologics, Office of Drug Research and Review, Food and Drug Administration, Rockville, MD.

[62] FDA. (1994). *Guide to Inspection of Bulk Pharamceutical Chemicals*. The U.S. Food and Drug Administration, Rockville, MD.

[63] FDA. (1997). *Guidance for Industry: Dissolution Testing of Immediate Release Solid Oral Dosage Forms*. The United States Food and Drug Administration, Rockville, MD.

[64] FDA. (1998). *Guidance for Industry: Stability Testing of Drug Substances and Drug Products (draft guidance)*. The United States Food and Drug Administration, Rockville, MD.

[65] FDA. (2000). *Guidance for Industry: Analytical Procedures and Methods Validation (draft guidance)*. The United States Food and Drug Administration, Rockville, MD.

[66] FDA. (2006). *FDA Enforcement Report*. The U.S. Food and Drug Administration, Rockville, MD, March 8, 2006.

[67] Federal Register. (1997). Prescription drug products: levothyroxine sodium. *Federal Register*, 62, (157), 43553–43537.

[68] Fuller, W.A. (1996). *Introduction to Statistical Time Series*, 2nd ed., Wiley, New York.

[69] Garnick, R.I., et al. (1982). Stability indicating high-pressure liquid chromatographic method for quality control of sodium liothyronine and sodium levothyroxine in tablet formulations. In *Hormone Drugs*, Gueriguian, J.L., Bransome, E.D., and Outschoorn, A.S., Eds. United States Pharmacdpedial Convention, pp. 504–516, Rockville, MD.

[70] Gill, J.L. (1978). *Design and Analysis of Experiments in Animal and Medical Sciences*, Vol. 1, Iowa State University Press, Ames.

[71] Gill, J.L. (1988). Repeated measurement: split-plot trend analysis versus analysis of first differences. *Biometrics*, 44, 289–297.

[72] Golden, M.H., Cooper, D., Riebe, M., and Carswell, K. (1996). A matrixed approach to long-term stability testing of pharmaceutical products. *Journal of Pharmaceutical Sciences*, 86, 240–244.

[73] Grimm, W. (1985). Storage conditions for stability testing. Part I: Long term testing and stress tests. *Drugs Made in Germany*, 28, 196–202.

[74] Grimm, W. (1986). Storage conditions for stability testing. Part II: Long term testing and stress tests. *Drugs Made in Germany*, 29, 39–47.

[75] Grimm, W. (1998). Extension of the International Conference on Harmonization tripartite guideline for stability testing of new drug substances and products to countries of climatic zones III and IV. *Drug Development and Industrial Pharmacy*, 24, 313–325.

[76] Grimes, J.A. and Foust, L.B. (1994). Establishing release limits with a random slope stability model. *Proceedings of the Biopharmaceutical Section of the American Statistical Association*, Alexandria, VA, pp. 498–502.

[77] Gumpertz, M. and Pantula, S.G. (1989). A simple approach to inference in random coefficient models. *American Statistician*, 43, 203–210.

[78] Habs, M. (1999). Herbal products: advances in preclinical and clinical development. *Drug Information Journal*, 33, 993–1001.

[79] Hahn, G.J. and Meeker, W.Q. (1991). *Statistical Intervals: A Guide for the Practitioner*. Wiley, New York.

[80] Halperin, M. (1970). On inverse estimation in linear regression. *Technometrics*, 12, 727–736.

[81] Haynes, J.D. (1971). Worldwide virtual temperatures for product stability testing. *Journal of Pharmaceutical Sciences*, 60, 927–929.

[82] Hedayat, A.S. (2001). Optimal Designs. Lecture notes, University of Illinois at Chicago.

[83] Hedayat, A.S., Yan, X., and Lin, L. (2006). Optimal designs in stability studies. *Journal of Biopharmaceutical Statistics*, 16, 35–59.

[84] Helboe, P. (1992). New design for stability testing problems: matrix or factorial designs, authorities' viewpoint on the predictive value of such studies. *Drug Information Journal*, 26, 629–634.

[85] Helboe, P. (1999). Matrixing and bracketing designs for stability studies: an overview from the European perspective. In *International Stability Testing*, Mazzo, D., Ed. Interpharm Press, Buffalo Grove, IL, pp. 135–160.

[86] Hettmansperger, T.P. (1984). *Statistical Inferences Based on Ranks*. John Wiley, New York.

[87] Hettmansperger, T.P. and McKean, J.W. (1998). *Robust Nonparametric Statistical Methods*. Atnold, New York.

[88] Hildreth, G. and Houck, J.R. (1968). Some estimators for a linear model with random coefficients. *Journal of the American Statistical Association*, 63, 584–595.

[89] Ho, C.H., Liu, J.P., and Chow, S.C. (1993). On analysis of stability data. *Proceedings of the Biopharmaceutical Section of the American Statistical Association*, Alexandria, VA, 198–203.

[90] ICH Q1A. (1993). *Stability Testing of New Drug Substances and Products*. Tripartite International Conference on Harmonization Guideline Q1A, Geneva.

[91] ICH Q1A. (R2) (2003). *Stability Testing of New Drug Substances and Products*. Tripartite International Conference on Harmonization Guideline Q1A (R2), Geneva.

[92] ICH Q2A. (1994). *Text on Validation of Analytical Procedures*. Tripartite International Conference on Harmonization Guideline Q2A, Geneva.

[93] ICH Q3A. (2003). *Impurities in New Drug Substances*. Tripartite International Conference on Harmonization Guideline Q3A, Geneva.

[94] ICH Q6A. (1999). *Specifications: Test Procedures and Acceptance Criteria for New Drug Substances and New Drug Products: Chemical Substances*. Tripartite International Conference on Harmonization Guideline Q6A, Geneva.

[95] ICH Q1B. (1996). *Photostability Testing of New Drug Substances and Products.* Tripartite International Conference on Harmonization Guideline Q1B, Geneva.

[96] ICH Q2B. (1996). *Validation of Analytical Procedures: Methodology.* Tripartite International Conference on Harmonization Guideline Q2B, Geneva.

[97] ICH Q3B (1996). *Impurities in New Drug Products.* Tripartite International Conference on Harmonization Guideline Q3B, Geneva.

[98] ICH Q3B (R). (2003). *Impurities in New Drug Products.* Tripartite International Conference on Harmonization Guideline Q3B (R), Geneva.

[99] ICH Q6B. (1999). *Specifications: Test Procedures and Acceptance Criteria for New Drug Substances and New Drug Products: Biotechnological/Biological Products.* Tripartite International Conference on Harmonization Guideline Q6B, Geneva.

[100] ICH Q1C. (1997). *Stability Testing of New Dosage Forms.* Tripartite International Conference on Harmonization Guideline Q1C, Geneva.

[101] ICH Q5C. (1995). *Quality of Biotechnological Products: Stability Testing of Biotechnological/Biological Products.* Tripartite International Conference on Harmonization Guideline Q5C, Geneva.

[102] ICH Q1D. (2003). *Guidance for Industry: Q1D Bracketing and Matrixing Designs for Stability Testing of New Drug Substances and Products.* Center for Drug Evaluation and Research and Center for Biologics Evaluation and Research, Food and Drug Administration, Rockville, MD.

[103] ICH Q1E. (2004) *Evaluation of Stability Data.* Tripartite International Conference on Harmonization Guideline Q1E, Geneva.

[104] ICH Q1F. (2004). *Stability Data Package for Registration Applications in Climatic Zones III and IV.* Tripartite International Conference on Harmonization Guideline Q1F, Geneva.

[105] Jaeckel, L.A. (1972). Estimating regression coefficients by minimizing the dispersion of residuals. *Annal. of Mathematical Statistics*, 43, 1449–1548.

[106] Johnson, R.A. and Wichern, D.W. (1998). *Applied Multivariate Statistical Analysis*, 3rd ed. Prentice Hall, Upper Saddle River, NJ.

[107] Ju, H.L. and Chow, S.C. (1995). On stability designs in drug shelf-life estimation. *Journal of Biopharmaceutical Statistics*, 5, 201–214.

[108] Ju, H.L. and Chow, S.C. (1996). An overview of stability studies. *Journal of Food and Drug Analysis*, 4, 99–106.

[109] Katori, N., Kaniwa, N., Aoyagi, N., and Kojima, S. (1998). A new acceptance plan for the official dissolution test. *JP Forum*, 7, 166–173.

[110] Kellicker, P.G. (2006). Drug expiration dates: how accurate are they? EBSCO Publishing, Ipswich, MA. (http://healthlibrary.epnet.com).

[111] Kervinen, L. and Yliruusi, J. (1993). Modeling S-shaped dissolution curves. *International Journal of Pharmaceutics*, 92, 115–122.

[112] Kiermeier, A., Jarrett, R.G., and Verbyla, A.P. (2004). A new approach to estimating shelf-life. *Pharmaceutical Statistics*, 3(1), 3–11.

[113] Kirkwood, T.B.L. (1977). Predicting the stability of biological standards and products. *Biometrics*, 33, 736–742.

[114] Kohberger, R.C. (1988). Manufacturing and quality control. In *Biopharmaceutical Statistics for Drug Development*, Peace, K., Ed. Marcel Dekker, New York.

[115] Kommanaboyina, B. and Rhodes, C.T. (1999). Trends in stability testing with emphasis on stability drug distribution and storage. *Drug Development and Industrial Pharmacy*, 25, 857–868.

[116] Krutchkoff, R.G. (1967). Classical and inverse methods of calibration. *Technometrics*, 9, 535–539.

[117] Kumkumian, C. (1994). International Council for Harmonization Stability Guidelines: Food and Drug Administration regulatory perspective. *Drug Information Journal*, 28, 635–640.

[118] Laird, N.M., Lange, N., and Stram, D. (1987). Maximum likelihood computations with repeated measures: application of the EM algorithm. *Journal of the American Statistical Association*, 82, 87–105.

[119] Laird, N.M. and Ware, J.H. (1982). Random effects models for longitudinal data. *Biometrics*, 38, 963–974.

[120] Lan, K.K.G. and DeMets, D. (1983). Discrete sequential boundaries for clinical trials. *Biometrika*, 70, 659–663.

[121] Langenbucher, F. (1972). Linearization of dissolution rate curves by the Weibull distribution. *Journal of Pharmacy and Pharmacology*, 24, 979–981.

[122] Lee, K.R. and Gagnon, R.C. (1994). Calibration and stability analysis with a simple mixed linear model when the experiment is of a split-plot type. *Proceedings of the Biopharmaceutical Section of the American Statistical Association*, Alexandria, VA, pp. 435–440.

[123] Leeson, L.J. (1995). *In vitro/in vivo* correlation. *Drug Information Journal*, 29, 903–915.

[124] Lin, J.S. and Chow, S.C. (1992). The estimating of drug shelf-life. *Proceedings of the 17th SUCI Conference*, Cary, North Carolina, pp. 1292–1297.

[125] Lin, K.K. (1990). Statistical analysis of stability study data. *Proceedings of the Biopharmaceutical Section of the American Statistical Association*, Alexandria, VA, pp. 210–215.

[126] Lin, K.K. and Lin, T.Y.D. (1993). Stability of drugs. Part A: Room temperature tests. In *Statisics in the Pharmaceutical Industry*,2nd ed. Buncher, C.R. and Tsay, J.Y., Eds. Marcel Dekker, New York, pp. 419–444.

[127] Lin, T.Y.D. (1994). Applicability of matrix and bracket approach to stability study design. *Proceedings of the Biopharmaceutical Section of the American Statistical Association*, Alexandria, VA, pp. 142–147.

[128] Lin, T.Y.D. (1999a). Statistical considerations in bracketing and matrixing. *Proceedings of IBC Bracketing and Matrixing Conference*, London.

[129] Lin, T.Y.D. (1999b). Study design, matrixing and bracketing. AAPS Workshop on Stability Practices in the Pharmaceutical Industry: Current Issues, Arlington, VA.

[130] Lin, T.Y.D. and Chen, C. W. (2003). Overview of stability study designs. *Journal of Biopharmaceutical Statistics*, 13, 337–354.

[131] Lin, T.Y.D. and Fairweather, W.R. (1997). Statistical design (bracketing and matrixing) and analysis of stability data for the US market. *Proceedings of IBC Bracketing and Matrixing Conference*, London.

[132] Lin, T.Y.D. and Tsong, Y. (1991). Determination of significance level for pooling data in stability studies. *Proceedings of the Biopharmaceutical Section of the American Statistical Association*, Alexandria, VA, pp. 195–201.

[133] Liu, J.P., Ma, M.C., and Chow, S.C. (1997). Statistical evaluation of similarity factor f_2 as a criterion for assessment of similarity between dissolution profiles. *Drug Information Journal*, 31, 1225–1271.

[134] Liu, J.P., Tung, S.C., and Pong, Y.M. (2006). An alternative approach to evaluation of poolability for stability studies. *Journal of Biopharmaceutical Statistics*, 16, 1–14.

[135] Lu, C.J. and Meeker, W.Q. (1993). Using degradation measures to estimate a time-to-failure distribution. *Technometrics*, 35, 161–174.

[136] Lyon, R.C., Taylor, J.S., Porter, D.A., Prasanna, H.R., and Hussain, A.S. (2006). Stability profiles of drug products extended beyond labeled expiration dates. *Journal of Pharmaceutical Sciences*, 95, 1549–1560.

[137] Ma, M.C., Lin, R.P., and Liu, J.P. (1999). Statistical evaluation of dissolution similarity. *Statistica Sinica*, 9, 1011–1028.

[138] Ma, M.C., Wang, B.B.C., Liu, J.P., and Tsong, Y. (2000). Assessment of similarity between dissolution profiles. *Journal of Biopharmaceutical Statistics*, 10, 229–249.

[139] Mazzo, D.J. (1998). *International Stability Testing*. Interpharm Press, Buffalo Grove, IL.

[140] Mellon, J.I. (1991). Design and Analysis Aspects of Drug Stability Studies When the Product is Stored at Several Temperatures. Presented at the 12th Annual Midwest Statistical Workshop, Muncie, IN.

[141] Min, Y. (2004). *Semiparametric inferences in stability design*. PhD thesis, Univeristy of Florida, Gainesville.

[142] Moore, J.W. and Flanner, H.H. (1996). Mathematical comparison of curves with an emphasis on dissolution profiles. *Pharmaceutical Technology*, 20, 64–74.

[143] Morris, J.W. (1992). A comparison of linear and exponential models for drug expiry estimation. *Journal of Biopharmaceutical Statistics*, 2, 83–90.

[144] Muirhead, R.J. (1982). *Aspects of Multivariate Statistical Theory*. John Wiley and Sons, New York.

[145] Murphy, J.R. (1996). Uniform matrix stability study designs. *Journal of Biopharmaceutical Statistics*, 6, 477–494.

[146] Murphy, J.R. (2000). Bracketing design. In *Encyclopedia of Biopharmaceutical Statistics*, Chow, S.C., Ed. Marcel Dekker, New York, p. 133.

[147] Murphy, J.R. and Weisman, D. (1990). Using random slopes for estimating shelf-life. *Proceedings of the Biopharmaceutical Section of the American Statistical Association*, Alexandria, VA, pp. 196–203.

[148] Nordbrock, E. (1989). Statistical Study Design. Presented at the National Stability Discussion Group, October, 1989.

[149] Nordbrock, E. (1991). Statistical Comparison of NDA Stability Study Designs. Presented at the Midwest Biopharmaceutical Statistics Workshop, Muncie, IN, May, 1991.

[150] Nordbrock, E. (1992). Statistical comparison of stability designs. *Journal of Biopharmaceutical Statistics*, 2, 91–113.

[151] Nordbrock, E. (1994). Design and analysis of stability studies. *Proceedings of the Biopharmaceutical Section of the American Statistical Association*, Alexandria, VA, pp. 291–294.

[152] Nordbrock, E. (2000). Stability matrix design. In *Encyclopedia of Biopharmaceutical Statistics*, S. Chow, Ed. Marcel Dekker, New York, pp. 487–492.

[153] Nordbrock, E. (2003). Stability matrix designs. In *Encyclopedia of Biopharmaceutical Statistics*. Ed. S.C. Chow, 2nd Edition, Marcel Dekker, New York, 934–939.

[154] Nordbrock, E. and Valvani, S. (1995). PhRMA Stability Working Group. In *Guideline for Matrix Designs of Drug Product Stability Protocols*, January, 1995.

[155] O'Brien, P.C. and Fleming, T.R. (1979). A multiple testing procedure for clinical trials. *Biometrics*, 35, 549–556.

[156] Pena Romero, A., Caramella, C., Ronchi, M., Ferrari, F., and Chulia, D. (1991). Water uptake and force development in an optimized prolonged release formulation. *International Journal of Pharmacy*, 73, 239–248.

[157] Pogany, J. (2006). Stability Studies Assessment Experience. Presented at WHO Training Workshop on Pharmaceutical Quality, GMP and Bioequivalence, China.

[158] Polli, J.E., Rekhi, G.S., Augsburger, L.L., and Shah, V.P. (1977). Methods to compare dissolution profiles and a rationale for wide dissolution specifications for metoprolol tartate tablets. *Journal of Pharmaceutical Sciences*, 86, 670–700.

[159] Pong, A. (2000). Expiration dating period. In *Encyclopedia of Biopharmaceutical Statistics*, Chow, S.C., Ed. Marcel Dekker, New York, pp. 336–342.

[160] Pong, A. and Raghavarao, D. (2000). Comparison of bracketing and matrixing designs for a two-year stability study. *Journal of Biopharmaceutical Statistics*, 10, 217–228.

[161] Pong, A. and Raghavarao, D. (2001). Shelf-life estimation for drug products with two components. *Proceedings of the Biopharmaceutical Section of the American Statistical Association*, Alexandria, VA.

[162] Pong, A. and Raghavarao, D. (2002). Comparing distributions of shelf lives for drug products with two components under different designs. *Journal of Biopharmaceutical Statistics*, 12, 277–293.

[163] Rahman, R.A. (1992). A comparison of expiration dating period methods. *Journal of Biopharmaceutical Statistics*, 2, 69–82.

[164] Rao, C.R. (1973). *Linear Statistical Inference and Its Applications. 2nd Edition*, Wiley, New York.

[165] Repeto, P. (2000). Some considerations about stability study design. *Journal of Biopharmaceutical Statistics*, 10, 73–82.

[166] Ruberg, S.J. and Hsu, J.C. (1990). Multiple comparison procedures for pooling batches in stability studies. *Proceedings of the Biopharmaceutical Section of the American Statistical Association*, Alexandria, VA, pp. 205–209.

[167] Ruberg, S.J. and Hsu, J. (1992). Multiple comparison procedures for pooling batches in stability studies. *Technometrics*, 34, 465–472.

[168] Ruberg, S.J. and Stegeman, J.W. (1991). Pooling data for stability studies: testing the equality of batch degradation slopes. *Biometrics*, 47, 1059–1069.

[169] Sathe, P., Tsong, Y., and Shah, V.P. (1996). *In vitro* dissolution profile comparison: statistics, and analysis, model dependent approach. *Pharmaceutical Research*, 13, 1799–1803.

[170] Schumacher, P. (1974). Aktuelle Fragen zur Haltbarkeit von Arzneimitteln [Current questions on drug stability]. *Pharmazeutische Zeitung*, 119, 321–324.

[171] Shah, V.P., Tsong, Y., Sathe, P., and Liu, J.P. (1998). In vitro dissolution profile comparison: statistics and analysis of the similarity factor f_2. *Pharmaceutical Research*, 15, 889–896.

[172] Shao, J. (1999). *Mathematical Statistics*. Springer, New York.

[173] Shao, J. and Chen, L. (1997). Prediction bounds for random shelf-lives. *Statistics in Medicine*, 16, 1167–1173.

[174] Shao, J. and Chow, S.C. (1994). Statistical inference in stability analysis. *Biometrics*, 50, 753–763.

[175] Shao, J. and Chow, S.C. (2001a). Two-phase shelf life estimation. *Statistics in Medicine*, 20, 1239–1248.

[176] Shao, J. and Chow, S.C. (2001b). Drug shelf life estimation. *Statistica Sinica*, 11, 737–745.

[177] Shao, J. and Tu, D. (1995). *The Jackknife and Bootstrap*. Springer, New York.

[178] Silverberg, A. (1997). A comparison of fixed effects model and mixed models for the calculation of shelf-life. *Proceedings of the Physical and Engineering Section of the American Statistical Association*, Alexandria, VA, pp. 153–162.

[179] Snedecor, G.W. and Cochran, W.G. (1980). *Statistical Methods, 7th Edition*, Iowa State University Press. Ames, Iowa.

[180] Stefan, G. and Chantal, B. (2005). The challenges of chemical stability testing of herbal extracts in finished products using state-of-the-art analytical methodologies. *Current Pharmaceutical Analysis*, 1, 203–215.

[181] Sun, Y., Chow, S.C., Li. G., and Chen, K.W. (1999). Assessing distributions of estimated drug shelf-lives in stability analysis. *Biometrics*, 55, 896–899.

[182] SUPAC-IR. (1995). The United States Food and Drug Administration Guideline *Immediate Release Solid Oral Dosage Forms: Scale-Up and Postapproval Changes: Chemistry, Manufacturing, and Controls, In Vitro Dissolution Testing, and In Vivo Bioequivalence Documentation* (SUPAC-IR). Rockville, MD.

[183] SUPAC-MR. (1997). The United States Food and Drug Administration Guideline *Modified Release Solid Oral Dosage Forms: Scale-Up and Postapproval Changes: Chemistry, Manufacturing, and Controls, In Vitro Dissolution Testing, and In Vivo Bioequivalence Documentation* (SUPAC-MR). Rockville, MD.

[184] SUPAC-SS. (1997). The United States Food and Drug Administration Guideline *Nonsterile Semisolid Oral Dosage Forms: Scale-Up and Postapproval Changes: Chemistry, Manufacturing, and Controls, In Vitro Dissolution Testing, and In Vivo Bioequivalence Documentation* (SUPAC-SS). Rockville, MD.

[185] Swarbrick, J. (1990). *Encyclopedia of Pharmaceutical Technology*. CRC Press, New York.

[186] Tapon, R. (1993). Estimating shelf-life using L_1 regression methods. *Journal of Pharm. Bm.*, 11, 843–845.

[187] Tonnesen, H.H. (2004). *Photostability of Drugs and Drug Formulations*. Taylor and Francis, New York.

[188] TPD (1997). *Stability Testing of Existing Drug Substances and Products*. Therapeutic Products Directorate, Ottawa, Canada.

[189] Tsong, Y. (2003). Recent issues in stability study introduction. *Journal of Biopharmaceutical Statistics*, 13, vii–ix.

[190] Tsong, Y.. Chen, W.J. and Chen, C.W. (2003a). ANOVA approach for shelf life analysis of stability study of multiple factor designs. *Journal of Biopharmaceutical Statistics*, 13, 375–393.

[191] Tsong, Y. Chen, W.J., Lin, T.Y.D., and Chen, C.W. (2003b). Shelf life determination based on equivalence assessment. *Journal of Biopharmaceutical Statistics*, 13, 431–449.

[192] Tsong, Y., Hammerstorm, T., Lin, K.K., and Ong, T.E. (1995). The dissolution testing sampling acceptance rules. *Journal of Biopharmaceutical Statistics*, 5, 171–184.

[193] Tsong, Y., Hammerstrom, T., Sathe, P., and Shah, V.P. (1996). Statistical assessment of mean differences between two dissolution data sets. *Drug Information Journal*, 30, 1105–1112.

[194] Tsong, Y., Shen, M., and Shah, V.P. (2004). Three-stage sequential statistical dissolution testing rules. *Journal of Biopharmaceutical Statistics*, 14, 757–779.

[195] USP/NF (2000). *The United States Pharmacopeia XXIV and the National Formulary XIX*. The United States Pharmacopeial Convention, Inc., Rockville, MD.

[196] Vonesh, E.F. and Carter, R.L. (1987). Efficient inference for random coefficient growth models with unbalanced data. *Biometrics*, 43, 617–628.

[197] Wang, H. (2007). Estimation of the probability of passing the USP content uniformity test and the dissolution test. *Journal of Biopharmaceutical Statistics*, 17, No. 3. In press.

[198] Wang, S.G. and Chow, S.C. (1994). *Advanced Linear Models*. Marcel Dekker, New York.

[199] WHO. (1996). Guideline for Stability Testing of Pharmaceutical Products Containing well established drug substances in conventional dosage forms. In *WHO Expert Committee on Specifications for Pharmaceutical Preparations*. Technical Report Series 863, World Health Organization, Geneva, pp. 65–69.

[200] WHO. (2006). WHO Working Document QAS/05.146: Stability Studies in a Global Environment. World Health Organization, Geneva.

[201] Won, C.M. (1992). Kinetics of degradation of levothyroxine in aqueous solution and in solid state. *Pharamceutical Research*, 9, 131–137.

[202] Wright, J. (1989). Use of factorial designs in stability testing. In *Proceedings of Stability Guidelines for Testing Pharmaceutical Products: Issues and Alternatives*. AAPS Meeting, December, 1989.

[203] Yoshioka, S. (1999). Current application in Japan of the ICH stability guideline: does Japanese registration require more than others do? In *International Stability Testing*, Mazzo, D., Ed. Interpharm Press, Buffalo Groove, IL, pp. 255–264.

[204] Yoshioka, S., Aso, Y., and Kojima, S. (1997). Assessment of shelf-life equivalence of pharmaceutical products. *Chemical and Pharmaceutical Bulletin*, 45, 1482–1484.

[205] Yoshioka, S. and Stella, V.J. (2000). *Stability of Drugs and Dosage Forms*. Springer, New York.

Index

Printed in the United States
by Baker & Taylor Publisher Services